Theory of Chemical Reaction Dynamics

Volume I

Editor
Michael Baer
Applied Mathematics
Soreq Nuclear Research Center
Yavne, Israel

CRC Press, Inc.
Boca Raton, Florida

Library of Congress Cataloging in Publication Data
Main entry under title:

Theory of chemical reaction dynamics.

Includes bibliographies and indexes.
1. Chemical reaction, Conditions and laws of —
Collected works. I. Baer, M. (Michael)
QD501.T58 1985 541.3′9 84-4237
ISBN 0-8493-6114-1 (v. 1)
ISBN 0-8493-6115-X (v. 2)
ISBN 0-8493-6116-8 (v. 3)
ISBN 0-8493-6117-6 (v. 4)

Direct all inquiries to CRC Press, Inc., 2000 orporate Blvd., N.W., Boca Raton, Florida 33431.

International Standard Book Number 0-8493-6114-1 (Volume 1)
International Standard Book Number 0-8493-6115-X (Volume 2)
International Standard Book Number 0-8493-6116-8 (Volume 3)
International Standard Book Number 0-8493-6117-6 (Volume 4)
Library of Congress Card Number 84-4237

Printed in the United States

PREFACE

In 1979 I was asked to talk at a forthcoming Gordon Conference on Few Body Problems on developments in the theory of reaction dynamics. This talk led to my writing a review of quantum mechanical approximate treatments of the three-body reactive system, which was published in *Advances in Chemical Physics,* Volume 49, 1982. While I was preparing this review (in 1980), I realized two things: (1) that it was almost 50 years since the field in its present form was founded and (2) that as far as I knew, no graduate level textbook on this subject was in existence. While I was weighing the pros and the cons of writing (or editing) such a book, I was contacted by CRC Press about the possibility of editing a book on this subject. It did not take much time for me to agree to their proposition.

I was determined that my book should cover the entire field of the theory of reaction dynamics, leaving no aspect untreated. Having decided on the main subjects, I contacted the best leading scientists in the field, assigned to each of them the subject he specialized in, and asked them to write comprehensive, self-contained reviews. In general, the response to my invitation was rapid and enthusiastic and it was later followed by illuminating chapters. I am very thankful to all the authors for their productive collaboration.

The volumes consist of 15 chapters, 13 of which are mainly concerned with atom-diatom reactive systems and the other 2 with more extended systems, i.e., reactions at solid surfaces and reactions in solutions.

INTRODUCTION

Of the four chapters in this first volume, the first two are devoted to the treatment of potential energy surfaces and the other two to general theories for the dynamics.

In Chapter 1 the authors provide a historical background, then discuss the *ab initio* characterizations of the reactive potential energy surface for an atom-diatom system, introducing among other things the concept of reaction path. Next they describe in detail some well-established methods such as the Hartree-Fock method and several multiconfiguration-self-consistent field methods (including the valence bond method, etc.) and a number of configuration interaction methods. They then present numerical results for three-body (H + HX; X = F,Cl Br, I, H + HC, and H + CO) and four-body (H_2 + D_2, CH + H_2) reactions.

Chapter 2 is mainly concerned with the problem of effectively combining empirical data with *ab initio* calculations in the construction of potential energy surfaces for uses in dynamical calculations. The methods are discussed under three main headings according to the amount of empiricism employed:

1. Small empirical adjustments to *ab initio* results.
2. Semiempirical methods involving diagonalization of an appropriate Hamiltonian matrix.
3. Mainly empirical methods.

Emphasis is placed on methods of the second kind, which leads to convenient and physically reasonable representations of the potential energy surface. With the increasing availability of good quality *ab initio* surfaces, it will be necessary in the future to turn more attention to the development of type 1 methods; it is suggested to use semiempirical methods to estimate in a systematic way the corrections to *ab initio* surfaces.

Chapter 3 is devoted to various treatments of reactive exchange collisions, employing the differential form of the Schrödinger equation. The first part of this chapter deals with the collinear system in which the three interacting particles are constrained to move along a line. The author derives the corresponding Schrödinger equation, represents it in terms of various relevant coordinates, and discusses the transition from one arrangement channel to the other. Next, he describes several numerical methods to solve the Schrödinger equation and discusses at length simplified models, mechanisms of vibrational excitation in exothermic reactions, and the resonance phenomenon. The second part of the chapter deals with the general three-dimensional case, gives the derivation of the Schrödinger equations for both the space-fixed and the body-fixed systems, and details the corresponding close-coupling treatment and the transformation from one arrangement channel to the other two, concluding with the results for the H + H_2 system.

Chapter 4 deals with the integral equation approach to reactive scattering and begins with an introductory discussion of a number of formal and computational features of the non-reactive Lippmann-Schwinger equation. The purpose of this discussion is to illustrate the similarities and differences between the integral equation methods appropriate to nonreactive and reactive scattering. The two primary noniterative solution methods which can be used are first illustrated within the context of nonreactive scattering. In addition, the basic complication of reactive scattering due to the role of the continuum in the uncoupled Lippmann-Schwinger equation is used to motivate the introduction of integral equations which are coupled in the arrangement channel. Some of the various alternative sets of coupled T-operator equations are then discussed, and some of their (formal) properties examined. The two types of equations on which most attention is focussed are those due to Faddeev and those due to Baer, Kouri, Levin, and Tobocman (BKLT). Next, Kouri gives a detailed discussion of how the BKLT equations can be made explicit in order to solve them for particular systems. This discussion is couched in terms of both simple collinear reactive scattering and the full three physical dimensional reactive scattering problems. A more brief discussion of the Faddeev equations then follows to conclude this chapter.

ACKNOWLEDGMENT

I would like to thank Professor Z. B. Alfassi of the Ben Gurion University in Beer Sheva, Israel and Professors D. J. Kouri, R. D. Levine, T. F. George, D. G. Truhlar, and R. E. Wyatt for their support and encouragement at the early stages of the project.

CONTRIBUTORS

Michael Baer
Applied Mathematics
Soreq Nuclear Research Center
Yavne, Israel

Thomas H. Dunning, Jr.
Theoretical Chemistry Group
Chemistry Division
Argonne National Laboratory
Argonne, Illinois

Donald J. Kouri
Department of Chemistry
and Department of Physics
University of Houston
University Park
Houston, Texas

P. J. Kuntz
Hahn-Meitner Institute
Department of Radiation Chemistry
Berlin, West Germany

Lawrence B. Harding
Theoretical Chemistry Group
Chemistry Division
Argonne National Laboratory
Argonne, Illinois

TABLE OF CONTENTS

THEORY OF CHEMICAL REACTION DYNAMICS

Michael Baer

Volume I

Ab Initio Determination of Potential Energy Surfaces for Chemical Reactions
Semiempirical Potential Energy Surfaces
The General Theory of Reactive Scattering: The Differential Equation Approach
The General Theory of Reactive Scattering: The Integral Equation Approach

Volume II

Approximate Treatments of Reactive Scattering: Infinite Order Sudden Approximation
Approximate Treatments of Reactive Scattering: The T-Matrix Approach
The General Theory of Reactive Scattering: A Many-Body Approach
The Theory of Electronic Nonadiabatic Transitions in Chemical Reactions

Volume III

The Classical Trajectory Approach to Reactive Scattering
Periodic Orbits and the Theory of Reactive Scattering
Semiclassical Reactive Scattering

Volume IV

Statistical Dynamics
Generalized Transition State Theory
Theory of Reactions at a Solid State Surface
The Theory of Reactions in Solution

Chapter 1

AB INITIO DETERMINATION OF POTENTIAL ENERGY SURFACES FOR CHEMICAL REACTIONS

Thom. H. Dunning, Jr. and Lawrence B. Harding

TABLE OF CONTENTS

I. INTRODUCTION

The Schrödinger equation for a molecular system composed of N nuclei and n electrons is

$$\mathcal{H}\Psi(\mathbf{r};\mathbf{R}) = E\Psi(\mathbf{r};\mathbf{R}) \tag{1}$$

In the above, **r** denotes the set of 3n electronic coordinates and **R** denotes the set of 3N nuclear coordinates. The total many-body Hamiltonian for the system $\mathcal{H}$ can be partitioned into two terms:

$$\mathcal{H} = \mathcal{H}_N(\mathbf{R}) + \mathcal{H}_e(\mathbf{r};\mathbf{R}) \tag{2}$$

where $\mathcal{H}_N(\mathbf{R})$ is the nuclear Hamiltonian (in atomic units):

$$\begin{aligned}\mathcal{H}_N(\mathbf{R}) &= T_N(\mathbf{R}) + V_{NN}(\mathbf{R}) \\ &= \sum_I -\frac{1}{2M_I}\nabla_I^2 + \sum_I\sum_J \frac{Z_IZ_J}{R_{IJ}}\end{aligned} \tag{3a}$$

and $\mathcal{H}_e(\mathbf{r};\mathbf{R})$is the electronic Hamiltonian (in atomic units):

$$\begin{aligned}\mathcal{H}_e(\mathbf{r};\mathbf{R}) &= T_e(\mathbf{r}) + V_{ee}(\mathbf{r}) + V_{eN}(\mathbf{r};\mathbf{R}) \\ &= \sum_i -\frac{1}{2}\nabla_i^2 + \sum\sum_{i>j}\frac{1}{r_{ij}} - \sum_i\sum_I \frac{Z_I}{r_{iI}}\end{aligned} \tag{3b}$$

Note that in Equation 3 we include the electron-nuclear attraction term $[V_{eN}(\mathbf{r};\mathbf{R}]$ in the electronic Hamiltonian.

Since the mass of a nucleus (M_I) greatly exceeds that of an electron (m_e) (e.g., $M_I > 1800\ m_e$), as a first approximation let us take M_I to be infinite. Then Equation 1 simply reduces to:

$$\mathcal{H}_e\psi_e(\mathbf{r};\mathbf{R}) = E_e(\mathbf{R})\psi(\mathbf{r};\mathbf{R}) \tag{4}$$

which is the Schrödinger equation describing only the motions of the electrons; the nuclear repulsion term has been omitted from Equation 4 since it does not depend on (**r**). As is evident from Equation 4, both the electronic energy and the electronic wavefunction depend parametrically on the nuclear coordinates (**R**) through the electron-nuclear attraction term $V_{eN}(\mathbf{r};\mathbf{R})$.

We now expand the solutions of the exact Hamiltonian (Equation 2) in terms of a complete set of solutions of Equation 4 [ψ_{ej} (**r**;**R**)] for each nuclear configuration

$$\Psi(\mathbf{r};\mathbf{R}) = \sum_j \chi_j(\mathbf{R})\psi_{ej}(\mathbf{r};\mathbf{R}) \tag{5}$$

with the expansion coefficients [$\chi_j(\mathbf{R})$] depending only on (**R**). Substituting Equation 5 into Equation 1 and using Equations 3 and 4, we obtain:

$$\sum_j[\psi_{ej}T_N\chi_j + \chi_jT_N\psi_{ej} + V_{NN}\chi_j\psi_{ej} + \chi_j\mathcal{H}_e\psi_{ej}] = E\sum_j \chi_j\psi_{ej}$$

To obtain an effective equation for the nuclear motion we premultiply the above equation by ψ_{ei}^* and integrate over the electronic coordinates:

$$[T_N + V_{NN} + E_{ei}(\mathbf{R})]\chi_i = E\chi_i - \sum_j \chi_j\Theta_{ji} \tag{6}$$

If the Θ_{ji} in Equation 6 are neglected, then this equation becomes:

$$[T_N + V_{ieff}(\mathbf{R})]\chi_i = E\chi_i \tag{7}$$

where

$$V_{ieff}(\mathbf{R}) = V_{NN}(\mathbf{R}) + E_{ei}(\mathbf{R}) \tag{8}$$

which is an equation for the motion of the nuclei on a potential energy surface defined by $V_{ieff}(\mathbf{R})$. Thus, by neglecting the Θ_{ji} we have uncoupled the calculation of the electronic and nuclear wavefunctions.

The procedure for solving the Schrödinger equation is to first solve Equation 4 for the electronic energy ($E_{ei}(\mathbf{R})$) and wavefunction ($\psi_{ei}(\mathbf{r};\mathbf{R})$) as a function of the nuclear configuration ($\mathbf{R}$). Then, using the effective potential energy surface obtained in the above step ($V_{ieff}(\mathbf{R})$) we solve Equation 7 for the nuclear wavefunction (χ_i) and total energy (E). This is the Born-Oppenheimer (BO) approximation (see, e.g., Born and Huang[1]). As is evident from the above, the concept of a potential energy surface describing the nuclear motion is a by-product of the BO approximation and its validity depends on the accuracy of this approximation.

The inclusion of the electron-nuclear coupling terms (Θ_{ji}) leads to a coupling between the various electronic states. Such effects are referred to as non-BO or nonadiabatic corrections. The Θ_{ji} terms involve derivatives of the electronic wavefunction with respect to the nuclear coordinates ($\mathbf{R}$). In general, the electronic wavefunction is only a slowly varying function of the nuclear coordinates and the non-BO corrections are small. If, however, two (or more) electronic states mix strongly in some region of nuclear configuration space, in that region small displacements in the nuclear coordinates can produce large changes in the electronic wavefunction. In this case, the appropriate Θ_{ji} terms must be included in the nuclear Equation 7.

In this chapter we will be concerned with the solution of the electronic Schrödinger equation (Equation 4) to obtain the potential energy surfaces ($V_{ieff}(\mathbf{R})$ for chemical reactions. While the methods used to calculate potential energy surfaces have changed little in the past few years, great strides have been taken to improve the efficiency of the computational techniques and algorithms. To date, only for the simplest reaction ($H + H_2$) has a potential energy surface of chemical accuracy (i.e., with errors of less than 1 kcal/mol) been reported, although the energetics of many more complex chemical reactions have been calculated with an accuracy of a few kilocalories per mole. In the near future, the application of these new techniques and algorithms for solving the electronic Schrödinger equation can be expected to greatly expand the number of chemical reactions for which accurate potential energy surfaces are known.

In past years, techniques for characterizing potential energy surfaces have also been rapidly evolving. Thus, general methods are now available for the location of saddle points and the determination of reaction paths for chemical reactions. These techniques provide a systematic means of determining the essential features of the potential energy surface for a chemical reaction and their exploitation can be expected to greatly increase our understanding of both the energetics and dynamics (using reaction path Hamiltonian approaches[2]) of chemical reactions.

II. *AB INITIO* CALCULATION OF POTENTIAL ENERGY SURFACES FOR CHEMICAL REACTIONS

In the Introduction it was shown that with the BO approximation, the problem of solving the Schrödinger equation for molecular systems can be separated into an electronic part (depending parametrically on the positions of the nuclei) and a nuclear part (depending on an effective potential energy surface obtained from the electronic calculation). In this section we discuss *ab initio* methods for solving the electronic Schrödinger equation. Most methods for solving this equation in use today are based, at least in part, on the variational principle. In a variational calculation a general form for the electronic wavefunction containing a number of parameters is assumed and then the values of the parameters are varied to obtain the lowest possible energy ($E(\mathbf{R})$) for the wavefunction ($\psi_e(\mathbf{r};\mathbf{R})$).

$$E(\mathbf{R}) = \int \psi_e(\mathbf{r};\mathbf{R})\mathcal{H}_e(\mathbf{r};\mathbf{R})\psi_e(\mathbf{r};\mathbf{R})d\mathbf{r}_1 \ldots d\mathbf{r}_n = \langle\psi_e|\mathcal{H}_e|\psi_e\rangle \tag{9}$$

In order to obtain physically meaningful results, the form of electronic wavefunctions must be constrained to satisfy the Pauli Principle. Since electrons are fermions, the Pauli Principle states that the only valid multielectron wavefunctions are those for which the sign of the wavefunction changes when any pair of electrons are interchanged. This antisymmetry constraint is usually imposed through the use of Slater determinants. A Slater determinant is the determinant of a matrix of one-electron functions where each column of the matrix corresponds to a different electron and each row of the matrix corresponds to a different one-electron function. For example, a two-electron Slater determinant can be written as follows:

$$\begin{aligned}\psi_e &= \frac{1}{\sqrt{2}}\begin{vmatrix}\phi_1(1)\alpha(1) & \phi_2(1)\beta(1)\\ \phi_1(2)\alpha(2) & \phi_2(2)\beta(2)\end{vmatrix}\\ &= \frac{1}{\sqrt{2}}[\phi_1(1)\alpha(1)\phi_2(2)\beta(2) - \phi_1(2)\alpha(2)\phi_2(1)\beta(1)]\\ &= \frac{1}{\sqrt{2}}[\phi_1\phi_2\alpha\beta - \phi_2\phi_1\beta\alpha] = \frac{1}{\sqrt{2}}(1 - e_{12})\phi_1\phi_2\alpha\beta\\ &= \mathcal{A}\,\phi_1\phi_2\alpha\beta\end{aligned} \tag{10}$$

where $\mathcal{A}$ is defined to be the antisymmetrizing operator and e_{12} is an operator which interchanges electrons 1 and 2. In the last two equations it is assumed that the functions are ordered by electron number.

An additional constraint that is often useful to build into variational wavefunctions is a requirement that the wavefunction be an eigenfunction of both the total spin angular momentum operator (S^2) and the spin angular momentum projection operator (S_z). This is usually accomplished by taking linear combinations of Slater determinants which, by construction, are eigenfunctions of both operators. For example, in the above case, while the determinant shown is an eigenfunction of S_z with an eigenvalue of 0, it is not an eigenfunction of S^2. However, by taking linear combinations of two Slater determinants as shown below,

$$\begin{aligned}\psi &= \frac{1}{\sqrt{2}}[\mathcal{A}\,\phi_1\phi_2\alpha\beta - \mathcal{A}\,\phi_1\phi_2\beta\alpha]\\ &= \frac{1}{\sqrt{2}}[\mathcal{A}\,\phi_1\phi_2(\alpha\beta - \beta\alpha)]\\ &= \frac{1}{2}(\phi_1\phi_2 + \phi_2\phi_1)(\alpha\beta - \beta\alpha)\end{aligned} \tag{11}$$

we obtain a wavefunction which is an eigenfunction of S^2, with eigenvalue $S(S+1) = 0$, i.e., $S = 0$ and Equation 11 corresponds to a singlet state. These spin-adapted combinations of Slater determinants are referred to as spin eigenfunctions (SEF) or configuration state functions (CSF). Note that the determinant in Equation 10 can be written as in Equation 12;

$$\mathscr{A}\ \phi_1\phi_2\alpha\beta = \frac{1}{\sqrt{2}}[\mathscr{A}\ \phi_1\phi_2(\alpha\beta - \beta\alpha) + \mathscr{A}\ \phi_1\phi_2(\alpha\beta + \beta\alpha)] \tag{12}$$

i.e., the Slater determinant (Equation 10) is a mixture of two SEFs, one corresponding to a singlet state and the other to a triplet state.

Slater determinants and SEFs form the building blocks of most commonly used electronic structure methods. In Hartree-Fock (HF) calculations, the form of the wavefunction is taken to be a single Slater determinant (or in some cases, a single SEF). The variational parameters in HF calculations are the one-electron functions or, as they are more commonly referred to, the orbitals (ϕ_i). In multiconfiguration self-consistent field (MCSCF) calculations, the form of the wavefunction is taken to be a sum of two or more SEFs in which both the orbitals and coefficients of the SEFs are variationally optimized. Typically, MCSCF wavefunctions consist of a relatively small number (several hundred) of SEFs chosen to ensure that the major features of the potential energy surface is described in (at least) a qualitatively correct manner. Finally, in configuration interaction (CI) calculations, the form of the wavefunction is again taken to be a sum of SEFs; however, only their coefficients are optimized; the orbitals are as obtained from an independent calculation. Typically, CI calculations involve a very large number of SEFs (the current record is about $1^1/_2$ million). Since no orbital optimization is done in CI calculations, some other method must be used to obtain the orbitals; typically, the CI orbital basis is determined from either a HF or a small MCSCF calculation.

In all of the above methods the orbitals are usually constructed from linear combinations of some finite set of basis functions. The choice of basis functions can have a pronounced influence on the accuracy of the calculated potential energy surface.

In the discussion which follows, application will be made to the $H + H_2$ reaction whenever possible. This reaction may be considered as a prototype three-body reaction and, while the conclusions drawn should not be generalized unquestioningly, they nonetheless provide one measure of the importance of a number of effects relevant to the computation of potential energy surfaces for chemical reactions.

A. The Hartree-Fock Method

Conceptually, the HF method presents an appealing model of the electronic structure of molecules in which each electron moves in the average field due to all of the other electrons. This is accomplished by assigning to each electron a spatial orbital. Spin orbitals are then formed by multiplying each spatial orbital by either an α- or β-spin function. The HF wavefunction is constructed by multiplying the occupied spin orbitals together and antisymmetrizing the resulting many-electron function, usually yielding a single Slater determinant. There are two variants of the HF method called restricted Hartree-Fock (RHF) and unrestricted Hartree-Fock (UHF). In RHF calculations the spatial orbitals associated with α-spin are required to be identical to the corresponding orbitals of β-spin. This restriction assures that the RHF wavefunctions are eigenfunctions of both S^2 and S_z. As an example, the two-electron singlet RHF wavefunction for the H_2 molecule is

$$\begin{aligned}\psi_e^{RHF}(H_2) &= \mathscr{A}\ \phi_1\phi_1\alpha\beta \\ &= \frac{1}{\sqrt{2}}[\phi_1\phi_1(\alpha\beta - \beta\alpha)]\end{aligned} \tag{13}$$

where ϕ_1 is of σ symmetry (the symmetry of the state is ${}^1\Sigma_g^+$). In UHF wavefunctions, no spatial equivalence (double occupancy) restrictions are imposed and as a result, while the total energies of UHF wavefunctions are equal to or lower than those of RHF wavefunctions, the UHF wavefunctions are not, in general, eigenfunctions of S_2, although they are eigenfunctions of S_z. The two-electron (S_z = O) UHF wavefunction for the H_2 molecule is

$$\psi_e^{UHF}(H_2) = \mathcal{A}\ \phi_1\phi_2\alpha\beta \tag{14}$$

As noted earlier, this wavefunction is a mixture of a singlet and a triplet spin-state wavefunction. Because of the problems associated with such spin contamination, we shall limit our discussion here to the RHF method.

The optimum orbital for the closed-shell RHF wavefunction (Equation 13) is a solution of:

$$(h + J_1)\phi_1 = \epsilon_1\phi_1 \tag{15a}$$

or equivalently,

$$(h + 2J_1 - K_1)\phi_1 = \epsilon_1\phi_1 \tag{15b}$$

where h contains the one-electron kinetic energy and nuclear attraction terms,

$$h = -\frac{1}{2}\nabla^2 - \sum_I \frac{Z_I}{r_I} \tag{16}$$

(J_1, K_1) are the coulomb and exchange operators for orbital ϕ_1,

$$J_1(\mathbf{r}_1) = \int d\mathbf{r}_2\phi_1{}^*(\mathbf{r}_2)\phi_1(\mathbf{r}_2)\frac{1}{\mathbf{r}_{12}} \tag{17a}$$

$$K_1(\mathbf{r}_1) = \int d\mathbf{r}_2\phi_1{}^*(\mathbf{r}_2)\frac{1}{\mathbf{r}_{12}}e_{12}\phi_1(\mathbf{r}_2) \tag{17b}$$

and ϵ_1 is the orbital energy of orbital ϕ_1. Thus, the RHF Hamiltonian for the electron in orbital ϕ_1 contains kinetic and nuclear attraction terms plus a Coulomb interaction with the other electron of opposite spin in the orbital. Equation 15b, which is the most common form of the closed-shell RHF equations, follows from Equations 15a and 17 since

$$(J_1 - K_1)\phi_1 = 0 \tag{18}$$

The RHF wavefunction for the H + H_2 system is

$$\psi_e^{RHF} = \mathcal{A}\ \phi_1\phi_1\phi_2\alpha\beta\alpha \tag{19}$$

In a collinear geometry both ϕ_1 and ϕ_2 are of σ symmetry (the symmetry of the state is ${}^2\Sigma^+$). Using the coordinate system depicted in Figure 1, when $R_{ab} = \infty$, ϕ_1 becomes the RHF orbital of the H_2 (H_bH_c) fragment, while ϕ_2 becomes the 1s hydrogen atom (H_a) orbital. Similar considerations apply when $R_{bc} = \infty$. Thus, the RHF wavefunction provides a conceptually consistent description of the H + H_2 reaction, at least for the region of the potential energy surface near the reaction path.

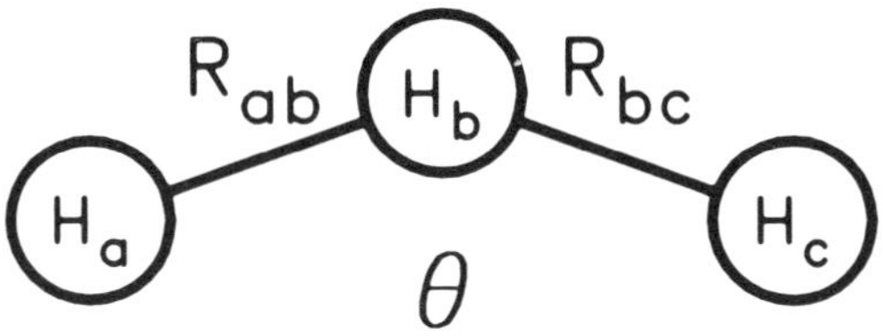

FIGURE 1. Coordinate system for the H + H_2 reaction.

The optimum orbitals for the open-shell RHF wavefunction (Equation 19) are the solutions of the following set of coupled equations,

$$[h + J_1 + {}^1/_2(2J_2 - K_2)]\phi_1 = \epsilon_1\phi_1 - \epsilon_{12}\phi_2 \tag{20a}$$

$$[h + (2J_1 - K_1)]\phi_2 = \epsilon_2\phi_2 - \epsilon_{21}\phi_1 \tag{20b}$$

i.e., in addition to the kinetic energy and nuclear attraction terms, the electron in orbital ϕ_1 has a Coulomb interaction with the other electron in ϕ_1 and an average of a Coulomb (J_2) and a Coulomb minus exchange ($J_2 - K_2$) interaction with the electron in ϕ_2, while the electron in ϕ_2 has a Coulomb interaction (J_1) with the electron with the opposite spin in ϕ_1 and a Coulomb minus exchange ($J_1 - K_1$) interaction with the electron in ϕ_1 with the same spin. Since the two orbitals are solutions of different Hamiltonian operators, Lagrange multipliers ($\epsilon_{12} = \epsilon_{21}$) must be introduced to ensure their orthogonality. A number of techniques have been developed to solve the closed-shell[3] (Equation 15) and open-shell[4-8] (Equation 20) RHF equations.

For regions of the potential surface far from the reaction path, e.g., when both R_{ab} and R_{bc} are large, the RHF wavefunction does not provide a proper description of the potential energy surface. Consider the case when R_{ab} is very large. Then to determine the behavior of the H_3 wavefunction far from the reaction path we need only consider the behavior of the wavefunction for the H_bH_c fragment at large R_{bc}. As $R_{bc} \rightarrow \infty$, the RHF wavefunction for this fragment becomes:

$$\begin{aligned} \psi_e^{RHF}(H_2) &= \mathscr{A}(1s_l + 1s_r)^2\alpha\beta \\ &= \frac{1}{\sqrt{2}}[(1s_l1s_r + 1s_r1s_l + 1s_r^2 + 1s_l^2)(\alpha\beta - \beta\alpha)] \end{aligned} \tag{21}$$

where $1s_l$ and $1s_r$ are orbitals on the left and right hydrogen atoms, respectively. It thus describes dissociation to a mixture of covalent and ionic states. The computed dissociation limit of this wavefunction lies over 160 kcal/mol above the energy of two hydrogen atoms. Also, as expected from the above, at large separations the RHF potential energy curve for H_2 is found to display a 1/R behavior caused by the spurious ionic terms in the wavefunction.

Accurate RHF calculations on the H + H_2 reaction[9] predict a barrier to reaction of ~24 kcal/mol at an HH separation ($R_{ab} = R_{bc}$) of 0.91 Å (for definitions of the barrier to reaction and the associated geometry, see Section III) (Table 1). The exact barrier height for this reaction is estimated to be 9.68 ± 0.12 kcal/mol.[10] The RHF method thus considerably overestimates the barrier to reaction. Electron correlation effects, which are neglected in the RHF wavefunction, are clearly important in the calculation of potential energy surfaces for chemical reactions and must be explicitly included.

B. Multiconfiguration Self-Consistent Field Methods

The goal in MCSCF calculations is to design wavefunctions that contain the most important

Table 1
CALCULATED SADDLE POINT GEOMETRIES AND BARRIER HEIGHTS FOR THE H + H_2 REACTION

	R_{sp} (HH) (Å)	$\Delta E_{barrier}$ (kcal/mol)
RHF	0.91	24.5
RHF + 1 + 2/IS		10.5
RHF + 1 + 2		10.3
SOGVB	0.96	17.1
GVB-CI	0.96	17.1
POL-CI	0.95	12.1
GVB + 1 + 2	0.93	10.0
Full CI	0.93	9.9

Note: Gaussian basis set is [3s2p1d] (Section III).

electron correlation effects in a relatively compact, comprehensible form. A general MCSCF wavefunction can be written as

$$\psi_e^{MCSCF} = \sum_i C_i \psi_{ei} \tag{22}$$

where each ψ_{ei} is a spin eigenfunction (SEF). Both the SEF coefficients (C_i) and the orbitals from which the SEFs are constructed are variationally optimized. The cost of MCSCF calculations, however, can be substantially greater than that of RHF calculations. For a discussion of recent developments in solving for MCSCF wavefunctions, see Olsen and Yeager.[11]

A crucial problem in MCSCF calculations is in the choice of configurations to be included in the wavefunctions. In the remainder of this section we will discuss two general procedures that have been developed to select configurations for MCSCF calculations of potential energy surfaces: the generalized valence bond (GVB) method and the fully optimized reaction space (FORS) or complete active space self-consistent field (CASSCF) method. These methods both provide a consistent description of the electronic structure of a reactive system over most of the potential energy surface of interest. A number of attempts have also been made to select for MCSCF calculations only the most important configurations; see, e.g., the optimized valence configuration (OVC) method of Wahl and Das.[12] This can be accomplished in several ways, including the use of preliminary CI calculations to determine which are the most important configurations. Care must be taken when using approaches of this type to calculate potential energy surfaces since obtaining a balanced description of the electronic structure of the system for all of the nuclear configurations of interest is fraught with difficulties.

1. The Generalized Valence Bond (GVB) Method

The GVB method exploits the relationship between valence bond (VB) and MCSCF wavefunctions to guide the choice of MCSCF configurations.[8,13] As a simple example, let us consider the VB wavefunction for H_2. In VB theory, wavefunctions are constructed from antisymmetrized products of atomic wavefunctions combined to give eigenfunctions of S^2. For H_2 the VB wavefunction is

$$\psi_e^{VB}(H_2) = \frac{1}{\sqrt{2(1 + S^2)}} \mathcal{A}\, \phi_l \phi_r (\alpha\beta - \beta\alpha) \tag{23}$$

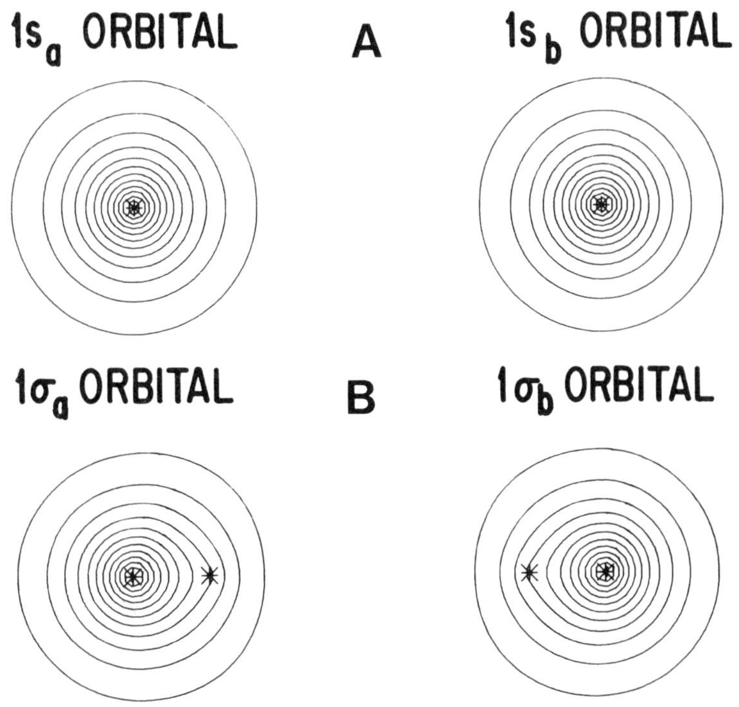

FIGURE 2. The GVB orbitals for the H_2 molecule (A) at large internuclear separations (15.0 a_0) and (B) near the equilibrium internuclear separation (1.42 a_0). The plots have uniformly spaced contour lines with increments of 0.05 a.u.; only the contour lines between -0.5 and $+0.5$ have been plotted.

where S is the overlap between the two nonorthogonal orbitals, $\phi_l = 1s_l$, $\phi_r = 1s_r$, and $1s_i$ is a 1s hydrogen atom orbital on either the left (i = l) or right (i = r) atom. In the GVB wavefunction, the orbitals in Equation 23 are variationally optimized to allow for any changes that occur upon bond formation. Thus, at $R_{HH} = \infty$, the GVB orbitals are just the 1s atomic orbitals; at shorter separations, however, the orbitals delocalize and hydridize to maximize the strength of the HH bond (see, e.g., Figure 2). The above wavefunction can be equivalently expressed as a two-determinant MCSCF wavefunction involving the orthogonal orbitals σ_g and σ_u,

$$\psi_e^{VB}(H_2) = C_1 \mathcal{A}\, \sigma_g\sigma_g\alpha\beta + C_2 \mathcal{A}\, \sigma_u\sigma_u\alpha\beta \tag{24}$$

where

$$\sigma_g = \frac{1}{\sqrt{2(1 + S)}} (1s_l + 1s_r) \tag{25a}$$

$$\sigma_u = \frac{1}{\sqrt{2(1 - S)}} (1s_l - 1s_r) \tag{25b}$$

and

$$C_1 = \frac{1 + S}{\sqrt{2(1 + S^2)}} \tag{26a}$$

$$C_2 = \frac{1 - S}{\sqrt{2(1 + S^2)}} \tag{26b}$$

Thus, variational optimization of the σ_g and σ_u orbitals and the coefficients (C_1, C_2), in Equation 24 is equivalent to optimizing the orbitals in the GVB wavefunction (Equation 23). The MCSCF wavefunction, which involves orthogonal orbitals, provides a computationally efficient scheme for optimizing the orbitals of the GVB wavefunction. The orbitals in Equation 24 are referred to as the GVB natural orbitals.

Note that the first term in Equation 24 is just the RHF wavefunction for H_2. At short internuclear distances $|C_1| >> |C_2|$, and the GVB and RHF wavefunctions differ little (although not negligibly). At large internuclear distances the energies of the two configurations in Equation 24 become equal (degenerate) and $|C_1| \rightarrow |C_2|$. Because of this, the GVB method is said to include near-degeneracy electron correlation effects; as can be seen in the present case, these effects are just those required to properly describe the dissociation of H_2 into two ground-state hydrogen atoms.

The GVB wavefunction for H_3 is constructed from the GVB wavefunction for H_2 and for the hydrogen atom,

$$\Psi_e^{GVB}(H_3) = \mathcal{A}\, \phi_{1\sigma_a}\phi_{1\sigma_b}\phi_{2\sigma}(C_1\chi_1 + C_2\chi_2) \tag{27}$$

where

$$\chi_1 = \frac{1}{\sqrt{2}}(\alpha\beta - \beta\alpha)\alpha \tag{28a}$$

$$\chi_2 = \frac{1}{\sqrt{6}}(2\alpha\alpha\beta - \alpha\beta\alpha - \beta\alpha\alpha) \tag{28b}$$

The first term in Equation 27 can be seen to arise directly from Equation 23; it is required to properly describe the dissociation of the H_3 species into $H + H_2$ or into three hydrogen atoms. The second term is required to properly describe the wavefunction in the vicinity of the saddle point and allows for the smooth transformation of the orbitals of the reactants into those of the products. Neglect of this term can lead to multiple solutions to the GVB equations in the saddle point region.[14]

As was the case for H_2, in Equation 27 all of the orbitals are nonorthogonal. However, it has been found that orthogonality can often be imposed between the orbitals in different pairs with little loss of accuracy; such is true in the present case. If $\phi_{2\sigma}$ is taken as orthogonal to $\phi_{1\sigma_a}$ and $\phi_{1\sigma_b}$ then,

$$\int \phi_{2\sigma}(\mathbf{r}_1)\phi_{1\sigma_a}(\mathbf{r}_1)d\mathbf{r}_1 = \int \phi_{2\sigma}(\mathbf{r}_1)\phi_{1\sigma_b}(\mathbf{r}_1)d\mathbf{r}_1 = 0 \tag{29}$$

or

$$\langle\phi_{2\sigma}|\phi_{1\sigma_a}\rangle = \langle\phi_{2\sigma}|\phi_{1\sigma_b}\rangle = 0$$

then Equation 27 can be rewritten as

$$\Psi_e^{GVB}(H_3) = C_1\,\mathcal{A}\,\phi_{1\sigma}\phi_{1\sigma}\phi_{2\sigma}\alpha\beta\alpha + C_2\,\mathcal{A}\,\phi_{3\sigma}\phi_{3\sigma}\phi_{2\sigma}\alpha\beta\alpha + + C_3\,\mathcal{A}\,\phi_{1\sigma}\phi_{3\sigma}\phi_{2\sigma}\left(\frac{2\alpha\alpha\beta - \alpha\beta\alpha - \beta\alpha\alpha}{\sqrt{6}}\right) \tag{30}$$

with definitions for the natural orbitals and CI coefficients similar to those for H_2 (Equations 25 and 26). Thus, the GVB wavefunction for H_3 (Equation 27) with the strong orthogonality constraint (Equation 29) is equivalent to a simple three-configuration MCSCF wavefunction (Equation 30). This wavefunction is referred to as the strongly orthogonal generalized valence bond (SOGVB) wavefunction. Note again that the first configuration in Equation 30 just corresponds to the RHF wavefunction for H_3.

The SOGVB orbitals for the H + H_2 calculation are plotted in Figure 3 at various points along the path leading from reactants to the saddle point to products;[9] for a definition of the reaction path and the saddle point, see Section IIIB. From this figure we see that:

1. The $1\sigma_a$ orbital, which correlates with the left $1s_H$-like bonding orbital in the reactants, changes only slightly on going from the reactants to the saddle point to the products, becoming the right $1s_H$-like bonding orbital in the products.
2. The $1\sigma_b$ orbital, which correlates with the right $1s_H$-like bonding orbital in the reactants, first delocalizes onto the attacking hydrogen atom on approaching the saddle point and then localizes on this atom when receding from the saddle point, becoming the left $1s_H$-like bonding orbital in the products.

In this way, the bonding orbitals maintain high overlap as the bond pair is transferred from the reactants to the products, e.g., at the saddle point the overlap of these two orbitals is 0.85, while in the reactants (or products) it is 0.80. It is for this reason that the barrier for the H + H_2 reaction is only a small fraction of the H_2 bond energy.

To remain orthogonal to the orbitals in the bond pair, the 2σ orbital, which correlates with the $1s_H$ orbital on the attacking atom in the reactants and on the leaving atom in the products, builds in a nodal plane in the bonding region. The resulting change in the phase of this orbital between reactants and products is the basis for the selection rules on chemical reactions proposed by Goddard.[15]

As can be seen, the GVB method provides a conceptually simple description of the changes in the electronic wavefunction during the course of the H + H_2 reaction. Furthermore, GVB calculations on the H_3 system (GVB,[16] SOGVB[9]) predict a barrier to reaction of ~17 kcal/mol (Table 1). This is more than 7 kcal/mol lower than that obtained from RHF calculations. The near-degeneracy effects included in the GVB wavefunction are clearly important in providing qualitatively correct descriptions of reactive potential energy surfaces. However, it is also clear that additional electron correlation effects must be included in the calculations to obtain accurate barrier heights.

The saddle point geometry determined from the SOGVB calculations is R_{HH} = 0.96 Å, which is 0.05 Å longer than that obtained in the RHF calculations.

In summmary, the advantages of the GVB method are

1. Configurations can be chosen based on simple VB concepts to emphasize those correlation effects liable to be most relevant to a particular potential energy surface.
2. The GVB MCSCF wavefunction can be transformed back to VB form, yielding a readily interpretable orbital model of the electronic structure of the system.
3. The calculations are sufficiently tractable that relatively large systems can be treated.

The disadvantages of the GVB method include:

1. For some molecules, more than one VB structure is possible, leading to a far more complicated form for the GVB wavefunction. For these molecules it has been found that projected or resonating GVB wavefunctions are required.[17]

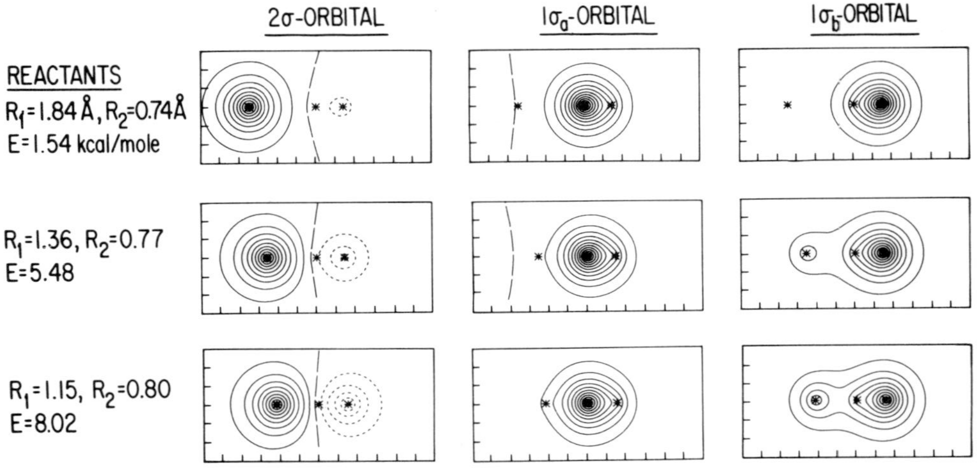

FIGURE 3. Contour plots of the SOGVB orbitals for the $H + H_2 \rightarrow H_2 + H$ reaction at selected points along the reaction path. Positive contour lines are traced in solid lines, negative contour lines in short dashed lines, and nodal (zero) lines in long dashed lines. Only contour lines between -0.5 and $+0.5$ have been plotted.

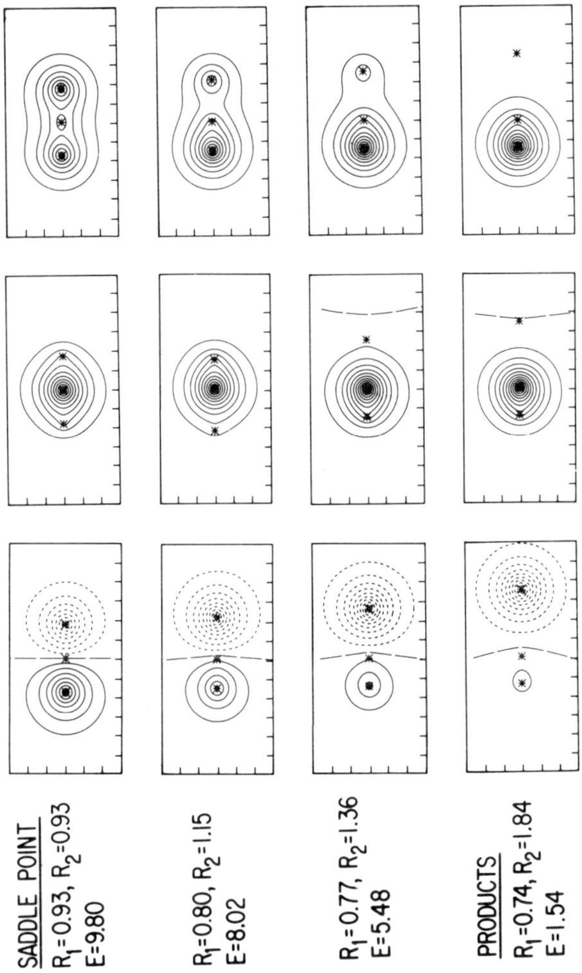

FIGURE 3 continued

2. Since GVB orbitals are usually localized, they often have a symmetry that is lower than that of the molecule, making it difficult to take full advantage of the any simplications due to symmetry.

2. The Fully Optimized Reaction Space/Complete Active Space Self-Consistent Field (FORS/CASSCF) Method

The approach is the FORS/CASSCF method is to place the emphasis on choosing orbitals to be included in the calculation rather than on choosing the configurations. The method divides the orbitals into inactive, active, and virtual sets. The configurations in a FORS/CASSCF calculation are then determined by requiring that the inactive orbitals be doubly occupied in all configurations, that the virtual orbitals be unoccupied in all configurations, and that all possible occupancies of the active orbitals be included in the wavefunction. This method was developed and applied by Ruedenberg et al.[18] (FORS) and by Siegbahn, Roos, and co-workers[19] (CASSCF). The main advantage of this method is that once the partitioning between active, inactive, and virtual orbitals is established, the wavefunction is completely specified; no further selection of configurations is required. The main disadvantage of this method is that the computational requirements of the calculation increases dramatically with the number of orbitals and electrons included in the active space. For examples of applications of the FORS/CASSCF method to reactive potential energy surfaces see Karlstrom and Roos[20] and Cheung et al.[18]

Often, FORS/CASSCF calculations are based on an orbital set whose composition, if not form, is identical to that used in the SOGVB calculations — this set is referred to as the valence orbital (VO) set and is composed of all of the molecular orbitals which can be formed from the valence shell atomic orbitals, i.e., the 1s orbital in hydrogen, the (2s, $2p_x$, $2p_y$, and $2p_z$) orbitals in the first row atoms (Li-Ne), etc. (the core orbitals are taken to be doubly occupied in all configurations). In this case, there is a close relationship between the two calculations, and the orbitals are found to be very similar even though many more configurations may be included in the FORS/CASSCF calculations (see below). In the H_3 system, e.g., at the (symmetric) saddle point the FORS/CASSCF wavefunction obtained using three σ orbitals, i.e., those correlating with the 1s orbitals of the three-hydrogen atoms, is identical to the SOGVB wavefunction. For nonsymmetric but collinear, geometries, the FORS/CASSCF wavefunction contains eight configurations rather than the three in the SOGVB wavefunction (Equation 30).

C. Configuration Interaction Methods

As we saw above, wavefunctions including near-degeneracy effects yield potential energy surfaces which have the correct qualitative shape. However, in order to obtain surfaces of quantitative accuracy, additional electron correlation effects must be included in the calculations. This can be accomplished with CI calculations.[21] The CI wavefunction has the same general form as an MCSCF wavefunction.

$$\psi_e^{CI} = \sum_i C_i \psi_{ei} \tag{31}$$

The difference is that in a CI wavefunction only the coefficients (C_i) of the SEFs are variationally optimized, the orbitals are not. This is accomplished by solving the matrix eigenvalue equation:

$$\mathbf{HC} = \mathbf{EC} \tag{32}$$

where $\mathbf{C}$ is the vector of CI coefficients in Equation 30 and the matrix elements of H are given by:

$$\mathcal{H}_{ij} = \langle\psi_{ei}|\mathcal{H}_e|\psi_{ej}\rangle = \int\psi_{ei}(\mathbf{r;R})\mathcal{H}_e(\mathbf{r;R})\psi_{ej}(\mathbf{r;R})d\mathbf{r}_1 \ldots d\mathbf{r}_n \quad (33)$$

CI matrix elements can be classified into three types: the diagonal elements, those between configurations differing by a single excitation, and those between configurations differing by a double excitation. It can be easily shown that since the electronic Hamiltonian has only terms involving, at most, two electrons, those matrix elements between configurations differing by more than a double excitation are zero. Also, only two-electron terms contribute to matrix elements between configurations that differ by a double excitation. Formulas are easily derived for calculating the Hamiltonian matrix elements between Slater determinants; these formulas may then be assembled into those between SEFs. While the above selection rule on nonzero matrix elements means that the CI matrix is quite sparse, many matrix elements must still be evaluated and the number of matrix elements which must be computed rises as N_c^2, where N_c is the number of configurations in the calculation. Great advances have, however, been made in recent years in techniques for solving the CI equations. Most notable among these are the direct CI techniques,[22,23] the graphical unitary group approach (GUGA),[24] the symbolic matrix method,[25] and the contracted CI method.[26] In the remainder of this section we confine ourselves to a brief discussion of some commonly used CI wavefunctions.

In order to uniquely define a CI wavefunction, it is necessary to specify both the orbitals used to construct the SEFs and the list of SEFs included in the calculation. Typically, orbitals for CI calculations are determined from RHF, SOGVB, or other MCSCF calculations. Configuration lists are most often specified with an order of excitation (singles, doubles, etc.) from some set of reference configurations together with possible restrictions on the occupations of certain orbitals, e.g., core orbitals are often required to be doubly occupied in all configurations. CI methods then, differ in the way the orbitals are defined, the choice of reference configurations, the order of excitation, and the restrictions placed on orbital occupancy. The following is a list of some commonly used CI methods.

1. Hartree-Fock (HF) Configuration Interaction Methods

Configuration interaction calculations based on a HF reference wavefunction are the most prevalent in the literature.

a. *HF + 1 + 2*

In a HF + 1 + 2 calculation the orbitals are determined in a HF calculation (either UHF or RHF) and then the configurations are selected by allowing all single and double excitations from the HF configuration.

An important variant of this method is the use of the interacting space (IS) restriction.[27,28] The IS is defined to be all configurations which have a nonzero matrix element with the reference configuration(s). As noted above, in order for a CI matrix element to be nonzero, the two configurations involved must differ by at most a double excitation. For a closed-shell RHF reference configuration there is no difference between a full RHF + 1 + 2 calculation and one using only the IS. For an open-shell RHF reference configuration, however, a full RHF + 1 + 2 calculation includes configurations which, due to differences in the spin functions, have zero matrix elements with the reference configuration. Elimination of these configurations represents a considerable savings in some cases.

The HF + 1 + 2 wavefunction includes the dominant electron correlation effects. Difficulties arise, however, in systems which are not well described by a single RHF configuration.

2. Generalized Valence Bond Configuration Interaction Methods

a. *GVB-CI*

Configurations in addition to those in the GVB wavefunction may be constructed from the SOGVB natural orbitals. The GVB-CI wavefunction[29,30] is obtained by including all single and double excitations from the SOGVB MCSCF configurations into other SOGVB natural orbitals which are unoccupied in those configurations. A full GVB-CI wavefunction includes all of the configurations which can be constructed from the SOGVB natural orbitals; this wavefunction is essentially identical to a similarly constructed FORS/CASSCF wavefunction (the orbitals are, of course, not optimum for the FORS/CASSCF wavefunctions).

The GVB-CI wavefunction includes any near-degeneracy effects not included in the simpler SOGVB wavefunction. In many cases, the GVB-CI wavefunction offers little improvement over the SOGVB wavefunction. In unsaturated systems, however, GVB-CI calculations do lead to significant improvements.[31] It should also be noted that GVB-CI has been used as a generic term to denote any CI based on SOGVB or GVB wavefunctions.

b. *POL-CI*

The polarization CI wavefunction[29,30] includes all single and double excitations from the SOGVB MCSCF wavefunction with the restriction that no more than one electron occupy a virtual orbital (i.e., a non-GVB natural orbital). In addition to the configurations included in the full GVB-CI wavefunction, the full POL-CI wavefunction contains all configurations which can be constructed by distributing (n-1) electrons in the SOGVB natural orbitals and one electron in the virtual orbitals. This latter method is closely related to the first-order CI (FO-CI) method (see below), the only difference resulting from the differences (usually small) in the orbitals.

The POL-CI wavefunction includes space- and spin-polarization effects which are not included in the GVB and GVB-CI wavefunctions. These effects are also referred to as semi-internal correlation effects.[32] As we shall see below, these effects must be included in the wavefunction to obtain a semiquantitative description of reactive potential energy surfaces. The RHF + 1 + 2 wavefunction also includes space and spin polarization effects.

A distinct advantage of the POL-CI (and FO-CI) method is that the size of the calculation increases only linearly with the number of external orbitals while the size of calculations such as RHF + 1 + 2, which allow two electrons in the virtual orbitals, is proportional to the square of the number of virtual orbitals. The POL-CI method is thus applicable to a wide range of chemical systems.

c. *GVB + 1 + 2*

A GVB + 1 + 2 wavefunction[30] includes all single and double excitations relative to the GVB wavefunction without restriction. In addition to the configurations included in the full POL-CI wavefunction, the full GVB + 1 + 2 wavefunction includes all possible arrangements of (n-2) electrons in the GVB natural orbitals coupled with all possible arrangements of two electrons in the virtual orbitals.

In addition to the effects included in the POL-CI wavefunction, the GVB + 1 + 2 wavefunction includes most of the remaining important electron correlation effects. The RHF + 1 + 2 includes most, but not all, of the correlation effects included in the GVB + 1 + 2 wavefunction.

3. General Multiconfiguration Configuration Interaction Methods

Valence CI — A valence CI calculation[33] relies on a partitioning of the orbitals into a core, a valence, and a virtual space; for an earlier, related method see Schaefer and Bender.[34] With this partitioning, all possible configurations which can be constructed from the set of VOs are included in the valence CI wavefunction. The core orbitals are doubly occupied and the virtual orbitals are unoccupied in all configurations.

FO-CI — In addition to the configurations included in the valence CI wavefunction, the FO-CI wavefunction[35] includes all configurations which have at most one electron in the external space.

SO-CI — In addition to the configurations included in the FO-CI wavefunction, the second-order CI (SO-CI) wavefunction[28,33] includes all configurations which have at most two electrons in the external space.

In many instances, the set of internal (valence) orbitals used in the valence CI, FO-CI, and SO-CI wavefunctions contains the same orbital types that are included in the SOGVB set. In these cases the valence-CI, FO-CI, and SO-CI wavefunctions are essentially identical to the full GVB-CI, full POL-CI, and full GVB + 1 + 2 wavefunctions.

MR + 1 + 2 — There are at least two variants of the simple multireference singles and doubles CI; both are designed to correct the problems of RHF + 1 + 2 calculations in regions where there is more than one dominant configuration in the CI wavefunction. The simplest procedure is to simply add more reference configurations to the RHF + 1 + 2 calculations (e.g., all configurations that are at most a double excitation from one of a set of reference configurations may be included). Alternatively, the orbitals can be obtained from a MCSCF calculation and then the single and double excitations taken relative to the MCSCF configurations. The latter procedure is preferable because any configurations that are important enough to be included as a reference configuration may also significantly affect the form of the orbitals, and therefore should be included in the orbital optimization step.

4. Correlation Effects in the $H + H_2$ Reaction

To obtain an appreciation of the various electron correlation effects included in the CI wavefunctions let us consider the results of the calculations of the barrier height in the $H + H_2$ reaction summarized in Table 1.

1. GVB-CI calculations offer no improvement over that obtained from the SOGVB calculations. In fact, it can readily be shown that for the H_3 system at the saddle point and in the reactants' and products' limits, the SOGVB, GVB-CI, full GVB-CI, and valence-CI wavefunctions are all identical.
2. The space- and spin-polarization effects included in the POL-CI calculation reduces the barrier to reaction from ~17 (SOGVB) to ~12 kcal/mol. In fact, the barrier height obtained from the POL-CI calculations is just 2 kcal/mol larger than the exact barrier estimated by Liu.[10] Likewise it can be shown that in the above regions the POL-CI, full POL-CI, and FO-CI wavefunctions for this system are also identical.

In general then, it is to be expected that explicit consideration of the near-degeneracy and polarization correlation effects included in the POL-CI wavefunction will be necessary to obtain barrier heights accurate to at least a few kilocalories per mole. Including the remaining electron correlation effects we find that:

3. The GVB + 1 + 2 wavefunction yields a barrier height only 0.1 kcal/mol higher than that obtained in the full CI calculations. This should be contrasted with the barrier obtained from the RHF + 1 + 2 calculation which is in error by 0.4 kcal/mol. This difference of 0.3 kcal/mol is a direct measure of the importance of the single and double excitations from the extra configurations included in the SOGVB wavefunction. The RHF + 1 + 2 calculation, of course, improves considerably on the RHF calculation.

Note that the elimination of configurations through use of the IS increases the error in the RHF + 1 + 2 calculation by only 0.2 kcal/mol.

As can be seen in Table 1, there is little variation in the calculated saddle point geometries. This provides some justification for the common technique of determining the saddle point geometry at a lower level of calculation than that used to determine the barrier height (at the saddle point geometry so obtained).

D. Basis Sets

The RHF equations are usually solved using finite basis set expansion techniques in which each of the orbitals (ϕ_i) is expressed as a linear combination of basis functions (χ_ν).

$$\phi_i = \sum_\nu \chi_\nu c_{\nu i} \tag{34}$$

The coefficients are then variationally optimized through an iterative procedure involving the solution of a series of eigenvalue equations. Using the notation of Roothaan,[3] the above equations can be written:

$$(F_{\mu\nu} - \epsilon_i S_{\mu\nu})c_{\nu i} = 0 \tag{35}$$

where $F_{\mu\nu}$ is the element of the RHF matrix (see below) involving basis functions χ_μ and χ_ν, $c_{\nu i}$ is the coefficient of the ν-th basis function in the i-th orbital, $S_{\mu\nu}$ is the overlap between basis functions χ_μ and χ_ν, and ϵ_i is the orbital energy of the i-th orbital. For a closed-shell wavefunction the RHF matrix has the form:

$$F_{\mu\nu} = h_{\mu\nu} + \sum_\lambda \sum_\gamma P_{\lambda\gamma}[(\mu\nu|\lambda\gamma) - {}^1\!/_2(\mu\lambda|\nu\gamma)] \tag{36}$$

The $h_{\mu\nu}$ in Equation 36 are integrals of the one-electron kinetic and nuclear attraction terms:

$$h_{\mu\nu} = <\chi_\mu|T + V_{eN}|\chi_\nu> = \int \chi_\mu(\mathbf{r}_1)\left(-\,{}^1\!/_2\,\nabla_1^2 - \sum_I \frac{Z_I}{r_{1I}}\right)\chi_\nu(\mathbf{r}_1)d\mathbf{r}_1 \tag{37}$$

The $P_{\lambda\gamma}$ are the elements of the density matrix:

$$P_{\lambda\gamma} = 2\sum_i c_{\lambda i}^* c_{\gamma i} \tag{38}$$

(the summation extends over the occupied orbitals only of which there are n/2) and, finally, the two-electron integrals have the form:

$$(\mu\nu|\lambda\gamma) = <\chi_\mu\chi_\nu|1/r_{12}|\chi_\lambda\chi_\gamma> = \int \chi_\mu(\mathbf{r}_1)\chi_\lambda(\mathbf{r}_1)\frac{1}{r_{12}}\chi_\nu(\mathbf{r}_2)\chi_\gamma(\mathbf{r}_2)d\mathbf{r}_1 d\mathbf{r}_2 \tag{39}$$

Note that subscripts μ, ν, λ, and γ refer to basis functions while the subscript i refers to an RHF orbital. The RHF operator ($F_{\mu\nu}$), then, includes both one-electron terms (kinetic energy and electron-nuclear attraction) and two-electron terms (electron-electron coulomb and exchange). The two-electron terms depend on the orbitals through the density matrix and it is for this reason that the equations must be solved iteratively to self-consistency, i.e., until the orbitals produced by solving Equation 35 are the same as those used to construct the F matrix. With these definitions the energy (E) of the above RHF wavefunction is

$$E^{RHF} = \sum_\mu \sum_\nu P_{\mu\nu}h_{\mu\nu} + {}^1\!/_2 \sum_\mu \sum_\nu \sum_\lambda \sum_\gamma P_{\mu\nu}P_{\lambda\gamma}\,[(\mu\nu|\lambda\gamma) - {}^1\!/_2\,(\mu\lambda|\nu\gamma)] + V_{NN} \tag{40}$$

One obvious choice of the basis functions (Equation 34) are the hydrogen atom-like functions:

$$r^{n-1}e^{-\zeta r}Y_{lm}(\theta,\phi) \tag{41}$$

where $Y_{lm}(\theta, \phi)$ are the spherical harmonics. The functions in Equation 41 are referred to as Slater functions. Such basis functions provide accurate, compact representations of atomic RHF wavefunctions. For molecular calculations, however, evaluation of the required many-centered, two-electron integrals (Equation 39) (of which there are $\sim 1/8N_{fn}^4$ where N_{fn} is the number of basis functions used in the expansion) is very time-consuming. One solution to this problem[36] is to use Gaussian functions of the form:

$$x^l y^m z^n e^{-\alpha r^2} \tag{42}$$

The needed molecular integrals involving Gaussian functions can be readily evaluated using simple formulas. However, Gaussian functions do not exhibit the correct behavior either at long- (exponential decay) or short-range (cusp condition). For these reasons, obtaining a wavefunction of a given level of accuracy requires many more Gaussian than Slater functions. Unfortunately, such an increase in the size of the basis set greatly increases the time required to solve the RHF, GVB, MCSCF, and CI equations. A compromise which has been found to be most effective is the use of contracted Gaussian functions which are simply linear combinations of Gaussian functions, the exponents and coefficients of which are chosen so that the contracted functions give both a compact and accurate description of the atomic and molecular wavefunctions. The use of contracted Gaussian functions does not reduce the number of integrals that must be computed; however, since only the integrals over the contracted functions are needed in subsequent steps of the calculation, considerable savings in these steps can be realized. The individual terms in the expansion of a contracted Gaussian are referred to as primitive Gaussian functions.

There are several standard contracted Gaussian basis sets in use today, including the STO-nG, 6-31G, and 6-311G basis sets.[37-39] Dunning's double-zeta and valence double-zeta contractions of Huzinaga's atomic Gaussian basis sets[40-42] and the powerful, general contraction scheme of Raffenetti[43] which greatly simplifies the derivation of contracted basis sets, especially for atoms beyond the first row of the periodic chart (the contraction coefficients are obtained directly from atomic RHF calculations). For a discussion and bibliography of contracted Gaussian basis sets see Dunning and Hay.[42]

The choice of basis functions for use in a calculation can have a profound effect on the accuracy of the results. To obtain an appreciation of the limitations imposed by the choice of basis functions, let us consider the results of full CI calculations on the $H + H_2$ reaction (Table 2). From the results summarized there, we see that:

1. The minimum basis set, a single 1s function on each atom denoted by [1s], results in a barrier to reaction greater than 23 kcal/mol which is in error by ~15 kcal/mol. A minimum basis set will clearly be of limited utility for the calculation of reactive potential energy surfaces.
2. The barrier height is reduced to 14 kcal/mol by doubling the size of the basis set to two 1s functions on each atom; this basis set is denoted by [2s].
3. Adding a 2p-polarization function to the [2s] basis set, denoted [2s1p], decreases the barrier to reaction to ~11 kcal/mol; this is in error by only ~1.5 kcal/mol.

For a general reactive system then, a polarized double-zeta basis set can be expected to be required to obtain a barrier height with an accuracy of a few kilocalories per mole. (The difference noted in the barrier height obtained with the Slater and Gaussian [2s1p] sets is not due as much to the inherent differences in the two basis sets as to the greater degree of exponent optimization in the Slater basis set.) Going beyond the [2s1p] basis set we see that:

Table 2
CALCULATED GEOMETRIES AND BARRIER HEIGHTS FOR THE BARRIER TO THE $H + H_2$ REACTION; FULL CI CALCULATIONS

	R_{sp} (HH) (Å)	$\Delta E_{barrier}$ (kcal/mol)
Slater basis sets		
[1s]	1.00	23.4
[2s]	0.95	14.0
[2s1p]	0.93	11.0
Gaussian basis sets		
[2s1p]	0.93	11.6
[3s2p]	0.93	10.4
[3s2p1d]	0.93	9.9
Slater basis		
[4s3p2d]	0.93	9.8

4. The error in the polarized valence double zeta basis set is reduced by a third upon expanding the (sp)-basis to [3s2p]; the calculated barrier height is now 10.4 kcal/mol.
5. There is a further reduction of ~0.5 kcal/mol upon adding a 3d-polarization function to the basis set. Furthermore, the result obtained with the [3s2p1d] basis set is within 0.1 kcal/mol of that obtained by Liu[10] with a much larger Slater basis set, [4s3p2d], and is within 0.22 ± 0.12 kcal/mol of the exact result as estimated by Liu.

Thus, to obtain a barrier height for the $H + H_2$ reaction to chemical accuracy, i.e., with an error less than 1 kcal/mol, requires a large ([3s2p] or [3s2p1d]) basis set.

III. *AB INITIO* CHARACTERIZATION OF POTENTIAL ENERGY SURFACES FOR CHEMICAL REACTIONS[43a]

Of the 3N nuclear coordinates in the nuclear Schrödinger equation, three describe the translational motion of the center of mass of the molecular system while another three describe overall rotation of the system (for a nonlinear molecule). Thus, the molecular potential energy surface (V_{ieff} (**R**)) is a parametric function of only 3N-6 internal nuclear coordinates (again for a nonlinear system) which, for convenience if not for clarity, we will also denote as **R**. A contour plot of the collinear $H + H_2$ potential energy surface is given in Figure 4; for the definition of the coordinate system, see Figure 1.

Because of the parametric dependence of the molecular potential energy surface on **R**, requiring the solution of the electronic Schrödinger equation at a number of points, the determination of such surfaces presents a formidable problem. To obtain an appreciation of the magnitude of the problem, note that if ten points are required to adequately represent the dependence of the energy on one of the nuclear parameters, specification of the potential energy surface for a triatomic system with 3N-6 = 3 degrees of freedom would require 1000 calculations, while for a tetratomic system with 3N-6 = 6 degrees of freedom one million calculations would be required. Clearly, the problem rapidly becomes intractable.

Fortunately, not all regions of the potential energy surface are of equal importance. For chemical reactions the regions of most importance are minima and saddle points and the paths which interconnect them. As stationary points on the potential energy surface, all of the first derivatives of the energy, i.e., the gradient, vanish at minima and saddle points.

$$q_i = \partial V/\partial R_i = 0 \qquad i = 1, \ldots, 3N - 6 \tag{43}$$

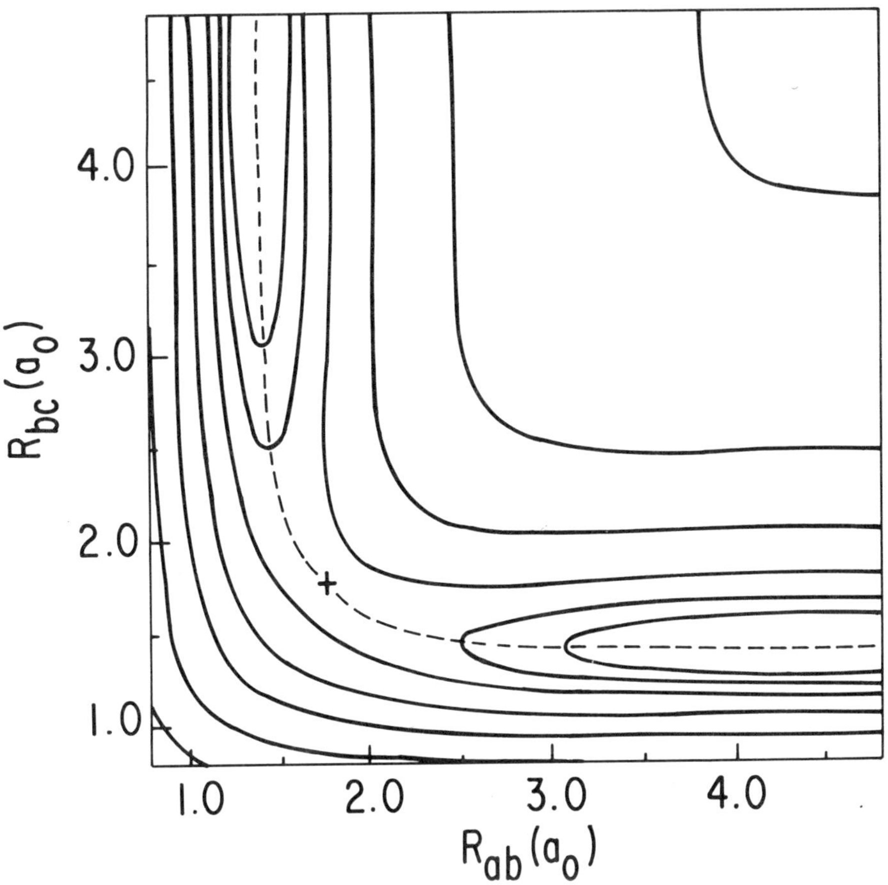

FIGURE 4. Contour plot of the collinear potential energy surface for the $H + H_2 \rightarrow H_2 + H$ reaction. Contour lines are drawn at 3, 6, 12, 24, 48, 96, and 192 kcal/mol. The steepest descent reaction path is given by the dashed line; the saddle point is indicated by an x.[45] This surface is an analytic representation of the calculations of Siegbahn and Liu.[46]

(Henceforth we will omit the "ieff" subscript from the potential energy function). Minima may correspond to either global or local minima on the potential energy surface and thus to either stable or quasistable molecular species; the reactants and products of a chemical reaction correspond to minima on the potential energy surface. For minima, the eigenvalues of the matrix of second derivatives (the Hessian) matrix:

$$k_{ij} = \partial^2 V/\partial R_i \partial R_j \tag{44}$$

must all be positive, corresponding to a "valley" or "well" on the potential surface. For simple saddle points there must be one and only one negative eigenvalue of the curvature matrix.[47] Saddle points correspond to "mountain passes" connecting one stable molecular species with another — it corresponds to the barrier which must be surmounted for the chemical reaction to occur. The saddle point for the $H + H_2$ reaction is indicated by an "x" in Figure 4.

The energy defect of a chemical reaction and the barrier height is defined by:

$$\Delta E_{rxn} = V_{prod} - V_{react} \tag{45a}$$

$$\Delta E_{barrier} = V_{sp} - V_{react} \tag{45b}$$

where V_{sp} is the electronic plus nuclear repulsion energy at the saddle point, and V_{react} and V_{prod} are the corresponding quantities for the reactants and products, respectively; $\Delta E_{barrier}$ is sometimes referred to as the "classical" barrier to reaction. For an exoergic chemical reaction V_{prod} will be more negative than V_{react} so that ΔE_{rxn} will be negative.

Perhaps the most important characteristics of both minima and saddle points are the vibrational frequencies, the normal modes describing the associated nuclear motion, and the zero point energy. The simplest procedure for determining the normal modes[48] employs the Hessian matrix (Equation 44) in cartesian coordinates, i.e., where the R_i and R_j in Equation 44 refer to all 3N cartesian coordinates of the atoms with respect to an (arbitrary) origin (usually taken to be the center of mass). The Hessian matrix in cartesian coordinates can be obtained from the Hessian matrix in internal coordinates (of which there are only 3N-6) by numerical differentiation. To obtain the normal modes, the atomic cartesian Hessian matrix is first transformed to mass-weighted coordinates, i.e., if R_{iA} is a cartesian coordinate of atom A and R_{jB} is a cartesian coordinate for atom B, then the mass-weighting involves dividing k_{iAjB} in Equation 44 by the square root of the product of the masses of atoms A and B. Diagonalization of this mass-weighted Hessian matrix then yields the normal modes.

For a nonlinear polyatomic molecule, six of the eigenvalues of the mass-weighted Hessian matrix in cartesian coordinates will be zero. These correspond to the translational and rotational modes of the system. The 3N-6 nonzero eigenvalues of the Hessian matrix are the squares of the normal mode vibrational frequencies (ω_i). Thus, for minima, all of the nonzero eigenvalues are positive and all of the frequencies are real. The eigenvectors for the nonzero eigenvalues are the corresponding normal modes. For saddle points, on the other hand, one of the eigenvalues will be negative and, therefore, the associated frequency will be imaginary (ω_{rxn}), and the corresponding normal mode is the reaction coordinate.

The zero point energy of the system at a given stationary point is simply defined as

$$E_{zpe} = {}^1\!/_2 \sum_i h\omega_i \tag{46}$$

where the sum runs over all of the (3N-6) modes (3N-5 real modes at the saddle point). The enthalpy change at 0 K and the vibrationally adiabatic threshold for a chemical reaction are defined by:

$$\Delta H_0 = \Delta E_{rxn} + \Delta E_{zpe}^{prod-react} \tag{47a}$$

$$\Delta E_{vat} = \Delta E_{barrier} + \Delta E_{zpe}^{sp-react} \tag{47b}$$

where

$$\Delta E_{zpe}^{i-j} = E_{zpe}^{i} - E_{zpe}^{j}$$

and E_{zpe}^{sp} is the zero point energy of the saddle point, and E_{zpe}^{react} and E_{zpe}^{prod} are the corresponding quantities for the reactants and products, respectively. The heat of the reaction at 0 K is ΔH_0. If the vibrational levels of the reactants continuously and smoothly evolve into those at the saddle point, the barrier to reaction is ΔE_{vat}.

An alternative procedure for evaluating the normal mode frequencies bypasses the atomic cartesian Hessian matrix and uses the internal coordinate Hessian matrix directly. In this method the mass factors are included in a second matrix referred to as the G matrix.[49] While this method minimizes the arithmetic involved in the calculation, it is at the expense of increased algebraic complexity. For this reason, use of the atomic cartesian Hessian matrix is the preferred method when a computer is available to handle the arithmetic.

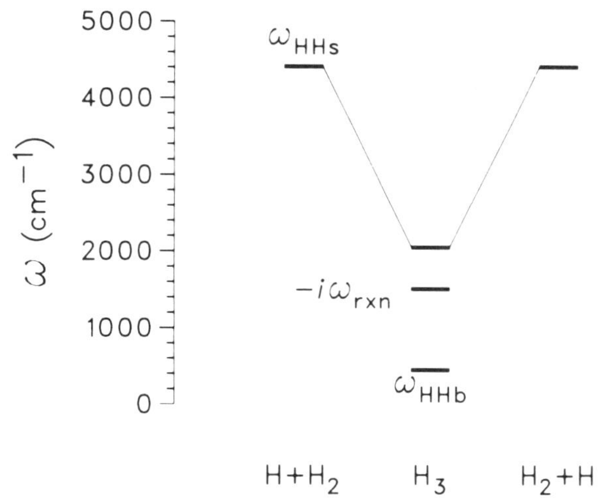

FIGURE 5. Vibrational energy levels of the H_3 system in the reactants ($H + H_2$) at the saddle point and in the products ($H_2 + H$).

Figure 5 shows the vibrational frequencies of the reactants, saddle point, and products for the $H + H_2$ reaction. For the reactants and products there is only one frequency, corresponding to the HH stretch of molecular hydrogen (4400 cm^{-1}). At the saddle point there are a total of four nonzero normal-mode frequencies. Two of these are the degenerate bending mode of the H_3 saddle point (455 cm^{-1}). The remaining two frequencies are the symmetric and asymmetric combinations of the two HH stretches. The symmetric stretch corresponds to motion along the top of the saddle point shown in Figure 4, and is found to have a frequency (2050 cm^{-1}) that is substantially less than that of the reactant H_2. The asymmetric stretch corresponds to motion over the barrier and is calculated to have a normal mode frequency of 1510.i cm^{-1}. The magnitude of the imaginary frequency is directly related to the width of the barrier and thus to the rate at which quantum mechanical tunneling can occur. The larger the frequency, the narrower the barrier.

The large decrease in the symmetric stretching frequency relative to that of the reactants results in a total zero point energy for the saddle point that is 2.1 kcal/mol less that of the reactants. Thus, for the $H + H_2$ reaction with $\Delta E_{barrier} \simeq 9.7$ kcal/mol, the calculated vibrationally adiabatic threshold is $\Delta E_{vat} \simeq 7.6$ kcal/mol.

A. Locating Stationary Points on Potential Energy Surfaces

A number of methods have been developed for locating minima and saddle points, the two most common techniques being energy mapping and gradient minimization methods. In the first method, an approximate location of the stationary point is estimated. The energy is calculated for a grid of points in the region of the approximate stationary point. The energy is then expanded as a polynomial function of the internal nuclear coordinates and a second approximation to the stationary point is determined by calculating the geometry for which all of the first derivatives (gradients) of the polynomial function are zero (Equation 43). If the new stationary point lies inside the presently determined grid of points, the search is stopped; otherwise, the procedure is reinitiated with the new guess for the stationary point. Of course, even if the new stationary point is located inside the grid, the grid may be refined and the procedure reinitiated. Usually, the polynomial function is limited to quadratic terms. This procedure works well for minima. The location of saddle points, however, requires a good initial guess for the stationary point; in fact, experience suggests that the saddle point must be located inside the grid for the method to be useful, i.e., extrapolation of the local polynomial representation of the surface in such cases is of limited usefulness. Determination of a suitable initial guess for the geometry of a saddle point can also be troublesome, although for some simple reactions a suitable initial guess can be obtained by varying only a subset of the **R**, thus simplifying the multidimensional search.

A simplified version of the above procedure often used to determine the geometry of minima cyclically optimizes the various internal nuclear coordinates. In this procedure, estimates are made for all but one of the internal coordinates. The optimum value of this coordinate is then determined by minimizing the energy along this coordinate (again, often using a polynomial expansion). Using this value as the estimate for the selected coordinate, a new coordinate is chosen and the procedure repeated. This continues until the changes observed in the internal coordinates fall below some threshold. This technique requires the energy to be a minimum for variations in all of the internal coordinates; it thus cannot be used to determine the geometry of saddle points.

In the past, saddle points have also been located by assuming that they lie along one of the internal coordinates. The energy along this coordinate is then computed with or without optimization of the remaining internal coordinates until a maximum is located. This point is then taken to be the saddle point. However, this procedure, referred to as the "distinguished reaction coordinate" method, while useful in some cases, often fails to locate a saddle point.

In the gradient minimization technique a multidimensional Newton-Raphson method is used to determine the zeros of the gradient.[50-54] The energy is expanded as a quadratic function of the internal nuclear coordinates about an approximation to the stationary point (R_O):

$$E(\mathbf{R}) = E_0 + \sum_i (\partial E/\partial R_i)_0 \Delta R_i + \sum_i \sum_j (\partial^2 E/\partial R_i \partial R_j)_0 \Delta R_i \Delta R_j \tag{48}$$

where

$$\Delta R_i = R_i - R_{i0}$$

In vector notation Equation 48 becomes:

$$E = E_0 + \mathbf{q} \cdot \mathbf{\Delta R} + {}^1/_2\, \mathbf{\Delta R} \cdot \mathbf{k} \cdot \Delta R \tag{49}$$

For a stationary point, Equation 43 must be satisfied, which leads to:

$$\mathbf{\Delta R} = -\mathbf{k}^{-1} \cdot \mathbf{q} \tag{50}$$

The gradient vector and the Hessian matrix is then computed at the new point, $R_O + \Delta R$) (the gradient will not be zero at the new point unless the surface was truly quadratic in the region) and the process is repeated until **q** and/or $\Delta\mathbf{R}$ falls below some preset threshold.

To increase the stability of the Newton-Raphson procedure, the length of the correction vector (Equation 50) can be adjusted by determining the point along the correction vector for which the norm of the gradient:

$$|\mathbf{q}|^2 = \mathbf{q} \cdot \mathbf{q} = \sum_i (\partial E/\partial R_i)^2$$

is a minimum. That is, $|\mathbf{q}|$ is evaluated at a number of points along the direction of the correction vector and the new point is taken to be the point at which $|\mathbf{q}|^2$ is a minimum. This procedure takes into account the nonquadratic nature of potential energy surfaces and ensures that the new point is the optimum point along the correction vector at which to initiate the next iteration. It does require, however, additional evaluations of the gradient. Other variants of this procedure, a combination of the Newton-Raphson method and a line search, are possible.

As with the energy mapping methods discussed above, gradient minimization methods also require an initial guess for the geometry of the stationary point, which can again prove troublesome for saddle points. Gradient minimization techniques can also lead to both stable and saddle points (and in fact also to other stationary points, e.g., double saddle points, maxima, etc.) and thus care must be exercised to ensure convergence to the desired stationary point. In general, for convergence to the desired stationary point the initial guess must be in a region of the surface for which the Hessian matrix has the proper number of positive and negative eigenvalues, i.e., all positive eigenvalues for a minimum and only one negative eigenvalue for a simple saddle point. This is a consequence of the step involving the minimization of the norm of the gradient, an inherently positive quantity.[53] Techniques have been suggested for forcing convergence to the desired stationary point even if the initial guess is outside the radius of convergence as defined above.[53]

In Equations 49 and 50, the computation of the matrix of second derivatives (k_{ij}) is time-consuming and therefore, in most applications, it is not computed directly; it is estimated from the changes in the gradient. Various methods have been developed to update the Hessian matrix based on the changes in the gradient.

With the development in recent years of methods for calculating the gradients of electronic wavefunctions analytically (Section III.C), the direct determination of stable and saddle points of potential energy surfaces using gradient techniques has become an active area of research (see, e.g., References 55 and 56).

More recently, techniques have been published for determining saddle points starting from the associated minima.[57] These techniques provide a systematic means of locating saddle points. However, they require the explicit evaluation of both the first and second derivatives of the energy at each point along the path leading from the minima to the saddle point. Application of these techniques will thus require efficient means of analytically evaluating both the first and second derivatives of the energy with respect to the internal nuclear coordinates for electronic wavefunctions (Section III.C).

Recent work on the general properties of saddle points[58-60] has shown that there are symmetry relationships connecting a saddle point and the two associated minima. For example, it can be shown that the symmetry of the transition state can be no greater than the symmetry of the two minima which it connects. Thus, linear saddle points must connect to linear minima, planar saddle points must connect to planar minima, etc. An exception arises when the two stable points are related by symmetry. Then the saddle point may have an additional symmetry element, the element which interchanges the two stable points. Examples of this include the $H + H_2$ exchange reaction (Figure 4). In this case, the saddle point possesses inversion symmetry, a symmetry not present in either reactants or products; the symmetry of the state at the saddle point is $^2\Sigma_u^+$. As the saddle point is also collinear, it can be characterized by only one internal nuclear coordinate, $R_{sp}(HH)$ [$= R_{sp}(H_aH_b) = R_{sp}(H_bH_c)$], the HH bond length.

While the above considerations place conditions on the minima to which saddle points are connected, they do little to aid in the location of saddle points given the two minima.

B. Characterization of Reaction Paths on Potential Energy Surfaces

Until recently, the most common technique for specifying the reaction path for a chemical reaction was to choose as the reaction coordinate a "distinguished" internal coordinate, a bond length, bond angle, etc. which connects the saddle point with the reactants and the products in some obvious physical manner (different coordinates were usually used in the two directions: reactants to saddle point, saddle point to products). Stepping along this pseudoreaction coordinate in the direction of the reactants or products, the remaining geometrical parameters were optimized at each step. The changes in these coordinates, along with the "distinguished" coordinate, then defined the reaction path for the reaction. Once the reaction path was known, the energy profile along this path could be computed; the barrier to reaction is then the highest energy point along the energy profile. We shall refer to this path as the "distinguished reaction coordinate" path.

There are a number of well–documented difficulties with this path. For example, it often has a "kink" at the saddle point. Consider, e.g., the three-atom abstraction reaction:

$$A + BC \rightarrow AB + C$$

If the reaction path is collinear, the energy can be written as a function of only two variables R_{AB} and R_{BC}, i.e., $E(R_{AB},R_{BC})$. If we denote the coordinates of the saddle point as $[R_{sp}(AB), R_{sp}(BC)]$, then the distinguished reaction coordinate (DRC) path is defined by:

Reactants' region:

$R_{AB} > R_{sp}(AB)$ $R_{BC} < R_{sp}(BC)$	The reaction path is taken to be (R_{AB},R_{BC}^{min}), where R_{BC}^{min} is chosen to minimize the energy along R_{BC}.

Products' region:

$R_{BC} > R_{sp}(BC)$ $R_{AB} < R_{sp}(AB)$	The reaction path is taken to be (R_{AB}^{min},R_{BC}), where R_{AB}^{min} is chosen to minimize the energy along R_{AB}.

McCullough and Silver[61] have shown for such a path that at the saddle point

$$(dR_{BC}/dR_{AB})_{spR} > (dR_{BC}/dR_{AB})_{spP}$$

where spR denotes the limit of the derivative at the saddle point in the reactants' region and spP denotes the corresponding limit in the products' region. That is, at the saddle point, the derivative of the DRC path is discontinuous, or the path possesses a kink at that point.

The above will be true whenever there is a discontinuous change in the internal coordinate used to characterize the DRC path in the reactants' and products' regions. For an addition reaction,

$$A + BC \rightarrow ABC$$

or an isomerization reaction,

$$ABC \rightarrow BCA$$

a single internal coordinate may well be appropriate in both the reactants' and products' regions, e.g., in the addition reaction, R_{AB} may suffice in both regions (from A + BC to the saddle point and from the saddle point to ABC). In these cases, the above arguments do not necessarily hold; however, as we shall see in the examples discussed in Section IV, other difficulties can arise for multidimensional surfaces.

The steepest descent reaction path starts at the saddle point and descends into the reactants' or products' regions following the steepest descent path. The steepest descent path follows the negative of the gradient of the potential. To see this, again consider the above collinear problem. Let the point (R_{AB0}, R_{BC0}) be a point on the reaction path arbitrarily close to, but not at, the saddle point. Expanding the energy about this point and retaining only the linear terms in the expansion:

$$E(R_{AB},R_{BC}) = E(R_{AB0},R_{BC0}) + (\partial E/\partial R_{AB})_0 \Delta R_{AB} + (\partial E/\partial R_{BC})_0 \Delta R_{BC} \tag{51}$$

we wish to take a step of fixed length:

$$\Delta R = \sqrt{\Delta R_{AB}^2 + \Delta R_{BC}^2}$$

such that

$$\Delta E = E(R_{ABO} + \Delta R_{AB}, R_{BCO} + \Delta R_{BC}) - E(R_{ABO},R_{BCO})$$

decreases by the maximum amount possible, i.e., we take the steepest descent path down from the saddle point. To determine the direction of this step, set:

$$\Delta R_{AB} = \Delta R \sin\theta$$

$$\Delta R_{BC} = \Delta R \cos\theta$$

Substituting the above into Equation 51 we obtain:

$$\Delta E = [(\partial E/\partial R_{AB})_0 \sin\theta + (\partial E/\partial R_{BC})_0 \cos\theta]\Delta R$$

For this to be a maximum the quantity in brackets must be a maximum, which yields:

$$\tan\theta_{sd} = (\partial E/\partial R_{AB})_0/(\partial E/\partial R_{BC})_0$$

Thus, the direction of the step

$$\frac{\Delta R_{AB}}{\Delta R_{BC}} = \frac{(\partial E/\partial R_{AB})_0}{(\partial E/\partial R_{BC})_0} \tag{52}$$

is proportional to the gradients in the two directions and the steepest descent path is in the direction of the negative of the gradient. The steepest descent path passes continuously through thc saddlc point from reactants to products,[61] i.e., it is "kinkless". At the saddle point the gradient, of course, vanishes. Thus, the above procedure does not define the direction of the step at the saddle point. However, expanding the energy as a quadratic function of (R_{AB}, R_{BC}) about [R_{sp}(AB), R_{sp}(BC)], it is straightforward to show that the initial direction of the steepest descent path is in the direction of the eigenvector of the Hessian matrix having the negative eigenvalue.

In summary, to determine the steepest descent path we first calculate the eigenvectors of the Hessian matrix at the saddle point and take a step in the direction of the eigenvector having the negative eigenvalue. At the new point we calculate the gradient and take another step in the direction of the negative of the gradient. This sequence of steps is continued until the reactants or products are reached. The resulting changes in the geometrical param-

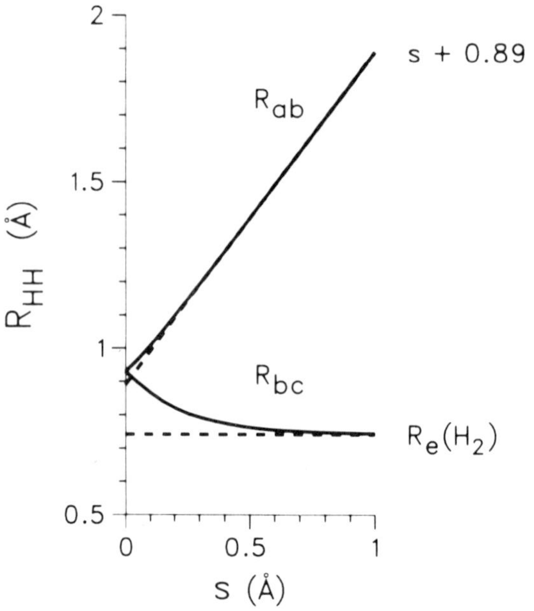

FIGURE 6. Steepest descent reaction path for the H + $H_2 \rightarrow H_2$ + H reaction. The coordinate system is given in Figure 1.[10] S is the distance along the reaction path starting at the saddle point.

eters define the reaction path; the energy profile along this reaction path is obtained as a by-product of this procedure.

The steepest descent path for the H + H_2 reaction is illustrated in Figures 4 and 6. The path plotted in Figure 4 is characteristic of simple thermoneutral abstraction or exchange reactions. At large R_{ab} (reactants' region) the reaction path is parallel to the R_{ab} axis and R_{bc} corresponds to the equilibrium H_2 bond distance. Below $R_{ab} \sim 1^1/_2$ Å, however, further shortening of the H_aH_b distance leads to a noticeable lengthening of the H_bH_c distance. This lengthening becomes more and more pronounced (increasing curvature of the reaction path) until the saddle point is reached at $R_{ab} = R_{bc} = 0.93$ Å. After passing the saddle point, R_{bc} continues to lengthen while R_{ab} continues to shorten. At large R_{bc} (products' region), the reaction path becomes parallel to the R_{bc} axis and R_{ab} approaches the H_2 bond length.

Another view of the reaction path is presented in Figure 6 where R_{ab} and R_{bc} are plotted as a function of s, the distance along the reaction path in the entrance (reactants') channel. The s is obtained as a line integral along the path with s = 0 at the saddle point. Here it is seen that R_{ab} approaches its asymptotic value (s + 0.89 Å) much faster than does R_{bc} [$R_e(H_2) = 0.74$ Å]. If there were no curvature in the reaction path R_{ab} would be simply given by s + 0.93 Å. Thus, the difference (0.93 − 0.89 = 0.04 Å) is one measure of the curvature in the reaction path.

The energy along the steepest descent path for the H + H_2 reaction is plotted in Figure 7. The calculated barrier height for this reaction, the maximum in the energy profile, is 9.8 kcal/mol. Given in the inset is the variation of the barrier height with bond angle $\theta = \theta H_aH_bH_c$. The angle $\theta = 180$ corresponds to the collinear configuration, and the barrier plotted in the inset is referenced to the barrier height at the collinear geometry. The dependence of the energy on angle was determined by locating for each angle (θ) the point in (R_{ab}, R_{bc}) space where the gradient vanished.

Although the above procedure for determining the steepest descent path is straightforward, the path obtained thereby tends to wobble around the "true" steepest descent path unless

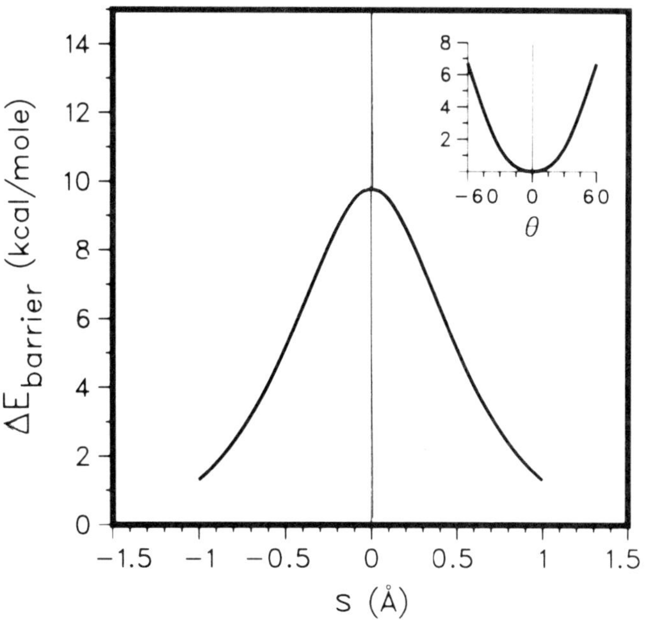

FIGURE 7. Energy profile along the steepest descent reaction path for the $H + H_2 \rightarrow H_2 + H$ reaction. The dependence of the barrier height on the bond angle (θ) is plotted in the inset; see the text.[10,46]

very small step sizes (△R) are used. The resulting error can be reduced by reducing the step size; however, this is often not practical. Ishida et al.[62] have proposed an alternate scheme for correcting this wobble, requiring additional energy (but not gradient) calculations at each step along the path.

While the steepest descent path provides a smooth, physically reasonable path connecting reactants to products in a chemical reaction, its definition depends on the coordinate system used. For example, for a three-atom reaction the steepest descent path in the (R_{AB}, R_{BC}, R_{CA}) coordinate system will not necessarily be equivalent to that in the (R_{AB}, R_{BC}, θ_{ABC}) coordinate system. In fact, it can be shown[63] that unless the two coordinate systems are related by a unitary transformation, the steepest descent paths will not be the same.

Fukui[64] (see also the review[65]) has proposed the use of the "intrinsic reaction coordinate" to describe the reaction path connecting reactants to products (it would be more consistent to refer to it as the "intrinsic reaction path" (IRP) which we shall do). The IRP is the steepest descent path connecting the transition state with the reactants and products in a mass-weighted coordinate system such as that used to define the vibrational energies and normal modes. In this coordinate system the kinetic energy is in diagonal form; this path describes the path that the reaction would take if the local speed were continuously damped to zero. It is thus intimately related to the dynamics of the reaction; in fact, one approach to treating the dynamics of a chemical reaction involves a description in terms of motion along the IRP coupled with harmonic motion along the degrees of freedom perpendicular to the IRP. A further advantage is that the IRP is independent of the choice of coordinate system.

In one of the first applications of this concept to a chemical reaction, Truhlar and Kuppermann[66] reported the IRP for the $H + H_2$ reaction using the surface calculated by Shavitt et al.[44] These authors also explored the dependence of the steepest descent path on the coordinate system used.

C. Analytic Derivatives of Electronic Wavefunctions

As is clear above, the derivatives of the energy with respect to the nuclear coordinates, especially the first derivative or gradient, are most useful in locating saddle points and in determining reaction paths. The simplest procedure for calculating the derivatives of the electronic energy is with numerical procedures involving the calculation of the energy at a grid of different nuclear geometries. The number of points needed to determine the first derivatives is proportional to the number of internal degrees of freedom (3N-6), while the number needed for second derivatives is proportional to $(3N\text{-}6)^2$. An alternative procedure is to differentiate the energy expression directly and thereby determine the derivatives from a calculation at a single geometry. It is now feasible to analytically determine both the first and second derivatives of HF, MCSCF, and certain CI wavefunctions, and it has been found that, for polyatomic systems, analytic differentiation is considerably more efficient and accurate than numerical differentiation. In the following discussion we follow the notation of Pople et al.[67]

Consider as an example the calculation of analytic first derivatives of a closed-shell RHF wavefunction. Differentiating the RHF energy expression with respect to a nuclear coordinate leads to the equation:

$$\frac{\partial E^{RHF}}{\partial R_i} = \sum_{\mu}\sum_{\nu} P_{\mu\nu}\frac{\partial h_{\mu\nu}}{\partial R_i} + {}^1\!/_2 \sum_{\mu}\sum_{\nu}\sum_{\lambda}\sum_{\sigma} P_{\mu\nu}P_{\lambda\sigma}\frac{\partial}{\partial R_i}\left[(\mu\nu|\lambda\sigma) - {}^1\!/_2(\mu\lambda|\nu\sigma)\right]$$

$$+ \frac{\partial V_{NN}}{\partial R_i} + \sum_{\mu}\sum_{\nu}\frac{\partial P_{\mu\nu}}{\partial R_i} h_{\mu\nu} + \sum_{\mu}\sum_{\nu}\sum_{\lambda}\sum_{\sigma}\frac{\partial P_{\mu\nu}}{\partial R_i} P_{\lambda\sigma} \quad (53)$$

$$\left[(\mu\nu|\lambda\sigma) - {}^1\!/_2(\mu\lambda|\nu\sigma)\right]$$

The first three parts of this equation involve derivatives of the one- and two-electron integrals and the nuclear-nuclear repulsion. The last two terms involve derivatives of the density matrix ($P_{\mu\nu}$). Taking the last two terms first, these can be rewritten in terms of RHF matrix elements ($F_{\mu\nu}$):

$$\sum_{\mu}\sum_{\nu}\sum_{j}\left(\frac{\partial C^*_{\mu j}}{\partial R_i}\right) h_{\mu\nu}C_{\nu j} + \sum_{\mu}\sum_{\nu}\sum_{\lambda}\sum_{\sigma}\sum_{j}\left(\frac{\partial C^*_{\mu j}}{\partial R_i}\right)P_{\lambda\sigma}[(\mu\nu|\lambda\sigma)$$

$$- {}^1\!/_2(\mu\lambda|\nu\sigma)]\, C_{\nu j} + \text{complex conjugate} \quad (54)$$

$$= \sum_{\mu}\sum_{\nu}\sum_{j}\left(\frac{\partial C^*_{\mu j}}{\partial R_i}\right) F_{\mu\nu}C_{\nu j} + \text{complex conjugate}$$

$$= \sum_{\mu}\sum_{\nu}\sum_{j}\left(\frac{\partial C^*_{\mu j}}{\partial R_i}\right) \epsilon_i S_{\mu\nu}C_{\nu j} + \text{complex conjugate}$$

where the closed-shell RHF equation (Equation 35) was used to replace the RHF matrix elements with the product of orbital eigenvalues and overlap matrix elements. As can be seen, Equation 54 requires the evaluation of the derivatives of the coefficients of the basis functions, which are difficult to calculate. However, the RHF orbitals must satisfy an orthonormality constraint which can be written as follows:

$$\sum_{\mu}\sum_{\nu} C^*_{\mu i}S_{\mu\nu}C_{\nu j} = \delta_{ij} \quad (55)$$

Differentiating and rearranging Equation 55 gives an expression for the derivatives of the orbital coefficients in terms of derivatives of the overlap matrix:

$$\sum_{\mu}\sum_{\nu}\left(\frac{\partial C^*_{\mu i}}{\partial R_i} S_{\mu\nu}C_{\nu j} + C^*_{\mu i}S_{\mu\nu}\frac{\partial C_{\nu j}}{\partial R_i}\right) = -\sum_{\mu}\sum_{\nu} C^*_{\mu i}\frac{\partial S_{\mu\nu}}{\partial R_i} C_{\nu j} \quad (56)$$

Substituting this expression into Equation 54 and combining with the rest of the terms in Equation 53 then leads to an expression for the derivative of the RHF energy which does not involve derivatives of the orbital coefficients.

$$\frac{\partial E^{RHF}}{\partial R_i} = \sum_{\mu}\sum_{\nu} P_{\mu\nu}\left(\frac{\partial h_{\mu\nu}}{\partial R_i}\right) + {}^1/_2 \sum_{\mu}\sum_{\nu}\sum_{\lambda}\sum_{\sigma} P_{\mu\nu}P_{\lambda\sigma}\frac{\partial}{\partial R_i}\left[(\mu\nu|\nu\sigma) - {}^1/_2(\mu\lambda|\nu\sigma)\right]$$
$$+ \frac{\partial V_{NN}}{\partial R_i} - \sum_{\mu}\sum_{\nu}\sum_{j} \epsilon_j C^*_{\mu j}\left(\frac{\partial S_{\mu\nu}}{\partial R_i}\right) C_{\nu j} \quad (57)$$

Thus, the calculation of the first derivative of the RHF energy reduces to the calculation of the derivatives of the one- and two-electron integrals. The ease of calculating the integral derivatives depends on the nature of the basis functions. If these functions are Gaussian, then the integral derivatives are easily evaluated since the derivative of a Gaussian function is simply a sum of Gaussian functions, one of higher angular momentum, i.e., higher $l + m + n$ in Equation 42, and one of lower angular momentum. Therefore, integral programs capable of evaluating integrals involving, say s, p, and d Gaussians, can be relatively easily modified to evaluate the necessary derivative integrals for s and p Gaussians.

In general, it has been found that the cost of calculating the first derivative (gradient) of the RHF energy is about equal to the cost of the original RHF calculation itself. In contrast, the cost of evaluating the RHF first derivative numerically is, as noted above, 3N-6 times the cost of one RHF calculation. However, the calculation of the first derivatives of the RHF energy is greatly simplified by the fact that the derivatives of the orbital coefficients are not needed. This simplification also applies to the calculation of the first derivatives of MCSCF energies. However, it does not apply to the calculation of the first derivatives of CI energies nor to the second (and higher) derivatives of HF, MCSCF, or CI energies. In these cases, expressions for the derivatives of the energy involve derivatives of the wavefunctions in addition to derivatives of the integrals. The problem of calculating these derivatives has been solved using coupled perturbed Hartree-Fock theory (CPHF). For a discussion of the application of CPHF to the evaluation of energy derivatives see Pople et al.[67]

IV. POTENTIAL ENERGY SURFACES FOR CHEMICAL REACTIONS: A SELECTION OF EXAMPLES

In this section we present the results of a selection of *ab initio* calculations of potential energy surfaces for chemical reactions. The examples were chosen to be representative of general classes of reactions; no attempt has been made to provide a comprehensive listing of surfaces which have been calculated to date. We apologize to any colleague whose favorite reaction has been omitted.

The examples chosen have also been limited to reactions involving three- and four-atom systems. While these reactions cannot be considered representative of chemistry as a whole, as we shall see, they provide a wide diversity of reaction types. In addition, dynamical studies beyond simple transition state theories (TST) are presently limited to few body systems.

A. Three-Body Reactions

1. The H + HX Reactions: Trends in a Series of Reactions

For the reactions of hydrogen atoms with the hydrogen halides, two reaction pathways (channels) are possible. These are abstraction,

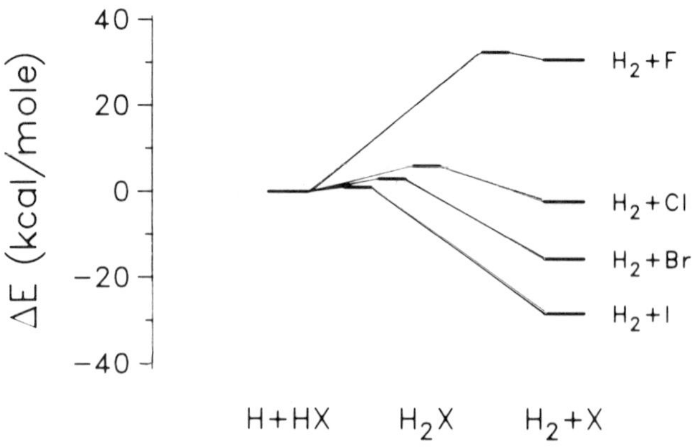

FIGURE 8. Schematic diagram of the energetics of the H + HX → H_2 + X abstraction reactions.

$$H + H'X \rightarrow HH' + X \tag{58a}$$

and exchange,

$$H + H'X \rightarrow HX + H' \tag{58b}$$

where the prime on the second hydrogen denotes the same or a different hydrogen isotope. The abstraction reactions vary from highly endoergic for reaction with hydrogen fluoride to highly exoergic for reaction with hydrogen iodide (Figure 8). The exchange reactions are essentially thermoneutral; they are, of course, exactly thermoneutral if the two hydrogen isotopes, H and H′, are identical.

Since the mid 1930s there have been numerous attempts to uncover relationships between the endo- or exoergicity of a reaction, the barrier to reaction, and the structure of the saddle point.[68-74] Dunning[75] has recently reported a series of calculations on the abstraction and exchange reactions for all of the hydrogen halide reactions for X = F − I. These calculations allow a detailed examination of the systematic trends observed in this prototypical series of homologous reactions. In this section we will examine the interrelationships found in these calculations.

The calculations of Dunning[75] used GVB and derived CI techniques with a polarized valence double-zeta basis set. We shall concentrate here on the POL-CI and GVB + 1 + 2 calculations. As we saw in Section II, for the H + H_2 reaction, the analogous calculations led to errors of 2 to 3 kcal/mol in the barrier height but yielded accurate saddle point geometries. For the H + HX reactions, since more electrons are involved, the errors would be expected to be larger, a fact borne out by the calculations. Although these calculations are not necessarily the most accurate calculations available (or possible) for these systems, they do provide a consistent description of all of the reactions of interest here. Other calculations on the H + H_2 reactions which should be noted include the abstraction and exchange pathways in H + HF,[17,76-80] the abstraction and exchange pathways in H + HCl,[17,79,81] and the exchange pathway in H + HBr.[82]

The orbital descriptions of the H + HX abstraction and exchange reactions are very much like that for the H + H_2 reaction. In the H + HX system there are, however, three distinct sets of orbitals:

1. The active orbitals, ($2\sigma_a$, $2\sigma_b$, 3σ) in Figures 9 to 11, which are those orbitals directly involved in the transfer of the bond pair from reactants to products; these orbitals are the counterparts of the ($1\sigma_a$, $1\sigma_b$, 2σ) orbitals in H + H_2.

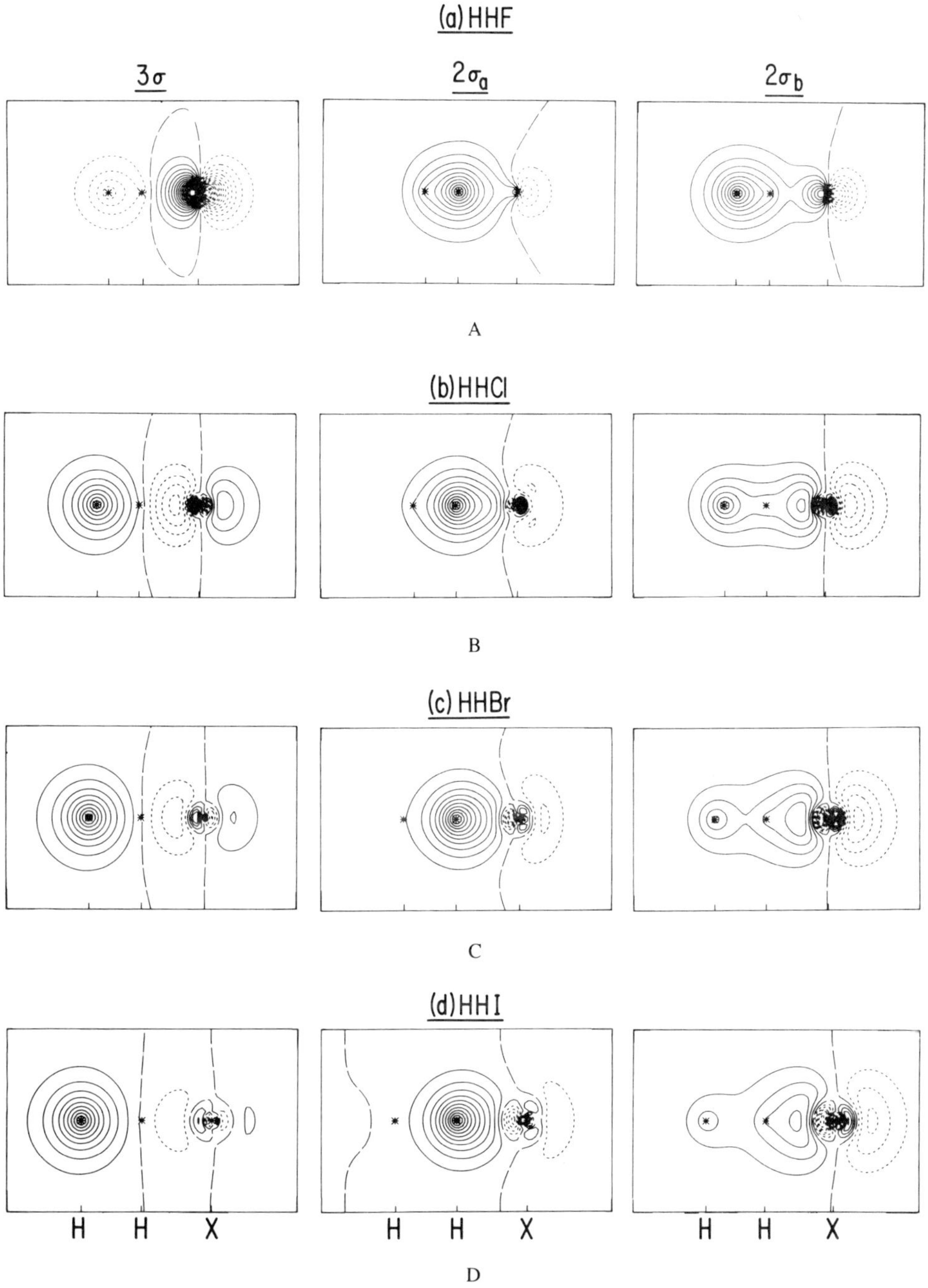

FIGURE 9. Active SOGVB orbitals for the H + HX abstraction reactions at the saddle point. For details on the plots, see the caption to Figure 3.

2. The semiactive orbitals, (1σ, 1π) in Figures 9 to 11 which are responsive to, but not directly involved in the bond transfer process; these are the lone pair orbitals on the halogen atoms.
3. The inactive orbitals, such as the inner shell (core) orbitals on the halogen, which have only a minor influence on the reaction energetics.

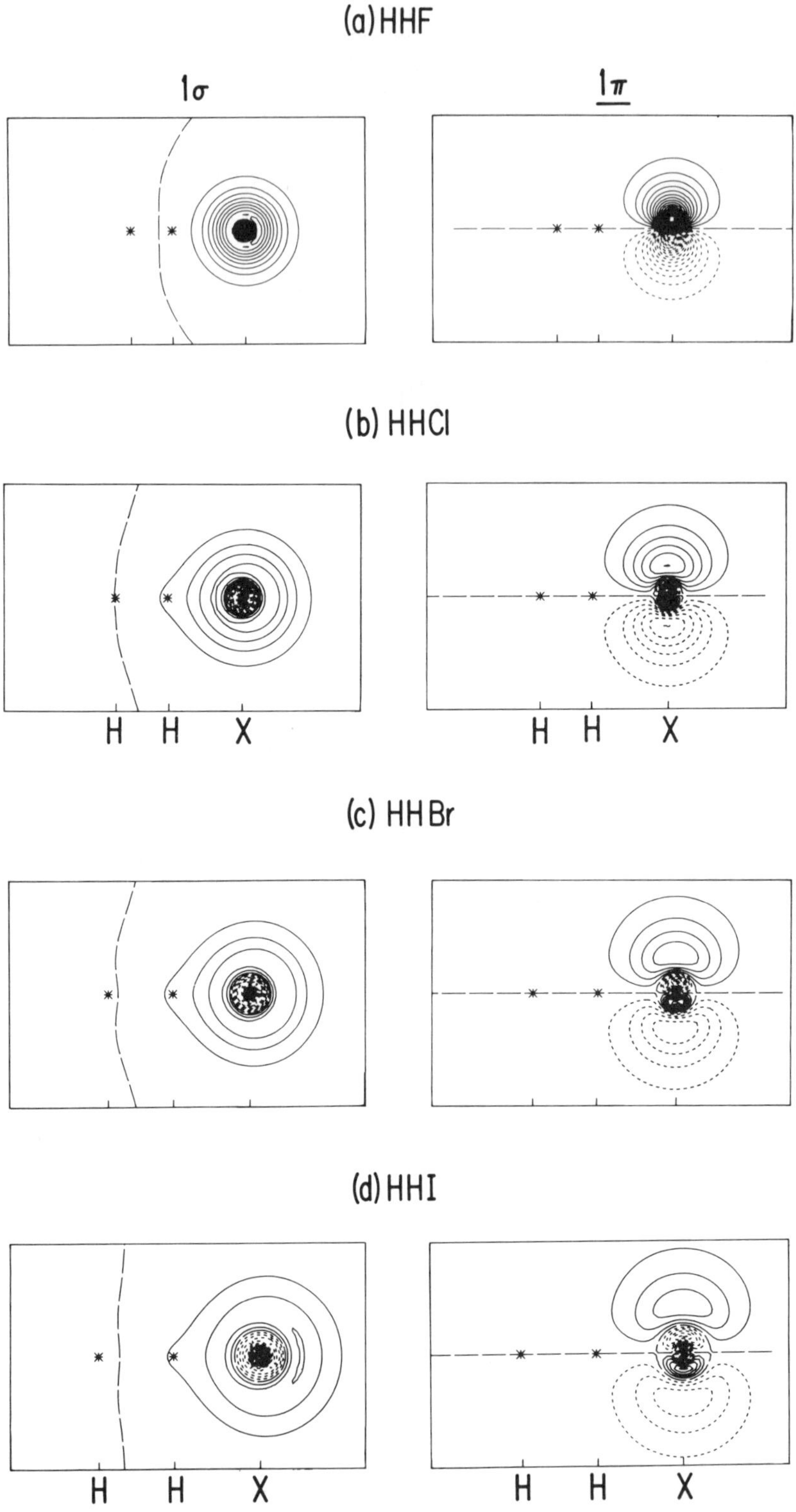

FIGURE 10. Semiactive SOGVB orbitals for the H + HX abstraction reaction at the saddle point. For details on the plots, see the caption to Figure 3.

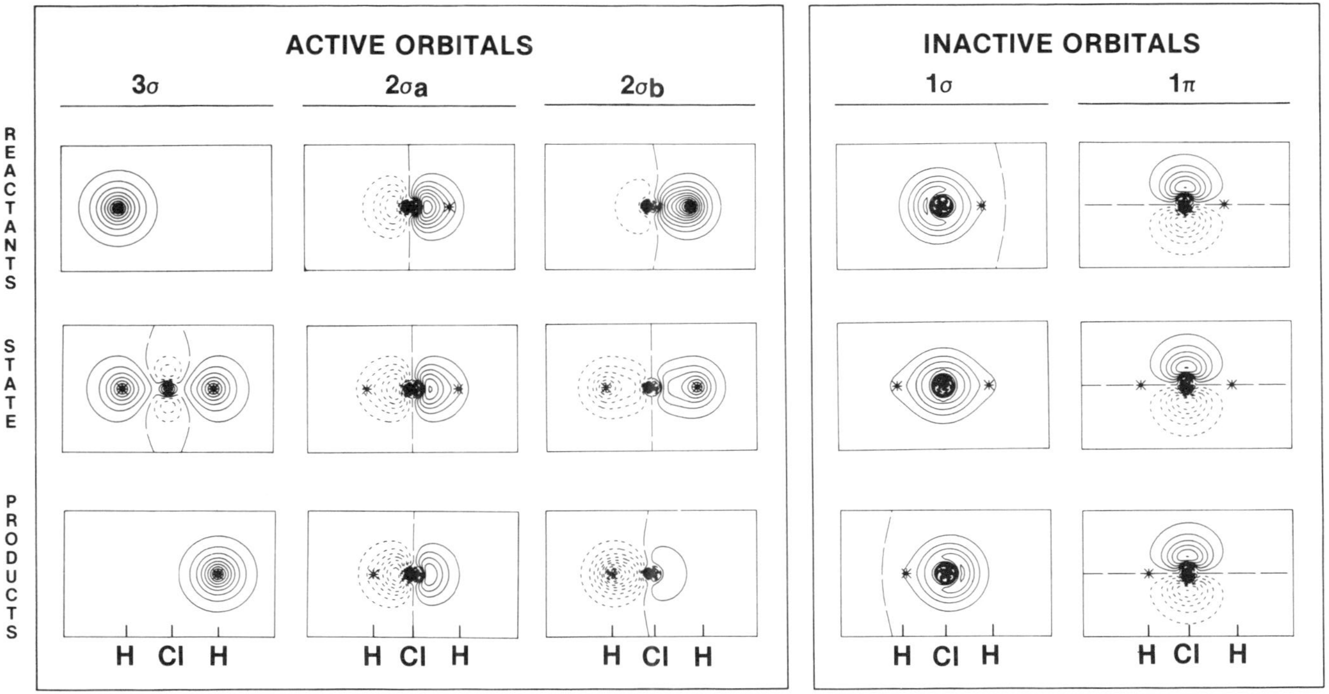

FIGURE 11. SOGVB orbitals for the H′ + H′X + H exchange reactions in the reactant (H′ + HX) saddle point (labeled transition state) and the product (H′X + H) regions. For details on the plots, see the caption to Figure 3.

Plots of the active and semiactive orbitals for the H + HX abstraction reactions at the saddle point are given in Figures 9 and 10. For the very endoergic H + HF reaction, the saddle point is in the products' channel and the singly occupied 3σ orbital resembles the $2p_z$ fluorine orbital and the ($2\sigma_a$, $2\sigma_b$) pair describes the HH bond. For the very exoergic H + HI reaction, on the other hand, the saddle point is in the reactants' channel and the 3σ orbital closely resembles a 1s hydrogen orbital and the ($2\sigma_a$, $2\sigma_b$) pair describes an HI bond. The active orbitals for the nearly thermoneutral H + HCl reaction and the exoergic H + HBr reaction are intermediate in character between these two extremes.

The semiactive orbitals for the H + HX abstraction reactions plotted in Figure 10 are seen to vary little from reaction to reaction, except for the changes expected with the halogen atom.

While the semiactive orbitals change only slightly during the course of the reaction, interactions between electrons in the active and semiactive orbitals can be energetically important. In the H + HX system, e.g., the differences in the energetics of the abstraction and exchange reactions can be largely attributed to interactions involving the active and semiactive orbitals. In the latter reaction the lone pair orbitals on the halogen atom (1σ and 1π in Figure 11) strongly interact with the incoming hydrogen atom. We would thus expect much higher barriers for exchange than for abstraction and, in fact, the calculations suggest differences of 10 to 20 kcal/mol. This interaction is expected to be most important for the 1σ orbital, which is seen to be polarized toward the attacking hydrogen atom in the reactants, to become symmetrical at the saddle point, and then to become polarized toward the leaving hydrogen atom in the products.

It should also be noted in Figure 11 that the 3σ orbital does not change phase between reactants and products. Rather the bonding orbitals $2\sigma_a$ and $2\sigma_b$ both change phase. This is a consequence of the asymmetric nature of the bonding orbital on the central (chlorine) atom (a $3p_z$-like orbital) and is still consistent with the Orbital Phase Continuity Principle.[15]

Let us consider first the calculated energy defects (ΔE_{rxn}) for the H + HX abstraction reactions:

$$\Delta E_{rxn} = D_e(HX) - D_e(H_2) \tag{59}$$

These are listed in Table 3 along with the experimental values obtained from the known bond energies (correction for spin-orbit effects is important here).[75] For the H + HX reactions we see that:

1. The POL-CI method provides the most consistent description of the overall energetics of the reactions with an average error of only 2.7 kcal/mol in the calculated energy defects.
2. The GVB + 1 + 2 method, although yielding much lower total energies and more accurate bond energies predicts energy defects in error by 4.2 kcal/mol on the average.

A review of the literature regarding the barrier heights inferred from experimental data is not possible in the limited space available here.[83] The values obtained in the calculations of Dunning[75] are consistent with errors of 3 to 4 kcal/mol in the barrier heights.

As can be seen in Table 3, there is a clear correlation between the barrier to reaction and the exoergicity of the reaction; namely, the barrier height decreases with increasing exoergicity. This correlation is quantified in Figure 12, where the dashed and solid lines are given by:

$$\Delta E_{barrier} = \Delta E_0 e^{-\alpha \Delta E_{rxn}} \tag{60}$$

Table 3
SADDLE POINT GEOMETRIES AND ENERGETICS OF THE H + HX ABSTRACTION REACTIONS AS OBTAINED FROM POL-CI AND GVB + 1 + 2 CALCULATIONS

	H + HF	H + HCl	H + HBr	H + HI
Saddle point geometries				
R_{sp} (HH)				
POL-CI	0.78 (0.02)	0.98 (0.21)	1.20 (0.44)	1.44 (0.68)
GVB + 1 + 2	0.79 (0.03)	1.01 (0.26)	1.26 (0.50)	1.46 (0.70)
R_{sp} (HX)				
POL-CI	1.60 (0.66)	1.46 (0.16)	1.52 (0.08)	1.68 (0.04)
GVB + 1 + 2	1.41 (0.49)	1.43 (0.14)	1.49 (0.06)	1.67 (0.04)
Barrier heights				
$\Delta E_{barrier}$				
POL-CI	36.5	8.2	4.7	2.8
GVB + 1 + 2	31.4	7.3	3.8	2.3
Energy defects				
ΔE_{rxn}				
POL-CI	33.6	−1.1	−12.6	−24.5
GVB + 1 + 2	24.0	−6.4	−19.0	−30.7
Exptl	30.8	−2.2	−15.6	−28.3

Note: Geometrics are in angstroms; energies are in kilocalories per mole; quantities in parentheses refer to bond extensions.

with

$$\Delta E_0 = 8.37 \text{ kcal/mol} \qquad \alpha = +0.0440 \text{ (kcal/mol)}^{-1}$$

from the POL-CI calculations, and

$$\Delta E_0 = 9.84 \text{ kcal/mol} \qquad \alpha = +0.0484 \text{ (kcal/mol)}^{-1}$$

from the GVB + 1 + 2 calculations. The average error in the analytic fit (Equation 60) is only 0.16 kcal/mol for the POL-CI calculations and 0.08 kcal/mol for the GVB + 1 + 2 calculations. While the simple exponential relationships in Equation 60 provide an excellent representation of the calculated barrier heights, it should be noted that this equation does not behave properly for very endoergic reactions. There the barrier height should approach the reaction endoergicity, while Equation 60 predicts an exponential increase in the barrier height. Also, it does not properly represent the energetics of the reverse (X + H_2) reactions. Nonetheless, there appears to be a propensity rule for chemical reactions which relates the barrier to reaction with the energy defect.

• **Rule 1** — Very exoergic reactions have small barriers to reaction with the barrier heights increasing with decreasing exoergicity.

The exponential relationship found here may be compared with the linear relationship postulated by Evans and Polyani,[68] the nonlinear relationship predicted by the bond energy-bond order (BEBO) method,[70] and the quadratic relationship put forth in the theory of electron transfer.[71]

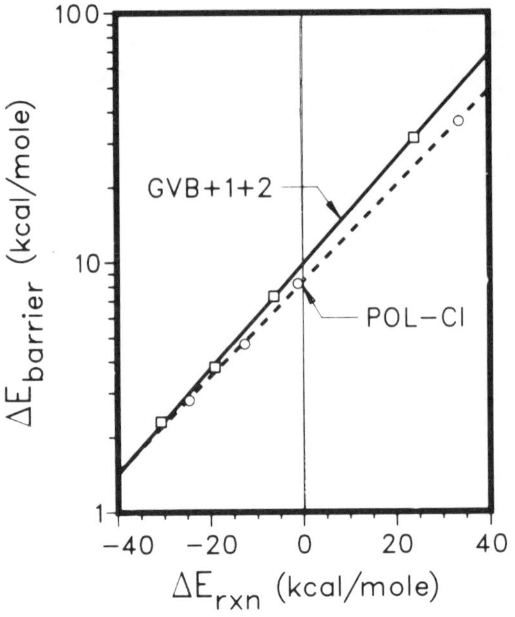

FIGURE 12. Variation of the calculated barrier heights ($\triangle E_{barrier}$) with the calculated energy defects ($\triangle E_{rxn}$) for the H + HX → H_2 + X abstraction reactions. The corresponding exponential fits (Equation 60), are also plotted; see the text.

It is interesting to note that the barrier for a "theoretical" thermoneutral H + HX abstraction reaction obtained from Equation 60, i.e., one for which $\triangle E_{rxn} = 0$ is 8 to 10 kcal/mol. This is comparable to the 11- to 12-kcal/mol barrier obtained from analogous calculations on the thermoneutral H + H_2 abstraction reaction.

The data in Table 3 also suggest two additional propensity rules for chemical reactions:

• **Rule 2** — Very exoergic reactions tend to have "early" saddle points, i.e., saddle points that resemble the reactants, with the saddle point moving to "later" positions with decreasing exoergicity.

• **Rule 3** — As the barrier to reaction increases, the saddle point moves to "later" positions.

Rule 2 is often referred to as Hammond's postulate.[69] In fact, there is also a quantitative correlation between the calculated bond extensions at the saddle point:

$$\Delta R_i = R_{sp}(i) - R_e(i) \tag{61}$$

namely,

$$e^{-\beta \Delta R_{HH}} + e^{-\beta \Delta R_{HX}} = 1 \tag{62}$$

where

$$\beta = 3.575 \text{ Å}^{-1}$$

for both the POL-CI and GVB + 1 + 2 calculations. Equation 62 is plotted in Figure 13. The average deviation of the constant in Equation 62 from unity is just 0.02. Note that in

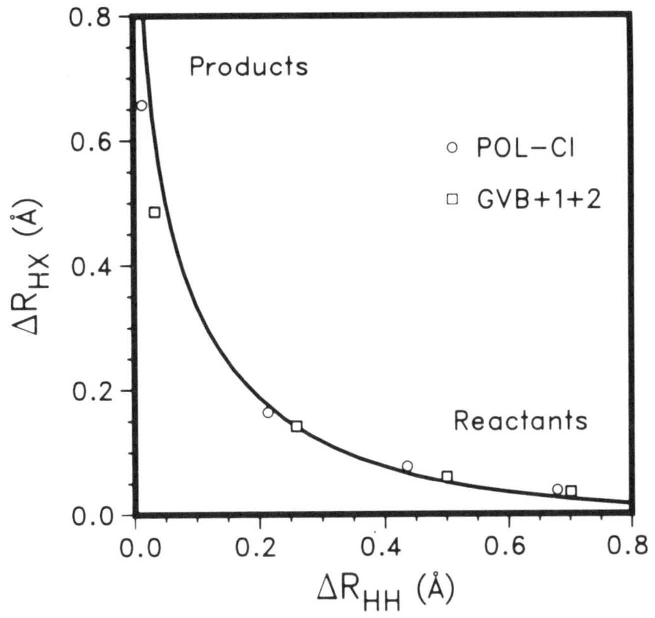

FIGURE 13. Relationship between the calculated bond extensions, $\triangle R_{HH}$ and $\triangle R_{HX}$, at the saddle point for the H + HX → H_2 + X abstraction reactions.[75] The solid curve corresponds to Equation 62; see the text.

this case the use of bond extensions rather than bond lengths is important because it compensates for the large variations in the diatomic bond lengths due to changes in the bond strengths and the halogen inner-shell (core) sizes.

The relationship in Equation 62 suggests that there is a generalized reaction path characterizing the H + HX abstraction reactions. Furthermore, we predict that the bond extension for a symmetric abstraction reaction, i.e., a reaction for which $\triangle R_{HH} = \triangle R_{HX}$, is 0.19 Å. This is identical to the 0.19-Å bond extension obtained in comparable calculations on the H + H_2 reaction[9] (Section II).

The reader should note that the errors in the calculated energy defects directly affect the barriers for the reverse X + H_2 reactions, i.e.,

$$\Delta E_{barrier}(X + H_2) = \Delta E_{barrier}(H + HX) - \Delta E_{rxn}$$

Thus, while the GVB + 1 + 2 calculations predict a barrier for the H + HF reaction of 31.4 kcal/mol in good agreement with recent estimates of 32 to 33 kcal/mol from experiment, the same calculations predict a barrier for the F + H_2 reaction of 7.4 kcal/mol, which is substantially larger than the 1 to 2 kcal/mol estimated from experiment.[83] In all cases, the GVB + 1 + 2 calculations appear to be biased against the X + H_2 limit, i.e., the correlation energy of the HHX complex at the saddle point more nearly equals that of the HX molecule than that of a halogen atom plus the hydrogen molecule. This appears to be so in spite of the fact that the saddle point for the F + H_2 reaction more closely resembles F + H_2 than H + HF.

The calculations on the H + HX exchange reactions are summarized in Table 4. As noted above, the barrier for a "theoretical" thermoneutral abstraction reaction is 8 to 10 kcal/mol. This is comparable to the barrier obtained for the H + HI exchange reaction (~10 kcal/mol). As noted earlier, the higher values obtained for the barriers in the other H + HX

Table 4
SADDLE POINT GEOMETRIES AND ENERGETICS OF THE H + HX EXCHANGE REACTIONS AS OBTAINED FROM POL-CI AND GVB + 1 + 2 CALCULATIONS

	H + HF	H + HCl	H + HBr	H + HI
Saddle point geometries				
R_{sp} (HX)				
POL-CI	1.19 (0.25)	1.52 (0.22)	1.66 (0.22)	1.84 (0.20)
GVB + 1 + 2	1.18 (0.25)	1.50 (0.21)	1.64 (0.21)	1.82 (0.19)
Barrier heights				
$\Delta E_{barrier}$				
POL-CI	46.6	22.8	17.3	10.2
GVB + 1 + 2	48.8	23.5	16.2	9.6
Bond energies				
D_e (HX)				
POL-CI	128.4	93.6	82.1	70.2
GVB + 1 + 2	129.3	99.0	86.4	74.7
Exptl	140.2	107.2	93.8	81.1

Note: Geometries are in angstroms; energies are in kilocalories per mole; quantities in parentheses are bond extensions.

reactions are in part due to the strong repulsive interaction of the attacking hydrogen atom with the electrons in the lone pair orbitals on the halogen atom. As expected, this interaction decreases in the series F > Cl > Br > I because of the increasingly more diffuse nature of the lone pair charge distribution in this same sequence.

While much of the early experimental data on the H + HX reactions were interpreted as indicating that the barrier for exchange was comparable to that for abstraction,[83] more recent experimental studies[84,85] are consistent with the high barriers predicted in the theoretical studies.

As the exchange reactions are thermoneutral, the observed variations in the barrier height cannot be correlated with changes in the reaction energy defects. A clear trend can, however, be established between the calculated barrier heights and the calculated HX bond energies, i.e., the energy of the bond being broken. This correlation is illustrated in Figure 14. The solid line in Figure 14 is given by:

$$\Delta E_{barrier} = (\gamma D_e - \delta) D_e \tag{63}$$

where

$$\gamma = 0.00451\ (\text{kcal/mol})^{-1} \qquad \delta = 0.206$$

for the GVB + 1 + 2 calculations. The average error for the quadratic fits (Equation 63) of the exchange barrier heights is 0.19 kcal/mol. An exponential relationship between $\Delta E_{barrier}$ and D_e(HX) such as was found to be appropriate for the abstraction reactions yields an error of ~1 kcal/mol (see the dashed line in Figure 14). With the form of Equation 63, $\Delta E_{barrier} \rightarrow 0$ as $D_e \rightarrow 0$, and $\Delta E_{barrier} \rightarrow \infty$ as $D_e \rightarrow \infty$, as expected intuitively.

The saddle point for the exchange reaction is symmetric with equal HX bond lengths. As can be seen from Table 4, the ΔR_{HX} remains relatively constant, varying only from 0.25 (H + HF) to 0.19 Å (H + HI).

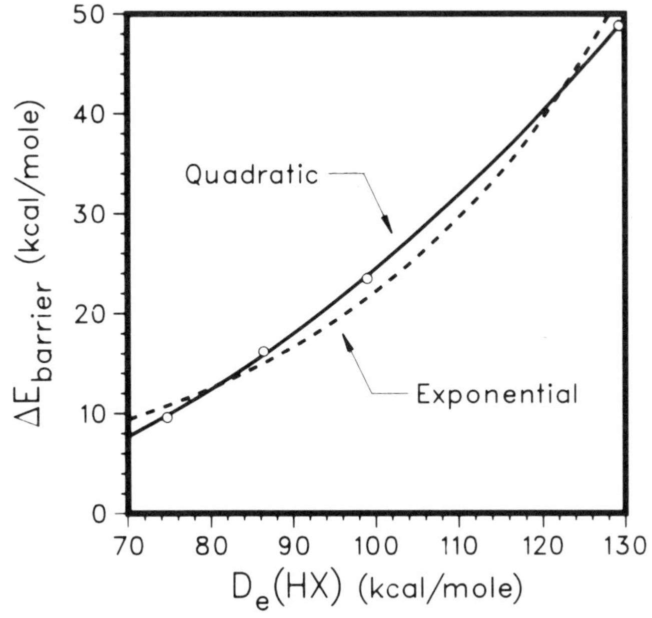

FIGURE 14. Variation of the calculated barrier heights for the H′ + HX → H′X + H exchange reactions with the calculated HX bond energy (D_e(HX)).[75] Both the quadratic (Equation 63) and exponential fits are also plotted.

2. The Li + HF Reaction: A Reaction Involving Highly Ionic Species

The reaction of lithium atoms with hydrogen fluoride

$$Li + FH \rightarrow LiF + H \tag{64}$$

is representative of a class of reactions which involve species with highly ionic bonds. Calculations on this reaction have been reported by Balint-Kurti and Yardley,[86] and Chen and Schaefer.[87] We present here a brief summary of the latter calculations.

These calculations[87] were based on the RHF method and included all single and double excitations from the HF configuration (RHF + 1 + 2 calculations). The basis set consisted of a polarized double-zeta basis set augmented with a diffuse set of fluorine 2p-functions to describe any fluorine negative ion (F^-) character in the wavefunction. Considering only the chemically active valence electrons, the Li + FH system is isoelectronic with the H + FH reaction considered in the last section. However, as we shall see below, there are marked differences in the potential energy surfaces for the two reactions.

The results of these calculations are summarized in Table 5; the structures of the species involved are plotted in Figure 15. As can be seen in Table 5, there is good agreement between the calculated and measured properties of the reactants (HF) and products (LiF). The energy defect for the reaction (Equation 64) obtained from experiment[88] is 3.50 (without the zero-point energy (ZPE) correction) and −0.95 kcal/mol (with the ZPE correction). These are to be compared to the calculated defects of 2.9 and −1.6 kcal/mol, respectively. Thus, overall the reaction is nearly thermoneutral.

From an examination of contour plots of the potential energy surface of reaction (Equation 64) as a function of (R_{HF}, R_{LiF}, θ_{LiFH},) Chen and Schaefer[87] drew the following conclusions about the reaction path:

Table 5
CALCULATED GEOMETRIES, VIBRATIONAL FREQUENCIES, ZERO POINT ENERGIES, AND RELATIVE ENERGIES FOR THE REACTANTS, SADDLE POINT, AND PRODUCTS FOR THE Li + F → LiF + H REACTION

	HF		LiFH	LiF-H	LiF	
	Calcd	Exptl	Calcd	Calcd	Calcd	Exptl
Geometries (Å)						
R_{HF}	0.93	0.917	0.94	1.29		
R_{LiF}			1.95	1.70	1.60	1.564
θ_{LiFH}			114	74		
Vibrational frequencies (cm^{-1})						
ω_{HFs}	4055.	4138.3	3940	1290.i		
ω_{LiFs}			305	670	860.	910.3
ω_{LiFHb}			400	880		
Zero point energies (kcal/mol)						
E_{zpe}	5.8	5.9	6.6	2.2	1.2	1.3
Relative energies (kcal/mol)						
ΔV_{eff}	0.0	0.0	−4.6	10.0	2.9	3.5
$\Delta V_{eff} + \Delta E_{zpe}$	0.0	0.0	−3.8	6.4	−1.6	−1.0

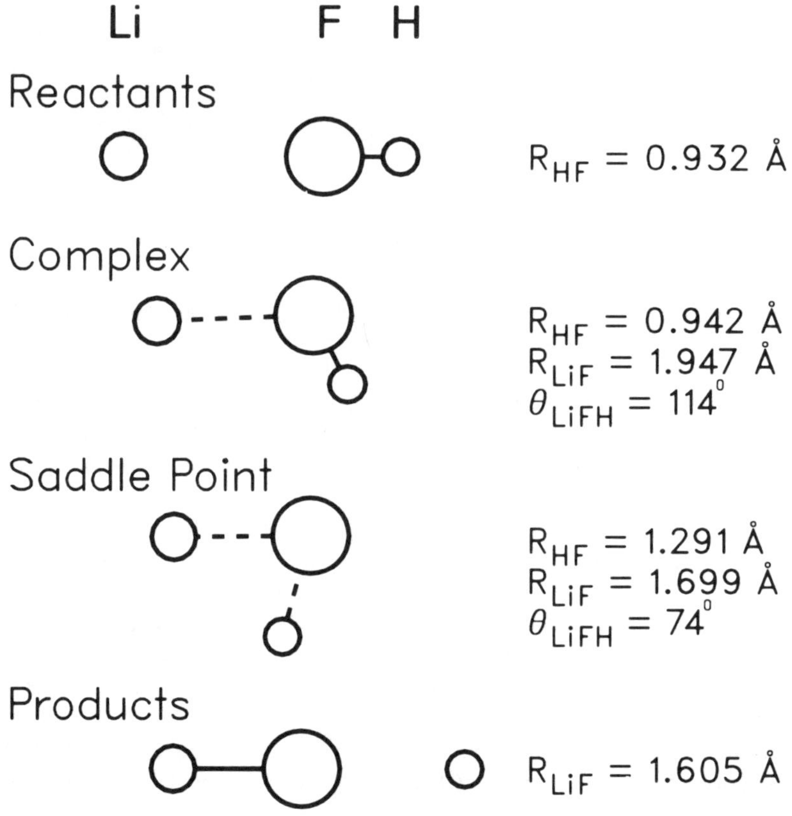

FIGURE 15. Structures of the reactants, complex, saddle point, and products in the Li + FH → LiF + H reaction.

1. Initially, the Li atom is attracted to the fluorine end of the HF, forming a bound complex ($\Delta E = -4.6$ kcal/mol) with an equilibrium geometry with $R_{LiF} = 1.95$ Å, $\theta_{LiFH} = 114°$, and $R_{HF} = 0.94$ Å.
2. Subsequently, the energy rises until the saddle point is reached, at $R_{LiF} = 1.70$ Å, $\theta_{LiFH} = 74°$, and $R_{HF} = 1.29$ Å, with an energy 10.0 kcal/mol above the reactants.
3. Finally, the energy drops until the products are reached, at 2.9 kcal/mol above the reactants.

Balint-Kurti and Yardley[86] have attributed the LiFH complex to an attractive dipole-induced dipole interaction between the lithium atom and hydrogen fluoride. Because of the presence of the LiFH complex, it should be noted that the saddle point in the table and figure corresponds to the reaction:

$$(\mathrm{LIFH})_{complex} \rightarrow \mathrm{LiF} + \mathrm{H} \qquad (65)$$

i.e., the saddle point does not directly connect the reactants and products in the Li + FH reaction. Equation 65 is calculated to be endoergic by 7.5 kcal/mol. As expected from the arguments given in the previous section, the saddle point is located in the products' region, e.g., $\Delta R_{LiFH} = 0.09$ Å while $\Delta R_{HF} = 0.36$ Å.

The Li + FH reaction is clearly quite different from the isoelectronic H + FH reaction discussed previously; not only is the barrier to reaction quite low (10 vs. nearly 50 kcal/mol in the H + FH reaction), but the saddle point for the reaction involves a "front side" configuration of the Li and H atoms rather than a "back side" configuration as in the analogous H reaction (Figure 15). The origin of the highly bent saddle point in the Li + FH reaction has not been clearly established, although various arguments have been advanced.[87] It is possible that a FLiH complex exists and that the saddle point leads to formation of this complex and not directly to products as assumed by Chen and Schaefer; a more comprehensive set of calculations on the potential energy surface for the Li + FH is needed to address this point.

A schematic energy diagram of the reaction is plotted in Figure 16 without (solid lines) and with (dashed lines) the ZPE correction. The changes in the ZPE for the LiFH system have a marked effect on the energetics of the reaction. The ZPE of HF is 5.8 kcal/mol. Because there is little decrease in the FH frequency upon complex formation (4055 to 3940 cm^{-1}), the ZPE of the system increases to 6.6 kcal/mol in the LiFH complex. The increase in ZPE in forming the complex is, of course, directly associated with the additional vibrational modes in the three-atom system (the LiF stretching, $\omega = 305$ cm^{-1}, and LiFH bending, $\omega = 400$ cm^{-1}, modes). At the saddle point, on the other hand, only the two low-frequency modes remain, the HF stretching mode now correlating with the reaction coordinate. Because of the loss of the HF stretching mode, the ZPE at the saddle point is only 2.2 kcal/mol. Finally, in the product, with loss of the LiFH bending mode, the ZPE decreases further to just 1.2 kcal/mol. In summary then, inclusion of the ZPE corrections:

1. Decreases the binding energy of the complex by 0.8 kcal/mol from 4.6 to 3.8 kcal/mol
2. Decreases the barrier height by 3.6 kcal/mol from 10.0 to 6.4 kcal/mol
3. Decreases the reaction energy defect by 4.5 kcal/mol from 2.9 (endoergic) to -1.6 kcal/mol (exoergic)

Clearly, changes in the ZPE must be taken into account in predictions of the overall energetics of chemical reactions.

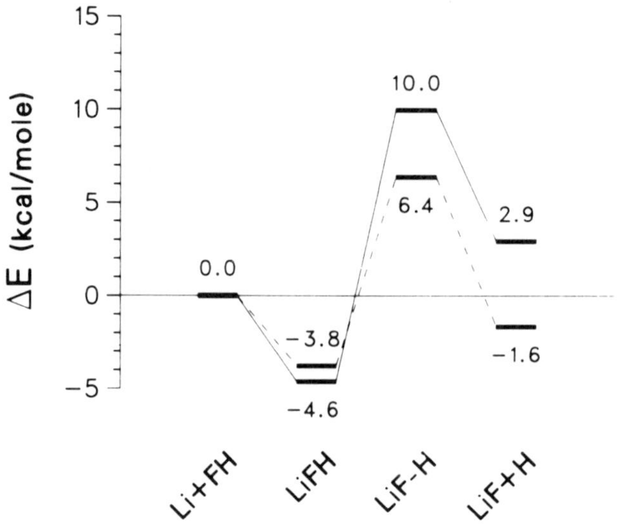

FIGURE 16. Schematic energy diagram for the Li + FH → LiF + H reaction without (solid lines) and with (dashed lines) the zero point energy correction.

From the work of Lee at al.[89] it is estimated that ΔE_{vat}, i.e., the barrier height with the ZPE correction, is less than 3 kcal/mol, implying an error of ~3 to 4 kcal/mol in the calculated barrier height.

The other reaction pathway for Li + FH, namely,

$$Li + HF \rightarrow LiH + F \tag{66}$$

is endoergic by nearly 80 kcal/mol (with ZPE corrections). It is thus not competitive with Equation 64.

Many reactions involving highly ionic species are expected to have one or more bound complexes in the entrance and/or exit channels. If there is a bound complex in the entrance channel and the barrier to reaction lies above the energy of the reactants, as in the present case, the reaction will be direct, but with complications. If, on the other hand, the barrier to reaction lies below the energy of the reactants, the reaction will be indirect, involving formation of the complex which can then either dissociate back to products or onto products. Of course, such reactions may also have a direct component.

3. The H + CH Reaction: An Abstraction Reaction Involving Two Reactive Species

The abstraction reaction between atomic hydrogen and the CH radical,

$$H(^2S) + CH(^2\Pi) \rightarrow H_2 + C(^3P) \tag{67}$$

presents some interesting mechanistic questions. In Section IV.A, examples of a series of hydrogen abstraction reactions involving one reactive (hydrogen) and one nonreactive species (hydrogen halide) were discussed in detail. It was found that these reactions proceed through a direct mechanism, i.e., the reaction path is collinear with a simple barrier separating reactants and products. Recent calculations[90] indicate that reactions involving two reactive species can be considerably more complex. A brief summary of these results are presented in this section.

The calculations of Harding[90] on Equation 67 include all single and double excitations from the RHF configuration (RHF + 1 + 2) employing a polarized double-zeta basis set. The RHF wavefunction for the x-component of the $^3\Pi$ state of linear HHC is

$$\mathcal{A}\ 1\sigma^2 2\sigma^2 3\sigma^2 4\sigma^1 1\pi_x^1 \alpha\beta\alpha\beta\alpha\beta\alpha\alpha \qquad (68)$$

wherein the reactants (1σ, 2σ, 3σ, and $1\pi_x$) correlate with the corresponding orbitals of CH and 4σ correlates with the 1s orbital on the hydrogen atom; in the products (1σ, 2σ, 4σ, and $1\pi_x$) correlate with the (1s, 2s, $3p_z$, and $2p_x$) orbitals of the carbon atom, and 3σ correlates with the bonding orbital of H_2. Thus, for this reaction the RHF configuration would be expected to be dominant throughout the course of the reaction and RHF + 1 + 2 calculations would provide an adequate description of the potential energy surface.

Preliminary calculations for a collinear geometry led to the prediction that the reaction is 29.5 kcal/mol exoergic with a barrier of 5.5 kcal/mol to formation of products. The barrier is located at a CH bond distance of 1.20 Å and an HH bond distance of 1.22 Å. These can be compared to a calculated CH distance in the reactant of 1.13 Å (i.e., ΔR_{CH} = 0.07 Å) and a calculated HH distance in the product of 0.74 Å (i.e., ΔR_{HH} = 0.48 Å). These results indicate an "early" barrier as expected for an exoergic reaction (Rule 2, Hammond's postulate).

As noted in Section II.A, for the collinear barrier to represent a saddle point of the abstraction reaction, there must be one and only one negative eigenvalue of the Hessian matrix (the matrix of second derivatives). Equivalently, there must be only one imaginary normal mode frequency. The calculated stretching frequencies are 1460 and 1810.i cm^{-1}, corresponding to symmetric and asymmetric combinations of the CH and HH stretches. As expected, the asymmetric combination correlates with the reaction coordinate. For linear molecules in a nondegenerate electronic state the bending vibrational modes are doubly degenerate. However, from Equation 68 the electronic wavefunction for the collinear HHC complex has a singly occupied σ orbital and a singly occupied π_x orbital. Since the electronic wavefunction is degenerate, the harmonic bending modes are nondegenerate. The calculated frequency for bending in the plane of the singly occupied π_x orbital (giving an electronic wavefunction of A′ symmetry) is 645 cm^{-1} while the frequency for bending in the plane perpendicular to the π_x orbital is 435.i cm^{-1}. Thus, the collinear barrier possesses two imaginary normal-mode frequencies and therefore cannot be the saddle point for the abstraction reaction.

Following the potential surface in the direction of the bending normal mode with the imaginary frequency is found to lead directly to the $^3B_1$ state of methylene (CH_2). In fact, a more careful search of the nonlinear potential energy surface shows that there is no barrier to the addition of atomic hydrogen to CH to form ground-state methylene, a reaction that is calculated to be 105 kcal/mol exoergic. Furthermore, the barrier for elimination of H_2 from methylene (see Section IV.A.5 below) is predicted to be 83 or 22 kcal/mol less than the energy of the reactants H + CH. This then provides an indirect, addition-elimination pathway leading to abstraction products which involves no net barrier relative to the reactants. The potential energy surface for Equation 67 is summarized schematically in Figure 17.

It is likely that many other radical-radical abstraction reactions will be found to proceed by similar mechanisms. For example, at low temperatures, the reaction of atomic hydrogen with HCO

$$H + HCO \rightarrow H_2 + CO \qquad (69)$$

almost certainly proceeds by an addition-elimination mechanism since calculations by Goddard and Schaefer[91] and Harding et al.[92] have shown that there is no barrier to addition-

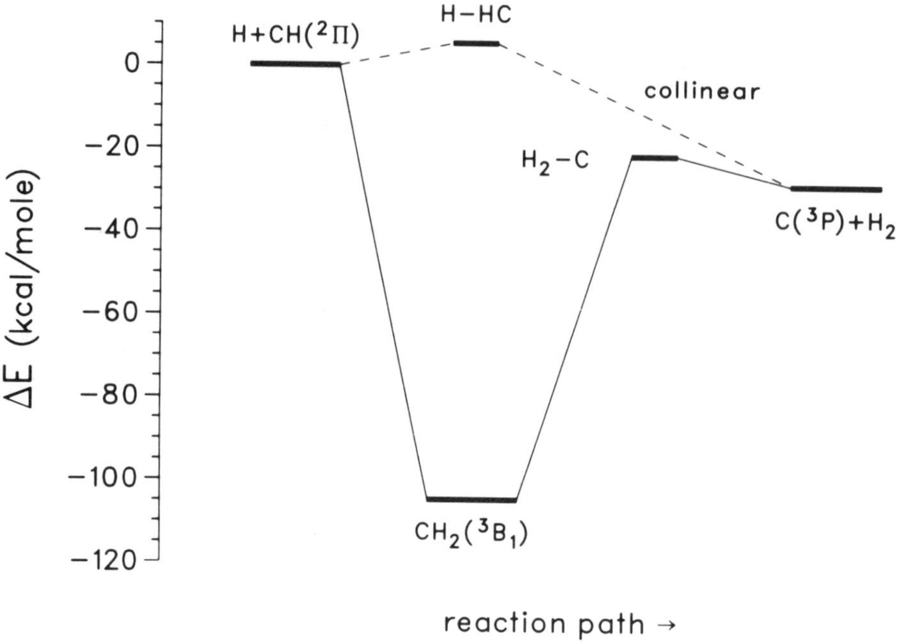

FIGURE 17. Schematic energy diagram for the H + HC → H_2 + C reaction without zero point corrections.

forming formaldehyde and that the barrier for elimination of H_2 from H_2CO is lower in energy than the reactants H + HCO. In this case, though, it is still not clear whether or not there is also a saddle point corresponding to a direct reaction which might play an increasingly important role at higher temperatures.

4. The H + CO Reactions: Addition and Migration Reactions

The potential energy surface for the addition of atomic hydrogen to carbon monoxide has been calculated by Dunning.[93] Dunning considered the two addition reactions,

$$\mathrm{H + CO \rightarrow HCO} \tag{70}$$

$$\mathrm{H + CO \rightarrow COH} \tag{71}$$

as well as the hydrogen migration reaction,

$$\mathrm{HCO \rightarrow COH} \tag{72}$$

The RHF + 1 + 2 method was used with a polarized valence double-zeta basis set. Again, the RHF configuration was found to be dominant over the regions of the potential energy surface of interest and so the RHF + 1 + 2 calculations are expected to adequately represent the reaction energetics. In this section some of the characteristics of the potential energy surfaces for Equations 70 to 72 will be discussed, the first two as prototypes of addition reactions, the last as a representative of (1,2)-hydrogen migration reactions.

A number of other theoretical studies have also been reported on the H + CO system. Bruna et al.[94] reported HF and CI calculations on the HCO and COH molecules. A detailed series of CI calculations exploring the cusp in the potential energy surface for the HCO molecule near the linear configuration has been reported.[95] Finally, Adams et al.[96] have reported many-body perturbation theory and coupled cluster calculations on Equation 70.

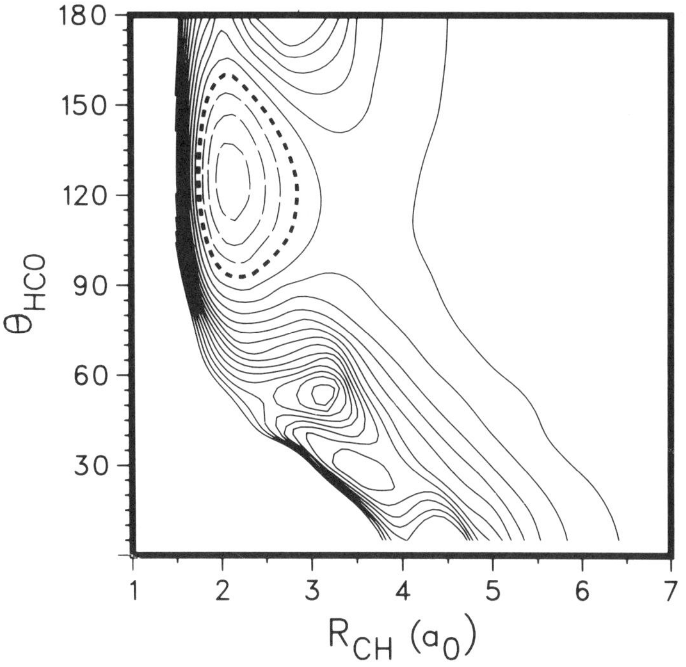

FIGURE 18. Contour plot of the potential energy surface for the H + CO system; for this plot the CO distance has been taken to be R_{CO} = 2.25 a_0 (1.19 Å), the calculated equilibrium value for HCO. Energies are relative to the H + CO limit. The contour spacing is 5 kcal/mol. Positive contour lines are traced in solid lines, negative contour lines in long dashed lines, and zero lines in short dashed lines.

While Equations 70 to 72 are discussed as separate reactions, all take place on the same potential energy surface. A contour plot of a portion of this surface is shown in Figure 18. Here the CO distance (R_{CO}) is kept constant at 2.25 a_O (1.19 Å) and the potential energy is plotted as a function of both the CH distance (R_{CH}) and the HCO angle (θ_{HCO}); R_{CO} = 2.25 a_O is the calculated equilibrium CO distance in HCO. The lowest (global) minimum on the surface corresponds to the HCO isomer while the second (local) minimum, at R_{CH} = 3.5 a_O (1.85 Å) and θ_{HCO} = 30° corresponds to the HOC isomer. The HOC minimum is actually somewhat deeper than is shown in Figure 18 because the optimum CO bond length for HOC is substantially longer than that used in the figure (Table 6). The saddle points for Equations 70 to 72 are also visible in this plot; for Equation 70 it appears at (R_{CH} = 3.3 a_O, θ_{HCO} = 120°), for Equation 71 it is found at (R_{CH} = 4.0 a_O, θ_{HCO} = 20°), and the third saddle point, for migration, occurs at (R_{CH} = 2.5 a_O, θ_{HCO} = 54°).

Let us now consider Equations 70 to 72 in detail. Concentrating on Equation 70 first, the calculated properties of the reactants, saddle point, and products are summarized in Table 6. From these results we see that the saddle point for the carbon addition reaction is very early, in accord with Rule 2 (Hammond's postulate); e.g., R_{CO} at the saddle point is only 0.01 Å longer than that of the CO molecule itself, while R_{CO} in the product is extended an additional 0.03 Å. The vibrational frequencies also exhibit trends expected of an exoergic reaction with an early transition state. The CO stretching frequency decreases only slightly (20 cm^{-1}) on going from the reactants to the saddle point and then somewhat more (~100 cm^{-1}) on going to product. The bending frequency is quite low at the saddle point (422 cm^{-1}), but increases dramatically as the CH distance shortens (to 1065 cm^{-1}). Because ω_{CO_s} decreases only slightly, the ZPE at the saddle point is slightly greater than that of the

Table 6
CALCULATED GEOMETRIES, VIBRATIONAL FREQUENCIES, ZERO POINT ENERGIES, AND RELATIVE ENERGIES FOR THE REACTANTS, SADDLE POINTS, AND PRODUCTS FOR THE H + CO REACTIONS

		C-addition		O-addition		
	H + CO	H-CO	HCO	H-OC	HOC	H-migration
Geometries (Å)						
R_{CO}	1.14	1.15	1.19	1.19	1.29	1.29
R_{XH}[a]		1.73	1.12	1.23	0.98	1.14
θ[b]		118	126	120	114	66
Vibrational frequencies (cm^{-1})						
ω_{COs}	2080	2060	1955	1885	1450	1454
ω_{XHs}[a]		1045.i	2920	3820.i	3815	2568
ω_b[b]		420	1165	1065	1080	2173.i
Zero point energies (kcal/mol)						
E_{zpe}	3.0	3.6	8.6	4.2	9.1	5.7
Relative energies (kcal/mol)						
ΔV_{eff}	0	7.6	−15.7	39.6	23.9	53.1
$\Delta V_{eff} + \Delta E_{zpe}$	0	8.2	−10.1	40.8	30	55.8

[a] For the carbon addition reaction, X = C; for the oxygen addition and hydrogen migration reactions, X = O.

[b] For the carbon addition reaction, $\theta = \theta_{HCO}$; for the oxygen addition and hydrogen migration reactions, $\theta = \theta_{COH}$.

reactants (by 0.6 kcal/mol); the vibrationally adiabatic barrier is thus larger than the classical barrier.

Equation 70 is calculated to be 10.1 kcal/mol exoergic; this is to be compared to a recent measurement of 14.4 kcal/mol for the dissociation energy of the H-CO bond.[98]

The calculated vibrationally adiabatic threshold for the carbon addition reaction is 8.2 kcal/mol. This is to be compared to a measured activation energy for the H + CO reaction of ~2 kcal/mol.[99] Thus, the error in the calculated barrier height for Equation 70 is ~6 kcal/mol, somewhat larger than found in previous examples. As noted earlier for the H + HX reactions, however, the correlation energy at the saddle point often resembles that of the more complex molecular species, even if the barrier is in the opposite reaction channel. Considering the reverse reaction,

$$HCO \rightarrow H + CO \tag{73}$$

the calculated barrier height is 10.1 + 8.2 = 18.3 kcal/mol. This is to be compared to the experimental estimate of 14.4 + 2.0 = 16.4 kcal/mol. The error in the barrier height for the dissociation reaction (Equation 73) is thus only 1.9 kcal/mol. In addition reactions it appears that the correlation energy at the saddle point more closely resembles that of the addition complex than that of the reactants.

At the saddle point the reaction coordinate for the carbon addition reaction closely resembles the CH stretching mode in the HCO molecule. Thus, the reaction path would be expected to be nearly parallel to the R_{CH} coordinate. The energy profile for Equation 70 along the R_{CH} coordinate is given in Figure 19. In computing the energies plotted in this figure, the CO distance was kept at the reactant value (1.14 Å) while θ_{HCO}, which is undefined in the reactants, was set to the product value (126°). The energy so obtained is a smooth function of R_{CH}. Complete optimization of the geometries at the saddle point and product

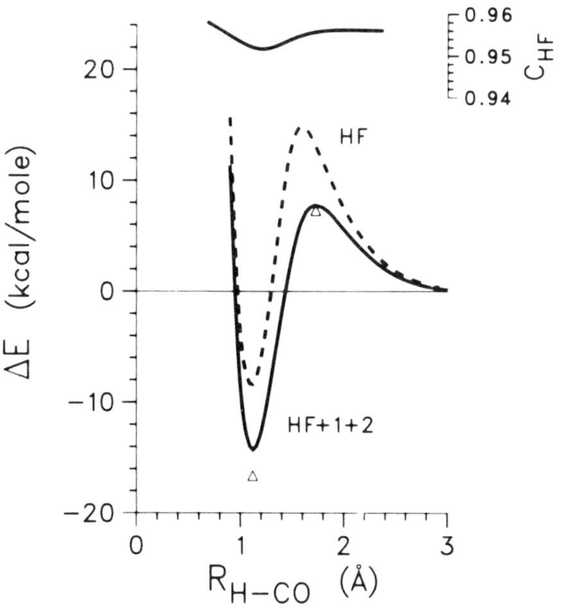

FIGURE 19. Energy profiles for the H + CO → HCO reaction as a function of the OH distance; for the curve the CO distance was taken to be R_{CO} = 1.14 Å, the calculated equilibrium value for the CO molecule, and the bond angle was taken to be θ_{HCO} = 126, the calculated equilibrium value for the HCO molecule. The triangles (△) correspond to fully optimized geometries for the saddle point and equilibrium structure. The CI coefficient for the RHF configuration is also plotted.[93,97]

leads to only minor changes (Figure 19). The energy profile plotted in the figure is expected to be close to that which would be obtained by following the steepest descent reaction path.

Also shown in Figure 19 is the energy profile from an RHF calculation and the variation in the magnitude of the coefficient of the RHF configuration in the CI calculation along the above approximate reaction path. Note that the RHF calculations lead to a barrier that is both too high ($\Delta E_{barrier} \simeq$ 15 kcal/mol) and too narrow ($\omega_{rxn} \simeq$ 1800.i). One reason for the poor behavior of the RHF wavefunction near the saddle point can be seen in the behavior of the CI coefficient of the configuration in this region. This coefficient exhibits a minimum very close to the saddle point geometry, indicating that the region of the saddle point is the least well described by a single configuration wavefunction of any along the reaction path. This is probably true of most saddle points since in saddle point regions the electronic wavefunction typically changes character and therefore, often involves more than one important configuration.

It is instructive to compare the exoergic addition to the carbon (Equation 70) to the endoergic addition to the oxygen (Equation 71). The calculated properties of saddle point and product for Equation 71 are also summarized in Table 6 and the energy profile is compared to that of Equation 70 in Figure 20. Here it is seen that the saddle point occurs much later in the reaction; compare, e.g., the ΔR_{CH} at the saddle point for Equation 70 (0.61 Å) with the saddle point ΔR_{OH} of 0.25 Å for Equation 71. A second characteristic of late addition saddle points is a large imaginary frequency for the reaction coordinate, an indication of a narrow barrier. For Equation 70 the calculated imaginary frequency is 1045.i cm^{-1} compared with a value of 3815.i cm^{-1} for Equation 71. Thus, the imaginary frequency for the endoergic addition reaction is almost four times that of the exoergic addition reaction. Note also that for Equation 71 there is a large change in the CO distance along the reaction path. As noted above for the carbon addition reaction, using the reactant CO distance at all

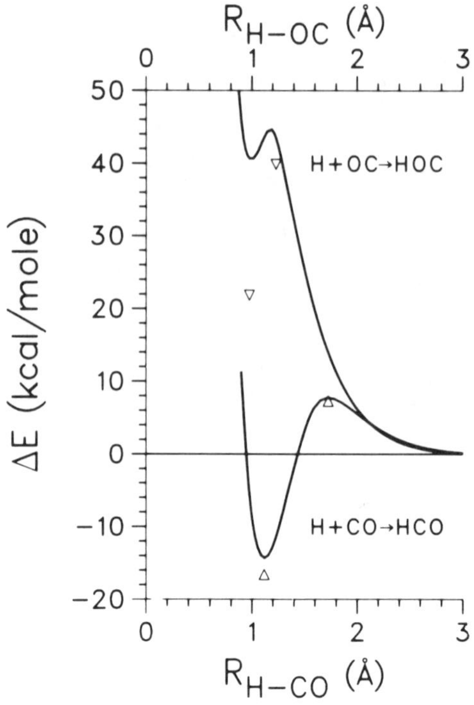

FIGURE 20. Energy profiles for the H + CO addition reactions as a function of R_{CH} (→ HCO; lower scale) and R_{OH} (→ COH; upper scale). Again, the CO distance was taken to be that appropriate for the CO molecule and the bond angle was taken to be that of the complex (θ_{HCO} = 126, θ_{COH} = 114).[93,97]

points along the path introduced an error of only a couple of kilocalories per mole. For the oxygen addition reaction the same approximation introduces an error of approximately 20 kcal/mol in the overall endoergicity of the reaction. In fact, at the saddle point for Equation 71 the reaction coordinate is a strong admixture of both the OH and CO stretching modes.

The energetics for all of the H + CO reactions are summarized in Figure 21. As can be seen there, the barrier for the (1,2)-hydrogen transfer reaction lies above that for the dissociation of COH to H + CO. This is in spite of the fact that isomerization is far more favorable energetically than is dissociation (23.9 vs. 39.6 kcal/mol). Hydrogen migration will therefore occur by a dissociation-recombination mechanism rather than by direct hydrogen transfer.

Similar features have been found in other radical addition reactions such as H + C_2H_2,[100,101] H + C_2H_4,[102,103] and H + HCN.[104]

5. The $C(^3P)$ + H_2 Addition Reaction: A Reaction Involving Multiple Potential Energy Surfaces

The addition of ground-state atomic carbon to molecular hydrogen

$$C(^3P) + H_2 \rightarrow CH_2(^3B_1) \qquad (74)$$

has been studied.[90,105,106] The reaction is found to proceed through a weakly avoided crossing between the potential energy surfaces of two states. This reaction is thus a simple example of a reaction involving more than one electronic state.

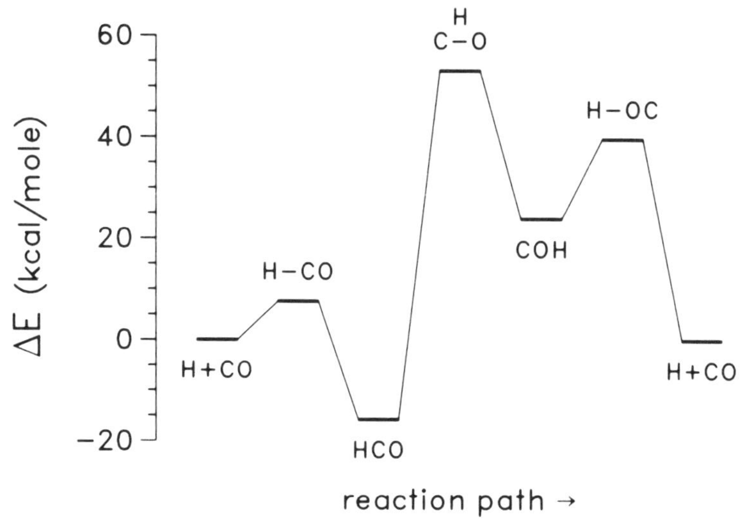

FIGURE 21. Schematic energy profile for the addition and migration reactions in the H + CO system without zero point corrections.[93]

Ground state atomic carbon has a threefold spatial degeneracy arising from its two singly occupied and one unoccupied 2p-orbital (the unoccupied orbital can point in any of three directions). If we define R_{cm} to be the distance between the carbon atom and the midpoint of the two hydrogen atoms, then at large R_{cm} the most favorable orientation of the carbon relative to the H_2 is to have the unoccupied 2p-orbital pointing toward the H_2, thereby minimizing repulsive interactions. In a C_{2v} configuration, this gives a state of $^3A_2$ symmetry. The ground state of the product, methylene, on the other hand, is of $^3B_1$ symmetry. At large distances, the $^3B_1$ state corresponds to orienting the carbon atom such that the unoccupied 2p-orbital is parallel to the H_2 bond. The third state is of $^3B_2$ symmetry, corresponding to placing both occupied 2p-orbitals in the plane of the three atoms. These orientations are depicted schematically in Figure 22.

In geometries with C_{2v} symmetry, these three states are of different symmetries and therefore do not interact. In non-C_{2v} geometries, however, the $^3A_2$ and $^3B_1$ states are both of A″ symmetry and therefore, these two states can interact. This interaction leads to an avoided crossing of the two states in non-C_{2v} geometries. The result is a lower surface with $^3A_2$ character at large R_{cm}, and $^3B_1$ character at short R_{cm}, which touchs an upper surface of opposite character along a line of C_{2v} geometries. The line of contact between the two surfaces forms a cusp in both surfaces which separates the regions of different character. It should be noted that it is artificial to treat these as two separate surfaces since in any given reactive collision there may be a high probability for jumping from one surface to the other at any point, i.e., the coupling term $(\theta_{^3A_2{}^3B_1})$ must be explicitly included in the dynamical calculations.

There are two equivalent saddle points for the addition reaction located on either side of the cusp. These saddle points necessarily lie at an energy lower than that of the cusp. The lowest energy intersection between the C_{2v} curves therefore represents an upper limit on the barrier for the reaction. Full CI calculations with a double-zeta basis set indicate that this lowest energy intersection occurs at an R_{cm} distance of approximately 2.15 a_O and an HH distance of 2.2 a_O. Potential curves in the region of intersection are shown in Figure 23. In this figure the HH distance is kept at 2.2 a_O and the energy is plotted as a function of the R_{cm} distance. The angle (θ) is defined to be the angle between the line joining the carbon atom to the midpoint of the two hydrogen atoms and the HH bond. The dotted lines are for

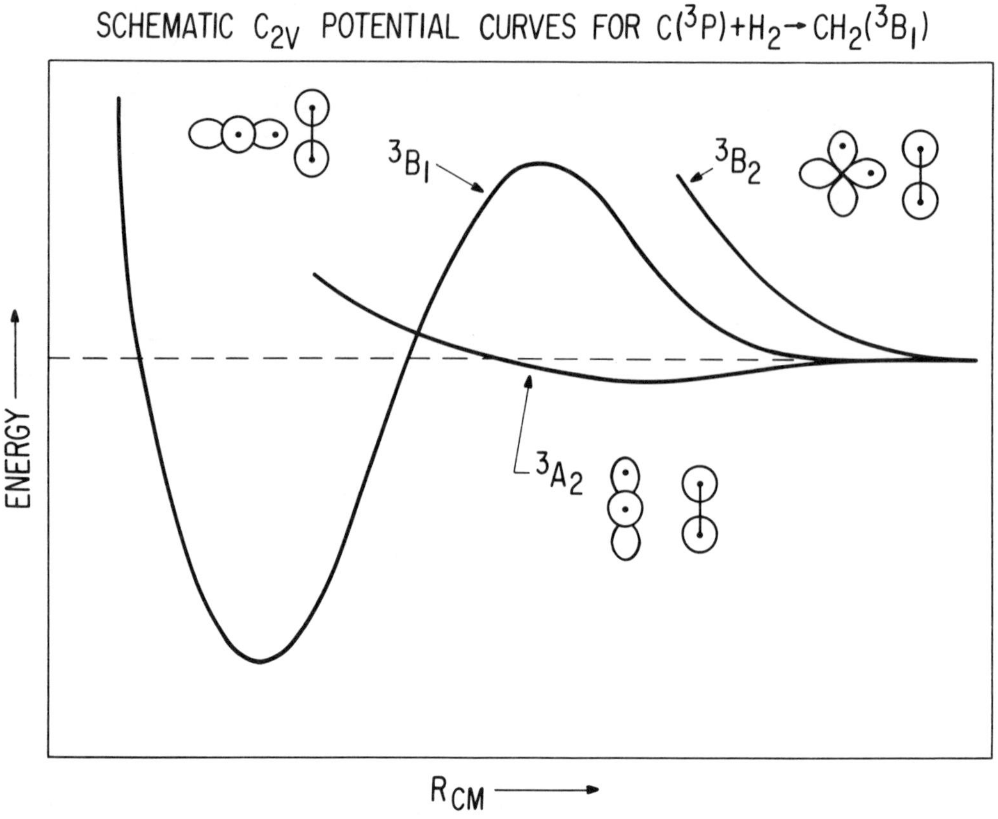

FIGURE 22. Schematic energy profile for the C + $H_2 \rightarrow CH_2$ reaction.

a C_{2v} approach (θ = 90) which shows a relatively flat $^3A_2$ curve, rising slowly as R_{cm} decreases, and a $^3B_1$ curve falling sharply as R_{cm} decreases. Making a small distortion from C_{2v} (θ = 85) yields the two dashed curves, the lower one having a maximum in the region of intersection and the upper one exhibiting a minimum. The two saddle points for this reaction occur at approximately θ = 80 and 100, corresponding to the solid curves on Figure 23. Thus, the saddle point lies only a few kilocalories per mole below the lowest C_{2v} intersection between the $^3A_2$ and $^3B_1$ states.

The energies of both the lowest C_{2v} intersection and the saddle points have been found to be very sensitive to the basis set employed. From Figure 23, one can see that with a double-zeta basis set, the intersection occurs 25 kcal/mol above the energy of the reactants. Addition of carbon 3d and hydrogen 2p polarization functions decreases this to 10 kcal/mol. Using a larger basis set [4s3p2d1f] on the carbon and [3s2p1d] on the hydrogen atoms decreases the intersection energy to 2.5 kcal/mol.[90] Of this 7.5-kcal/mol decrease, 2.5 can be traced to the addition of carbon 4f functions and 0.8 is due to the addition of hydrogen 3d functions. The remaining 4 kcal/mol is due to the additional s and p functions. The larger basis sets also shift the position of the intersection to shorter R_{cm}. This implies that the larger basis sets stabilize the $^3A_2$ state more than the $^3B_1$ state leading to the lower intersection energy.

In summary, the most accurate calculations on Equation 73 predict that the lowest C_{2v} intersection between the $^3A_2$ and $^3B_1$ states occurs 2.5 kcal/mol above the reactants. If the saddle point lies ~1 kcal/mol below the intersection, as indicated by polarized double-zeta basis set calculations, then the actual barrier for the reaction must be near zero. A global

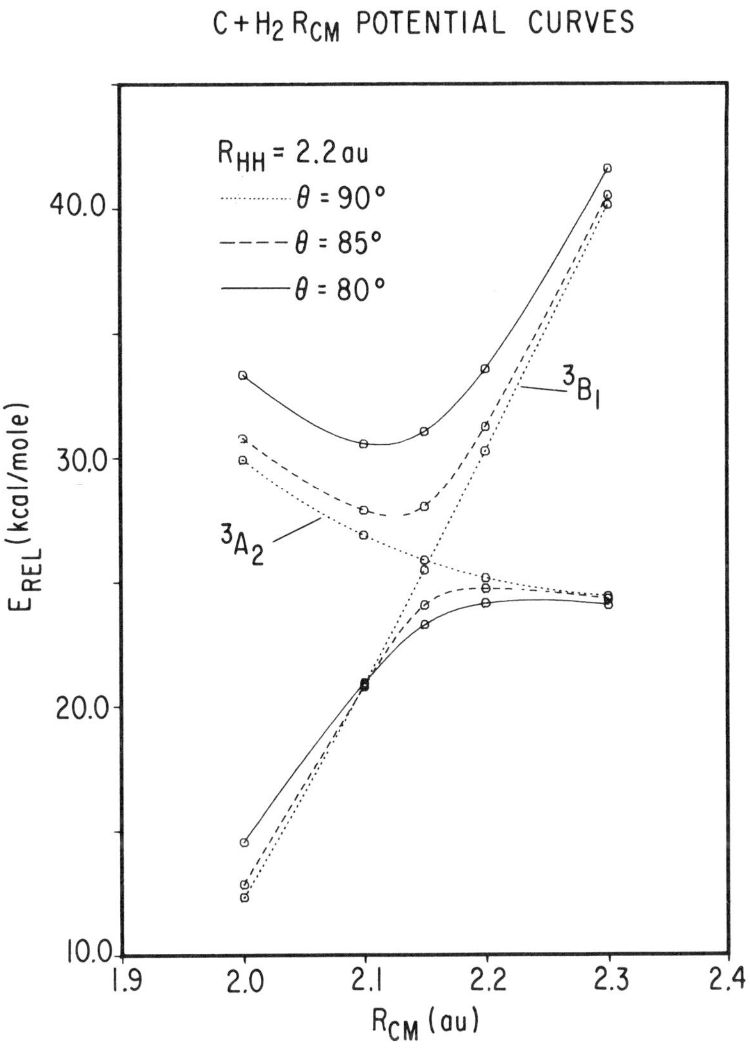

FIGURE 23. Energy profiles for the $^3A_2$ and $^3B_1$ states (θ = 90) and two lowest $^3A''$ states (θ = 80, 85) of the CH_2 species. R_{HH} has been fixed at 2.2 a_0 (1.16 Å).[90]

fit for this potential energy surface has been reported recently; however, the saddle point region for this reaction appears to be very poorly represented by the surface, probably due to the neglect of the $^3A_2$ state.

B. Four-Body Reactions

1. The OH + H_2 Reaction: An Atom-Diatom Reaction with Complications

Many reactions involving two diatomic species may simply be regarded as atom-diatom reactions with complications. The reaction of OH with H_2,

$$OH + H_2 \rightarrow H_2O + H \tag{75}$$

is a case in point. In Equation 75 the hydroxyl radical may be considered a pseudohalogen atom. The potential energy surface for Equation 75 has been studied in some detail[108] with both SOGVB and POL-CI calculations using a slightly better than polarized valence double-zeta basis set (O[3s3p1d] and H[3s1p]) as well as selected calculations with a more accurate O[4s3p2d] and H[3s2p] basis set.[109]

The orbital description of the OH + H_2 abstraction reaction is very similar to that for the H + H_2 and X + H_2 abstraction reactions (the reverse of which were discussed above). There are again a set of three active orbitals which are directly involved in the transfer of the bond pair: the singly occupied hydroxyl radical orbital and the two GVB orbitals comprising the H_2 bond, and three sets of semiactive orbitals composed of the two lone pair orbitals and the OH bonding orbitals of the hydroxyl radical. The semiactive orbitals remain well localized throughout the reaction; the active orbitals are plotted in Figure 24. Note that at the saddle point the active orbitals resemble those of the reactants more than those of the products, as expected of an exoergic reaction ($\Delta E_{rxn} = -16.0$ kcal/mol), but that otherwise the active orbitals of the OH + H_2 system resemble those of the H + H_2 system plotted in Figure 2 and the X + H_2 reactions plotted in Figures 9 and 10.

The results of the calculations of Walch and Dunning are summarized in Table 7. As is evident there, the agreement between the calculated and measured properties of the reactants and products is, in general, quite satisfactory: bond lengths are in error by no more than 0.02 Å, vibrational frequencies differ by less than 10% (of which half is due to the neglect of anharmonic effects in the calculations), and the reaction exoergicity is accurate to ~1 kcal/mol. In addition, transition state theory (TST) calculations using the calculated barrier height and vibrational frequencies led to substantial agreement with the measured rate constant for this reaction over a temperature range of 300 to 2000 K.[110]

As noted above, the OH + H_2 reaction closely resembles the X + H_2 reactions discussed previously. The orbital descriptions of the two reactions are clearly similar. Furthermore, from Table 7 we see that at the saddle point the oxygen atom and the two hydrogen atoms are nearly collinear ($\theta_{OH'H''} = 165$) and the two H_2 rocking frequencies ($\omega_{ipr} = 690$, $\omega_{opr} = 440$ cm^{-1}) are similar (they are degenerate in the X + H_2 systems). In fact, the collinear saddle point geometry of the OH + H_2 reaction ($\Delta R_{OH} = 0.34$ Å, $\Delta R_{HH} = 0.10$ Å) lies on the curve plotted in Figure 13 for the X + H_2 reactions. Finally, from the trend in the barrier heights for the X + H_2 reactions evident in Table 3, we would predict a barrier height of 4 to 5 kcal/mol for a reaction which is exoergic by ~15 kcal/mol, in reasonable agreement with the calculated value of 7.4 kcal/mol.

In spite of the similarities noted in the OH + H_2 and X + H_2 reactions above, differences do exist. Thus, in the three–body system there is no analog of the HOH′ bending motion; this mode contributes 1.8 kcal/mol to the ZPE at the saddle point. Furthermore, while the deviation of the oxygen and 2 hydrogen atoms from collinearity is small, the energy of the system is raised by nearly 0.5 kcal/mol if these 3 atoms are forced to be collinear; the nonlinearity of the OH′H″ moiety is a direct consequence of the asymmetry of the interaction between the active orbitals and semiactive orbitals on the OH moiety.

The vibrational energy levels of the reactants, saddle point, and products for Equation 75 are plotted in Figure 25; the associated normal modes are plotted in Figure 26. For the reactants, the high frequency mode corresponds to the stretching mode of the H_2 molecule. At the saddle point, on the other hand, the highest frequency mode corresponds to the stretching mode of the hydroxyl radical moiety, its energy being only slightly less than that in the reactants. The HH stretching mode strongly couples with the reaction coordinate (compare, e.g., the HH normal modes in Figure 26 with the reaction coordinate in Figure 27) and as a result, the frequency of this mode decreases from 4120 cm^{-1} in H_2 to only 1940 cm^{-1} at the saddle point. There is thus a crossing in the vibrational energy levels associated with the OH and HH stretching modes along the reaction path (more correctly, a near-crossing since the abiabatic vibrational energy levels obey a noncrossing rule). Proper correlation of the vibrational energy levels of the reactants and the saddle point is most important in understanding the effect of vibrational excitation of the reactants upon the rate

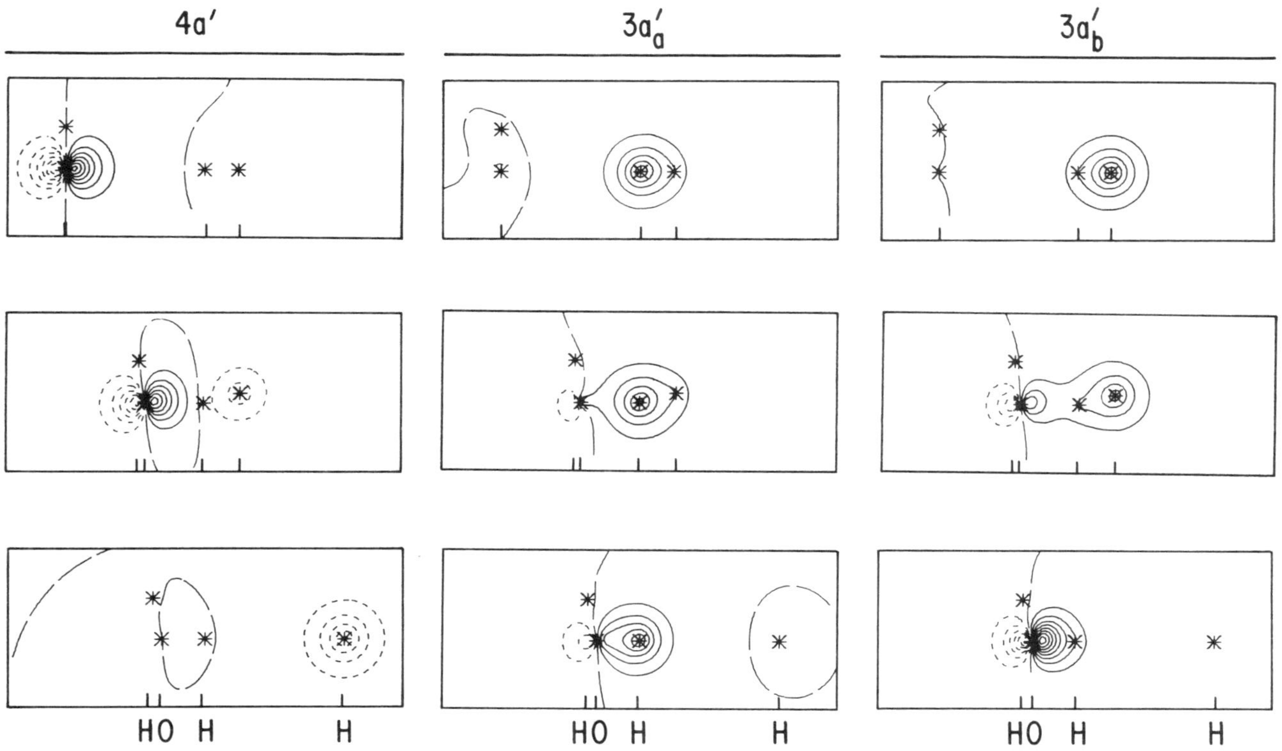

FIGURE 24. SOGVB active orbitals for the OH + H_2 → H_2O + H abstraction reaction in the reactants' (OH + H_2) saddle point (labeled transition state) and products' (H_2O + H) regions. For details on the plots, see the caption to Figure 3. Top row: reactants; middle row: state; bottom row: products.

Table 7
CALCULATED GEOMETRIES, VIBRATIONAL FREQUENCIES, ZERO POINT ENERGIES, AND RELATIVE ENERGIES FOR THE REACTANTS, SADDLE POINT, AND PRODUCTS FOR THE $OH + H_2 \rightarrow H_2O + H$ REACTION

	$OH + H_2$		HOH_2	$H + H_2O$	
	Calcd	Exptl	Calcd	Calcd	Exptl
Geometries (Å)					
R_{OH}	0.99	0.970	0.99	0.98	0.957
R_{HH}	0.76	0.741	0.86		
R_{OH}'			1.34	0.98	0.957
θ_{HOH}'			98	102	104.5
$\theta_{OH'H''}$			165		
Vibrational frequencies (cm^{-1})					
ω_{OHs}	3690	3737.8	3370	3650[a]	3657.0[a]
ω_{HHs}	4120	4401.2	1940	3765[a]	3755.8[a]
$\omega_{HOH'b}$			1250	1660	1594.8
ω_{ipr}			690		
ω_{opr}			440		
Zero point energies (kcal/mol)					
E_{zpe}	11.2	11.6	11.0	13.0	12.9
Relative energies (kcal/mol)					
ΔV_{eff}	0.0	0.0	7.4(6.2[b])	−15.2(−16.7[b])	−16.1
$\Delta V_{eff} + \Delta E_{zpe}$	0.0	0.0	7.1	−13.4	−14.8

[a] In H_2O the first two stretching frequencies correspond to the OH symmetric and antisymmetric modes (Figure 26).
[b] Obtained from calculations with an O[4s3p2d] and H[3s2p] basis set.

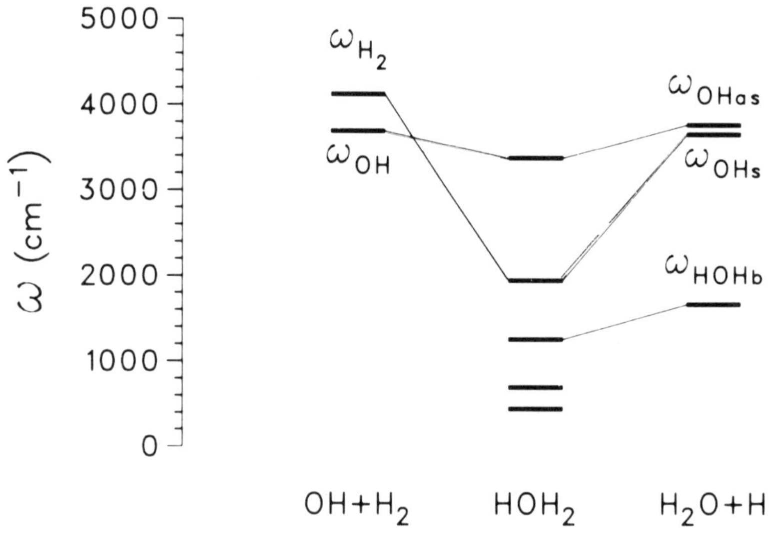

FIGURE 25. Vibrational energy levels of the HOH_2 system in the reactants ($OH + H_2$), at the saddle point and in the products ($H_2O + H$).

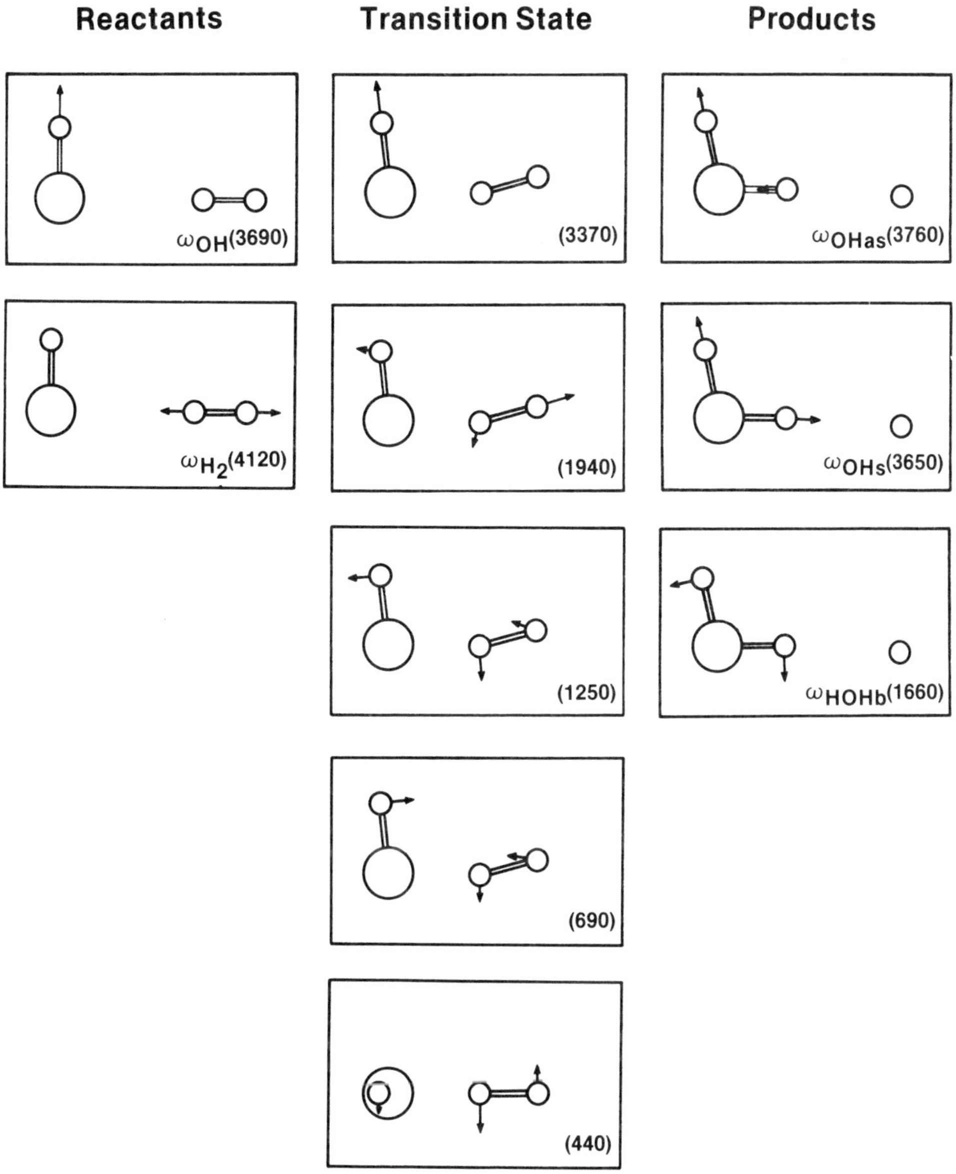

FIGURE 26. Normal modes for the bound vibrational energy levels of the HOH_2 system in the reactants ($OH + H_2$), at the saddle point and in the products ($H_2O + H$).

of the reaction.[110a] The OH and H_2 stretching modes correlate with the OH asymmetric and symmetric stretching modes in the product (H_2O).

The three lowest saddle point frequencies correspond to bending or rocking motions. The highest of these three frequencies corresponds to the HOH′ bending mode in the H_2O molecule. Even though the saddle point is well into the entrance channel, the HOH′ bending frequency is just 400 cm^{-1} less than that in water. The lowest two frequency modes are best described as rocking motion of the H_2 and OH fragments relative to each other. In the reactants or products the latter two modes correspond to rotational modes, as does the HOH′ bending mode in the reactants.

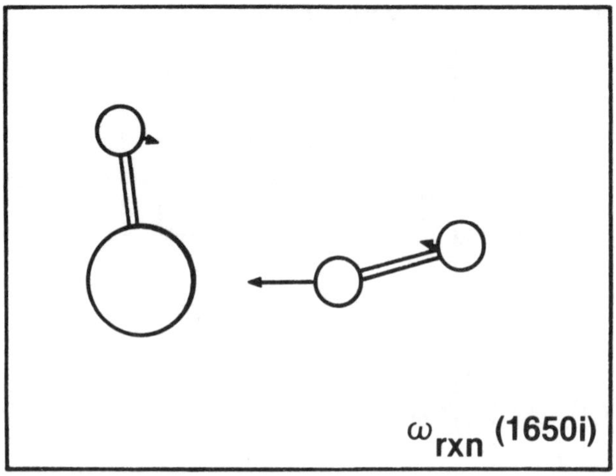

FIGURE 27. Reaction coordinate for the OH + H_2 → H_2O + H abstraction reaction at the saddle point.

The calculated, imaginary, reaction coordinate frequency is depicted in Figure 27. The reaction coordinate at the saddle point corresponds primarily to motion of the central hydrogen atom roughly along the OHH′ axis.

In summary, the OH + H_2 reaction is found to be very similar to both the H + H_2 and the X + H_2 abstraction reactions, the main difference being the presence of an extra bond (that of the OH moiety) which plays a largely "spectator" role in the reaction.

2. The H_2 + D_2 Reaction: A Four-Center Exchange Reaction?

The exchange reaction between H_2 and D_2

$$H_2 + D_2 \rightarrow 2HD \qquad (76)$$

was long considered to be the prototype for four-center exchange reactions. Shock tube studies[112] derived an activation energy of ~42 kcal/mol for Equation 76, a result which was confirmed shortly thereafter.[113] Beginning in the late 1960s, a number of calculations were reported on the potential energy surface for this reaction.[114-117] These calculations all failed to locate a saddle point for Equation 76 which lies below the energy required to dissociate the H_2 molecule, i.e., below ~109 kcal/mol. At this energy the exchange reaction could occur by the sequence of reactions

$$D_2(H_2) + M \rightarrow 2D(2H)$$

$$D + H_2 \rightarrow HD + H$$

$$H + D_2 \rightarrow HD + D$$

where the first step is a collision-induced dissociation reaction involving the buffer gas (M). We shall briefly review here the calculations of Wilson and Goddard.[114]

In 1969, they reported calculations on a number of possible saddle point structures for Equation 76. They used a minimum basis set of Slater functions and included all possible configurations which could be constructed from these functions in the calculations (full CI). They first considered a square configuration of the four hydrogen atoms, the most logical saddle point configuration for Equation 76. Optimizing the bond length of the square, they obtained an energy of 148 kcal/mol above that of two H_2 molecules. Furthermore, they found that the square configuration was unstable with respect to dissociation yielding H_2 + 2H. Many other structures were then considered: a tetrahedron, a centered equilateral triangle, a rhombus, a regular trapezoid, a kite-shaped molecule, and a linear molecule. The energies for all of these structures, except that of linear H_4, was above that for H_2 dissociation. The energy of the linear H_4 configuration with equal bond lengths was found to be only 52 kcal/mol above the energy of two H_2 molecules. However, it is difficult to see how this configuration could correspond to the saddle point for Equation 76, which requires exchange of two atoms.

Rubinstein and Shavitt,[115] Bender and Schaefer,[116] and Silver and Stevens[117] reported results in substantial agreement with those of Wilson and Goddard.[114]

Before proceeding, it should be noted that in none of the above calculations was the Hessian or curvature matrix computed and diagonalized. Therefore, it is not known which, if any, of the proposed structures correspond to actual saddle points, although in some cases, e.g., the square structure, it was shown that it was unstable with respect to dissociation to H_2 + 2H and therefore could not be the saddle point for the reaction.

Arguments based on both the Conservation of Orbital Symmetry (Woodard-Hoffmann rules)[118] and the Orbital Phase Continuity Principle[15] have been advanced to explain the high barrier for the H_2 + D_2 exchange reaction. Consider first the arguments of Hoffmann. For the present purpose let us assume that the reaction proceeds through a square saddle point:

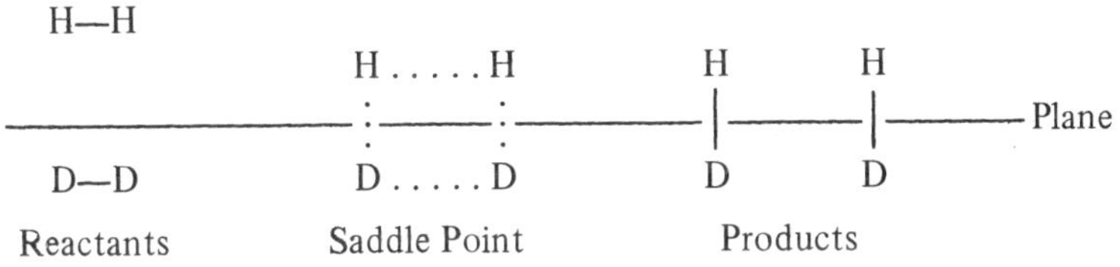

With respect to reflection in the plane indicated above, the HF configuration of the reactants is

$$a^2b^2$$

where a indicates an orbital which is symmetric with respect to reflection in the plane and b indicates an orbital which is antisymmetric with respect to reflection in the plane. On the other hand, the HF configuration of the products is

$$1a^22a^2$$

The orbital correlation diagram for this reaction is given in Figure 28. As can be seen in the figure, the reactant configuration correlates with a highly excited state of the product configuration and vice versa. The energy of this excited configuration, which corresponds to promoting two of the electrons to an antibonding orbital in the product, is very high, i.e., much greater than the bond energy of H_2 (~109 kcal/mol). Of course, a multiconfiguration wavefunction composed of both of the above configurations would properly describe the evolution of the wavefunction from reactants through the saddle point to products. However, as the above calculations indicate, the energy at the square configuration (indicated by the maximum in Figure 28) is still above that for breaking one of the H_2 bonds.

Finally, let us consider the arguments put forward by Goddard.[15] The GVB orbitals of the reactants and saddle point (square configuration) in the H_2 + D_2 reaction are represented schematically in Figure 29. Ignore for the moment the lower right hydrogen atom. The

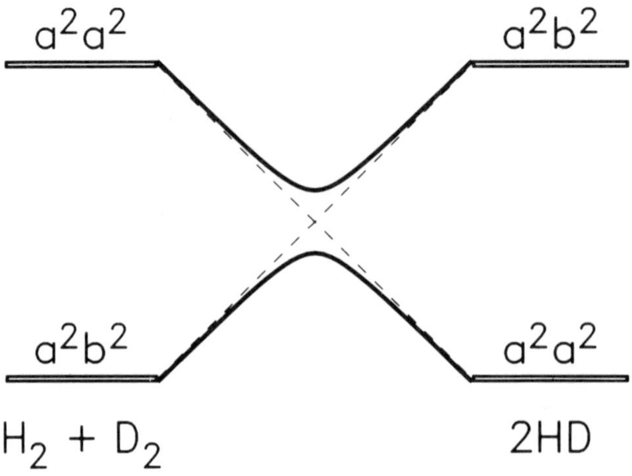

FIGURE 28. Correlation diagram for the RHF wavefunction of the H_2 + D_2 → 2HD exchange reaction.

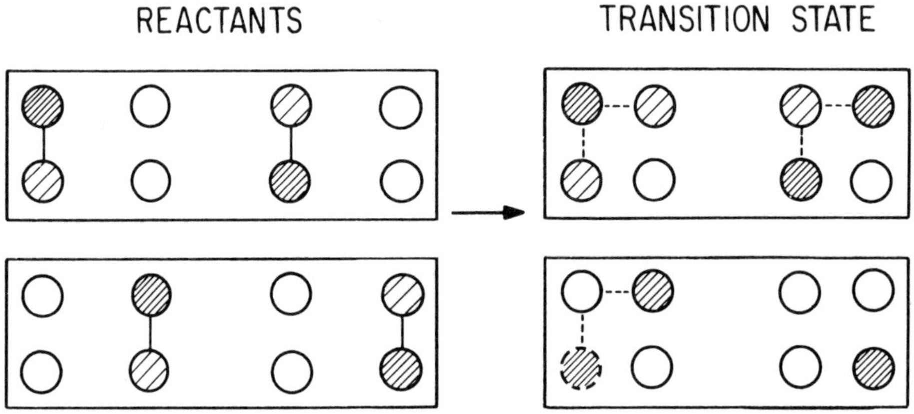

FIGURE 29. Schematic representation of the GVB orbitals of the H_2 + D_2 → 2HD exchange reaction in the reactants and at the saddle point.

orbitals in the resulting H_3 system will behave as indicated earlier in Section III.A. At the saddle point the upper left orbital (ϕ_{1a}) will delocalize slightly onto the upper right and lower left atoms (as did the $1\sigma_a$ orbital in H + H_2), while the lower left orbital (ϕ_{1b}) will delocalize onto the upper right orbital (as did the $1\sigma_b$ orbital in H + H_2). In response to this, the orbital originally on the upper right atom (ϕ_{2a}) will delocalize onto the lower left atom with a change of phase (i.e., in an antibonding fashion). With this, however, the ϕ_{2a} orbital becomes orthogonal to the orbital on the lower right atom (ϕ_{2b}). Thus, at the square configuration the ($2\phi_{2a}$, $2\phi_{2b}$) orbital pair overlap vanishes and, as a result, the corresponding bond is broken. Given this and the other unfavorable electronic interactions at the saddle point, it is not surprising that the energy of the square configuration is higher than that for H_2 + 2H. Of course, in the GVB wavefunction, the orbitals delocalize as much as possible to minimize the unfavorable interactions; however, the barrier still remains above that for dissociation of H_2.[114]

Both of the above arguments can be generalized to the other saddle point structures. In all cases high barriers are predicted.

While the source of the disagreement between theory and experiment remains unresolved at the present time, it appears unlikely that a low energy pathway will be found for the $H_2 + D_2$ exchange reaction.

3. The CH + H_2 Reaction: A Reaction with a Nonleast Motion Reaction Path

As was evident in the previous sections, the general features of the potential energy surfaces for chemical reactions are largely determined by the orbital structure of the reactants. This is nowhere more evident than in the reaction of methylidene (CH) with molecular hydrogen. Reaction of CH with H_2 occurs by two routes:

1. Abstraction of one of the hydrogens by the singly occupied π-orbital of CH, directly forming $CH_2 + H$.
2. Insertion of the lone pair orbitals into the HH-bond forming CH_3, which can then dissociate to form $CH_2 + H$.

It is the latter process,

$$CH + H_2 \rightarrow [CH_3]^{\dagger} \rightarrow CH_2 + H \tag{77}$$

which is the subject of the present section. In fact, here we will examine only the initial step in this reaction: the formation of the CH_3 complex. Calculations have been reported on this reaction by Brooks and Schaefer;[119] Dunning[120] has also carried out calculations on Equation 77.

As we saw in previous sections, a low energy pathway for a reaction will occur only if the overlaps between all of the electron pairs can be maintained during the course of the reaction, i.e., if the new pairs can be formed at the same time that the old pairs are broken. For Equation 77 the active orbitals are the carbon lone pair orbitals, $\phi_{2a'_a}$ and $\phi_{2a'_b}$ and the hydrogen bond orbitals, $\phi_{3a'_a}$ and $\phi_{3a'_b}$ at the limit of the reactant in Figure 30. Consider first a least-motion approach, i.e., an approach in which the CH bond axis is perpendicular to the HH bond axis:

```
        H
H—C     |
        H
```

An analysis of the changes in the orbitals on going from reactants to the saddle point, similar to that given earlier for the $H_2 + D_2$ reaction, shows that for this pathway, one of the pair overlaps must go to zero at the saddle point and thus the barrier to reaction will be high. Identical conclusions can be drawn from the Woodward-Hoffmann rules.[118] Brooks and Schaefer,[119] who carried out HF + 1 + 2 calculations on Equation 77 with a polarized double-zeta basis set reported a barrier of 76.6 kcal/mol for the least-motion pathway.

Consider now an approach in which the CH and HH bond axis are nearly parallel. The active orbitals for this pathway, i.e., the carbon lone pair orbitals and the HH bond orbitals, are plotted in Figure 30. As is evident in this figure, the CH + H_2 reaction may now be viewed as an attack by one of the CH lone pair orbitals on the HH bond; compare, e.g., the behavior of the $2a_b'$, $3a_a'$, and $3a_b'$ orbitals of CH + H_2 at the limit of the reactant and the saddle point with the 2σ, $1\sigma_a$, and $1\sigma_b$ orbitals of H + H_2 at the equivalent points. In this configuration, the overlap between the $2a_a'$ and $2a_b'$ orbitals can be maintained (as can that between $3a_a'$ and $3a_b'$). Thus, at the saddle point, the overlap between the $2a_a'$ and $2a_b'$ orbitals is 0.69, exactly the same as that between the CH lone pair orbitals in the reactants. Thus, a path involving an initially parallel approach of the CH and H_2 species will have a low barrier. Of course, in the product (CH_3) the CH and HH bond axis are again perpendicular (Figure 30).

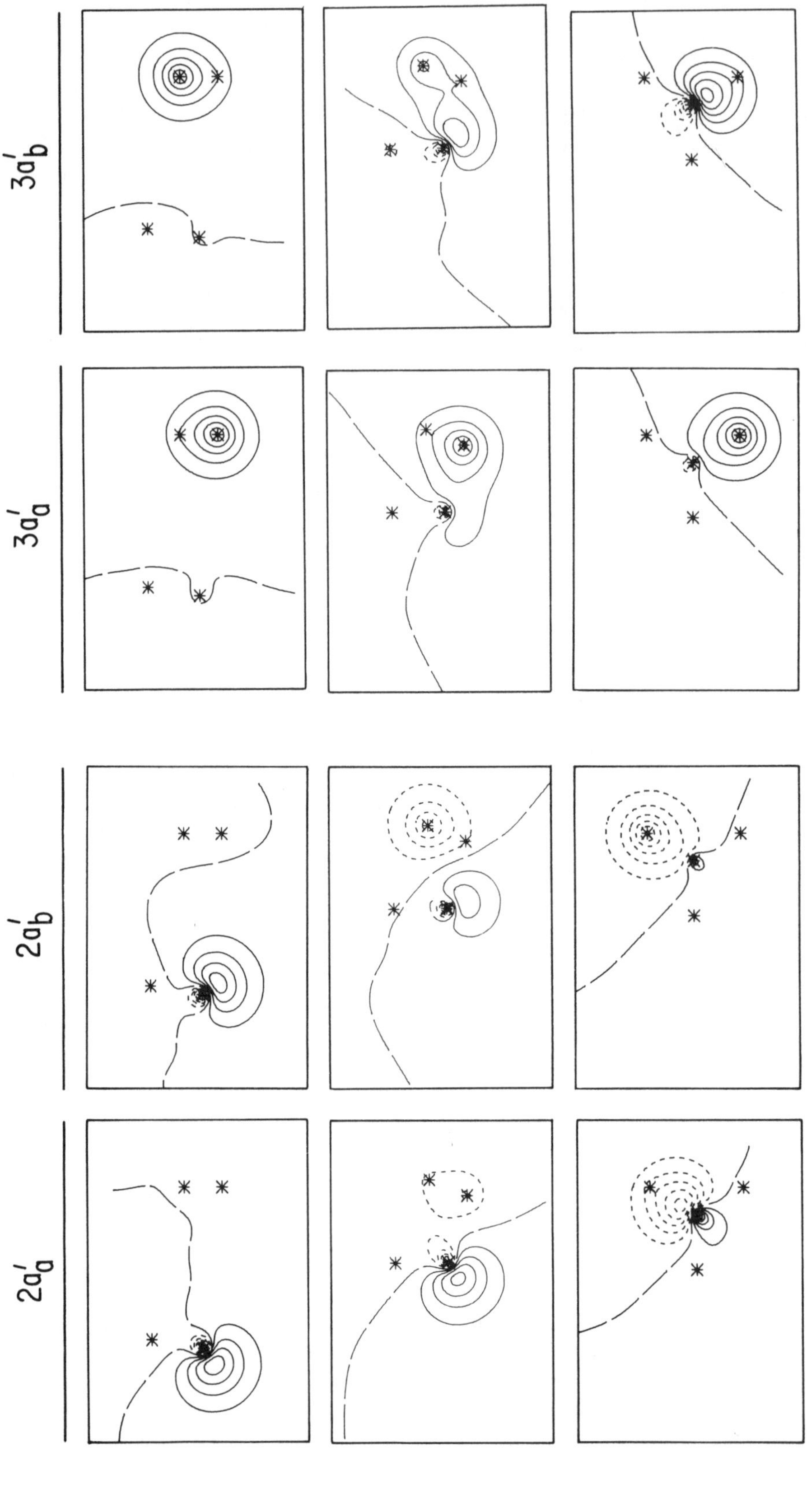

FIGURE 30. SOGVB active orbitals for the $CH + H_2 \rightarrow CH_3$ addition reaction in the reactants' ($CH + H_2$) saddle point (labeled transition state) and products' (CH_3) regions. For details on the plots, see the caption to Figure 3. Top row: reactants; middle row: state; bottom row: products.

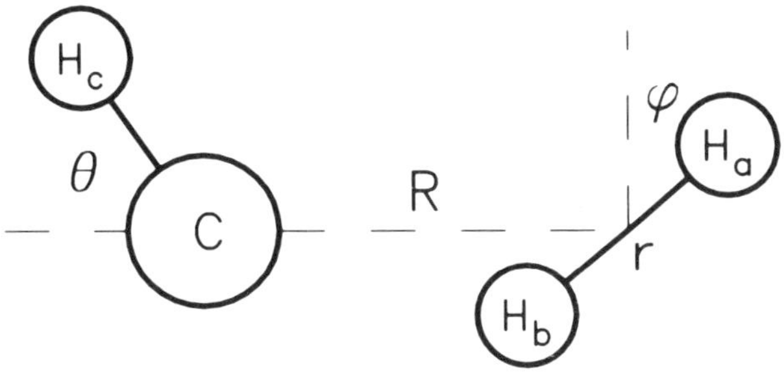

FIGURE 31. Internal coordinates used in the calculations on the (planar) $CH + H_2 \rightarrow CH_3$ addition reaction.

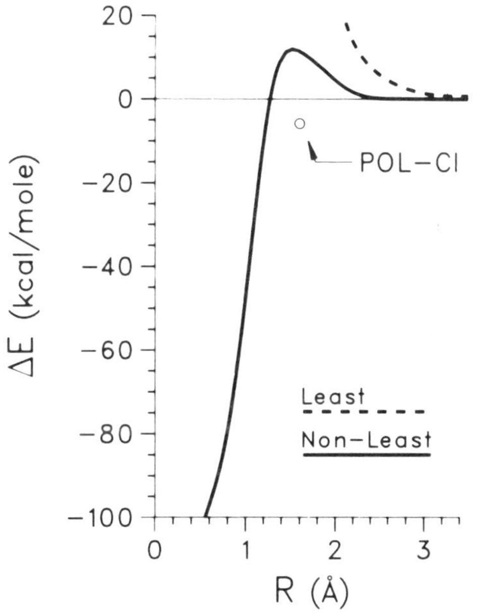

FIGURE 32. Energy profile for the $CH + H_2 \rightarrow CH_3$ addition reaction as a function of R; r and Θ were optimized for each value of R.[120]

The saddle point geometry illustrated in Figure 30 was determined from SOGVB calculations with a [3s3p1d] basis on the carbon atom and a [3s1p] basis set on the hydrogen atoms.[120] For a four-atom system there are six internal degrees of freedom; in the SOGVB calculations on Equation 77 only 4 degrees of freedom were explored. The internal coordinates used are illustrated in Figure 31. The SOGVB calculations predict that the reaction is 97.4 kcal/mol exoergic; Brooks and Schaefer reported an exoergicity of 107.6 kcal/mol. The saddle point geometry obtained in the GVB calculations was (R = 1.60 Å, r = 0.81 Å, θ = 90, ϕ = 22); the barrier height was 8.1 kcal/mol. Brooks and Schaefer, who fixed θ = 90 and varied only R and r, reported (R = 1.46 Å, r = 0.82 Å) and a barrier of 4.0 kcal/mol. In both cases the barrier occurs early, as would be expected for such an exoergic reaction. Considering the differences in the two calculations, the agreement is satisfactory.

At the saddle point the dominant component of the reaction coordinate corresponds to R (with a smaller admixture of r). To determine an approximate energy profile for Equation 77, R was taken as the "distinguished reaction coordinate" and (r, θ) were optimized at each value of R; this then defines the "R" reaction path. The energy profile so obtained is plotted in Figure 32. The barrier obtained from these calculations is ~12 kcal/mol, somewhat

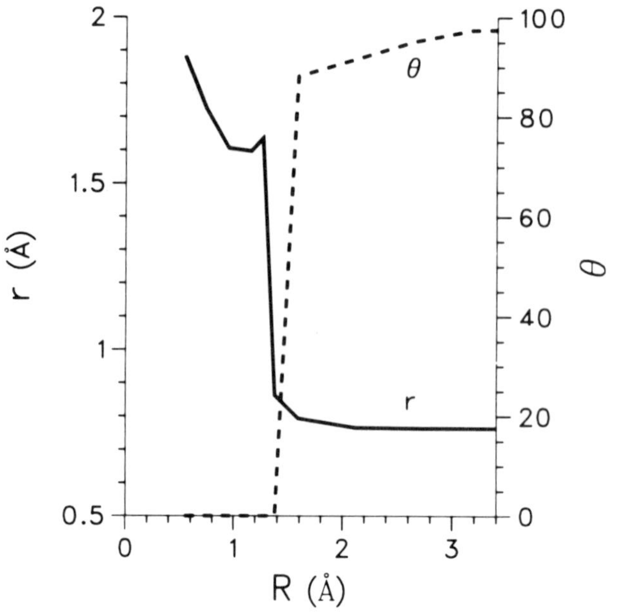

FIGURE 33. Dependence of r and θ on R for the CH + $H_2 \rightarrow CH_3$ addition reaction (for a definition of the coordinates see Figure 31).

larger than that obtained in the more complete calculations above. However, as can be seen in Figure 32, POL-CI calculations at the reactant and saddle point geometries eliminates the barrier to reaction. Although not definitive, these results strongly suggest that there is no barrier to the formation of the CH_3 complex. This is consistent with the results of Berman and Lin.[121]

While the energy obtained in the above calculations appears to be a smooth and continuous function of R, this is not the case for the other geometrical parameters, r and θ. These quantities are plotted in Figure 33. This behavior is a consequence of the curvature of the reaction path: for R shorter than the saddle point, the reaction path becomes a strong admixture of R, r, and θ. Any attempt to describe the reaction path in terms of a single ''distinguished reaction coordinate'' results in discontinuities in the other coordinates which are strongly coupled to the ''distinguished coordinate''.

ACKNOWLEDGMENT

We wish to thank our colleagues Dr. Albert F. Wagner, Dr. Andrew Komornicki, and Prof. Donald G. Truhlar for their numerous helpful comments on the manuscript. Work was performed under the auspices of the Office of Basic Energy Sciences, U.S. Department of Energy, under contract W-31-109-ENG-38.

REFERENCES

1. **Born, M. and Huang, K.,** *The Dynamic Theory of Crystal Lattices,* Oxford University Press, Oxford, 1954.
2. **Miller, W. H., Handy, N. C., and Adams, J. E.,** Reaction path Hamiltonian for polyatomic molecules, *J. Chem. Phys.,* 72, 99, 1980.
3. **Roothaan, C. C. J.,** New developments in molecular orbital theory, *Rev. Mod. Phys.,* 23, 69, 1951.
4. **Roothaan, C. C. J.,** Self-consistent field theory for open shells of electronic systems, *Rev. Mod. Phys.,* 32, 179, 1960.
5. **Huzinaga, S.,** Analytical methods in Hartree-Fock self-consistent theory, *Phys. Rev.,* 122, 131, 1961.
6. **Hunt, W. J., Goddard, W. A., III, and Dunning, T. H., Jr.,** The incorporation of quadratic convergence into open-shell self-consistent field equations, *Chem. Phys. Lett.,* 6, 147, 1970.
7. **Davidson, E. R.,** Spin-restricted open-shell self-consistent-field theory, *Chem. Phys. Lett.,* 21, 565, 1973.
8. **Bobrowicz, F. W. and Goddard, W. A., III,** The self-consistent field equations for generalized valence bond and open-shell Hartree-Fock wavefunctions, in *Methods of Electronic Structure Theory,* Schaefer, H. F., III, Ed., Plenum Press, N.Y., 1977, chap. 4.
9. **Dunning, T. H., Jr.,** unpublished.
10. **Liu, B.,** *Ab initio* potential energy surface for linear H_3, *J. Chem. Phys.,* 58, 1925, 1973.
11. **Olsen, J. and Yeager, D. L.,** Optimization and characterization of a multiconfigurational self-consistent field (MCSCF) state, in *Advances in Chemical Physics,* Prigogine, I. and Rice, S. A., Eds., John Wiley & Sons, N.Y., 1983, 1.
12. **Wahl, A. C. and Das, G.,** The multiconfigurational self-consistent field method, in *Methods in Electronic Structure Theory,* Schaefer, H. F., III, Ed., Plenum Press, N.Y., 1977, chap. 3.
13. **Goddard, W. A., III, Dunning, T. H., Jr., Hunt, W. J., and Hay, P. J.,** Generalized valence bond description of bonding in low-lying states of molecules, *Acc. Chem. Res.,* 6, 368, 1973.
14. **Baybutt, P., Bobrowicz, F. B., Kahn, L. R., and Truhlar, D. G.,** Generalized valence bond investigation of the reaction $H + Br_2 \rightarrow HBr + Br$, *J. Chem. Phys.,* 68, 4809, 1978.
15. **Goddard, W. A., III,** Selection rules for chemical reactions using the orbital phase continuity principle, *J. Am. Chem. Soc.,* 94, 793, 1972.
16. **Goddard, W. A., III and Ladner, R. C.,** The optimum orbitals for the $H_2 + D \rightarrow H + HD$ exchange reaction, *Int. J. Quantum Chem.,* IIIS, 63, 1969.
17. **Voter, A. F. and Goddard, W. A., III,** The generalized resonating valence bond method; barrier heights in the HF + D and HCl + D exchange reactions, *J. Chem. Phys.,* 75, 3638, 1981.
18. **Cheung, L. M., Sundberg, K. R., and Rudenberg, K.,** Electronic rearrangements during chemical reactions. II. Planar dissociation of ethylene, *Int. J. Quantum Chem.,* 16, 1103, 1979.
19. **Roos, B. O., Taylor, P. R., and Siegbahn, P. E. M.,** A complete active space SCF method (CASSCF) using a density matrix formulated super-CI method, *Chem. Phys.,* 48, 157, 1980.
20. **Karlstrom, G. and Roos, B. O.,** A multiconfigurational SCF study of the transformation of methylene peroxide to dioxirane. Intermediates in the ozonolysis of ethylene, *Chem. Phys. Lett.,* 79, 416, 1981.
21. **Shavitt, I.,** The method of configuration interaction, in *Methods of Electronic Structure Theory,* Schaefer, H. F., III, Ed., Plenum Press, N.Y., 1977, chap. 6.
22. **Roos, B. O. and Siegbahn, P. E. M.,** The direct configuration interaction method from molecular integrals, in *Methods of Electronic Structure Theory,* Schaefer, H. F., III, Ed., Plenum Press, N.Y., 1977, chap. 7.
23. **Saunders, V. R. and van Lenthe, J. H.,** The direct CI method. A detailed analysis, *Mol. Phys.,* 48, 923, 1983.
24. **Shavitt, I.,** The graphical unitary group approach and its application to direct configuration interaction calculations, in *The Unitary Group for the Evaluation of Electronic Energy Matrix Elements,* Lecture Notes in Chemistry, Vol. 22, Hinze, J., Ed., Springer-Verlag, Berlin, 1981; Lischka, H., Shepard, R., Brown, F. B., and Shavitt, I., New implementation of the graphical unitary group approach for multireference direct configuration interaction calculations, *Int. J. Quantum Chem.,* 15, 91, 1981.
25. **Liu, B. and Yoshimine, M.,** The Alchemy configuration interaction method. I. The symbolic matrix method for determining elements of matrix operators, *J. Chem. Phys.,* 74, 612, 1981.
26. **Siegbahn, P. E. M.,** Direct configuration interaction with a reference state composed of many reference configurations, *Int. J. Quantum Chem.,* 18, 1229, 1980.
27. **Bunge, A.,** Electronic wavefunctions for atoms. III. Partition of degenerate spaces and ground state of C, *J. Chem. Phys.,* 53, 20, 1970.
28. **McLean, A. D. and Liu, B.,** Classification of configurations and the determination of interacting and non-interacting spaces in configuration interaction, *J. Chem. Phys.,* 58, 1066, 1973.
29. **Hay, P. J. and Dunning, T. H., Jr.,** Polarization CI wavefunctions: the valence state of the NH radical, *J. Chem. Phys.,* 64, 5077, 1976.

30. **Dunning, T. H., Jr.,** The low-lying states of hydrogen fluoride: potential energy curves for the $X^1\Sigma^+$, $^3\Sigma^+$, $^3\Pi$, and $^1\Pi$ states, *J. Chem. Phys.*, 65, 3854, 1976.
31. **Dunning, T. H., Jr., Cartwright, D. C., Hunt, W. J., Hay, P. J., and Bobrowicz, F. W.,** Generalized valence bond calculations on the ground state ($X^1\Sigma_g^+$) of nitrogen, *J. Chem. Phys.*, 64, 4755, 1976.
32. **Silverstone, H. J. and Sinanoglu, O.,** Many-electron theory of nonclosed-shell atoms and molecules. I. Orbital wavefunctions and perturbation theory, *J. Chem. Phys.*, 44, 1899, 1966.
33. **Lie, G. C., Hinze, J., and Liu, B.,** Valence excited states of CH. I. Potential curves, *J. Chem. Phys.*, 59, 1872, 1973.
34. **Schaefer, H. F., III and Bender, C. F.,** Multiconfiguration wavefunctions for the water molecule, *J. Chem. Phys.*, 55, 1720, 1971.
35. **Schaefer, H. F., III, Klemm, R. A., and Harris, F. E.,** First-order wavefunctions, orbital correlation energies, and electron affinities of first-row atoms, *J. Chem. Phys.*, 51, 4643, 1969.
36. **Boys, S. F.,** Electronic wavefunctions. I. A general method of calculation for the stationary states of any molecular system, *Proc. R. Soc. London Ser. A.*, 200, 542, 1950.
37. **Hehre, W. J., Stewart, R. F., and Pople, J. A.,** Self-consistent molecular-orbital methods. I. Use of gaussian expansions of Slater-type atomic orbitals, *J. Chem. Phys.*, 51, 2657, 1969.
38. **Hariharan, P. C. and Pople, J. A.,** Influence of polarization functions on molecular orbital hydrogenation energies, *Theor. Chim. Acta.*, 28, 213, 1973.
39. **Krishnan, R., Binkley, J. S., Seeger, R., and Pople, J. A.,** Self-consistent molecular orbital methods. XX. A basis set for correlated wavefunctions, *J. Chem. Phys.*, 72, 650, 1980.
40. **Huzinaga, S.,** Gaussian-type functions for polyatomic systems. I., *J. Chem. Phys.*, 42, 1293, 1965.
41. **Dunning, T. H., Jr.,** Gaussian basis functions for use in molecular calculations. I. Contraction of (9s,5p) atomic basis sets for first row atoms, *J. Chem. Phys.*, 53, 2823, 1970.
42. **Dunning, T. H., Jr. and Hay, P. J.,** Gaussian basis sets for molecular calculations, in *Methods of Electronic Structure Theory*, Schaefer, H. F., III, Ed., Plenum Press, N.Y., 1977.
43. **Raffenetti, R. C.,** General contraction of gaussian atomic orbitals; core, valence, polarization, and diffuse basis sets; molecular integral evaluation, *J. Chem. Phys.*, 58, 4452, 1973.
43a. **Flanagan, M. C., Komornicki, A., and McIver, J. W., Jr.,** Ground-state potential surfaces and thermochemistry, in *Semiempirical Methods of Electronic Structure Calculation, Part B: Applications*, Segal, G. A., Ed., Plenum Press, N.Y., 1977.
44. **Shavitt, I., Stevens, R. M., Minn, F. L., and Karplus, M.,** Potential energy surface for H_3, *J. Chem. Phys.*, 48, 2700, 1968.
45. **Truhlar, D. G. and Horowitz, C. J.,** Functional representation of Liu and Siegbahn's accurate *ab initio* potential energy calculations for $H + H_2$, *J. Chem. Phys.*, 68, 2466, 1978.
46. **Siegbahn, P. and Liu, B.,** An accurate three-dimensional potential energy surface for H_3, *J. Chem. Phys.*, 68, 2457, 1978.
47. **Murrell, J. N. and Laidler, K. J.,** Symmetry of activated complexes, *Trans. Faraday Soc.*, 64, 371, 1968.
48. **Gwinn, W. D.,** Normal coordinates: general theory, redundant coordinates, and general analysis using electronic computers, *J. Chem. Phys.*, 55, 477, 1971.
49. **Wilson, E. B., Decius, J. E., and Cross, P. E.,** *Molecular Vibrations*, McGraw-Hill, N.Y., 1955.
50. **Pulay, P.,** *Ab initio* calculation of force constants and equilibrium geometries in polyatomic molecules. I. Theory, *Mol. Phys.*, 17, 197, 1969.
51. **McIver, J. W., Jr. and Komornicki, A.,** Rapid geometry optimization for semi-empirical molecular orbital methods, *Chem. Phys. Lett.*, 10, 303, 1971.
52. **McIver, J. W., Jr. and Komornicki, A.,** Structure of transition states in organic reactions. General theory and an application to the cyclobutene-butadiene isomerization using a semiempirical molecular orbital method, *J. Am. Chem. Soc.*, 94, 2625, 1972.
53. **Poppinger, D.,** On the calculation of transition states, *Chem. Phys. Lett.*, 35, 550, 1975.
54. **Komornicki, A., Ishida, K., Morokuma, K., Ditchfield, R., and Conrad, M.,** Efficient determination and characterization of transition states using *ab initio* methods, *Chem. Phys. Lett.*, 45, 595, 1977.
55. **Bell, S., Crighton, J. S., and Fletcher, R.,** A new efficient method for locating saddle points, *Chem. Phys. Lett.*, 82, 122, 1981.
56. **Schlegel, H. B.,** Optimization of equilibrium geometries and transition structures, *J. Comput. Chem.*, 3, 214, 1982.
57. **Cerjan, C. J. and Miller, W. H.,** On finding transition states, *J. Chem. Phys.*, 75, 2800, 1981.
58. **Metiu, H., Ross, J., Silbey, R., and George, T. F.,** On symmetry properties of reaction coordinates, *J. Chem. Phys.*, 61, 3200, 1971; McIver, J. W., Jr., The structure of transition states: are they symmetric, *Acc. Chem. Res.*, 7, 72, 1974.
59. **Stanton, R. E. and McIver, J. W., Jr.,** Group theoretical selection rules for the transition states of chemical reactions, *J. Am. Chem. Soc.*, 97, 3632, 1975.

60. **Pechukas, P.,** On simple saddle points of a potential energy surface, the conservation of nuclear symmetry along paths of steepest descent, and the symmetry of transition states, *J. Chem. Phys.*, 64, 1516, 1976.
61. **McCullough, E. A., Jr. and Silver, D. M.,** Reaction path properties at critical points on potential surfaces, *J. Chem. Phys.*, 62, 4050, 1975.
62. **Ishida, K., Morokuma, K., and Komornicki, A.,** The intrinsic reaction coordinate. An *ab initio* calculation for HNC → HCN and $H^- + CH_4 \rightarrow CH_4 + H^-$, *J. Chem. Phys.*, 66, 2153, 1977.
63. **Garrett, B. C. and Truhlar, D. G.,** Generalized transition state theory. Quantum effects for collinear reactions of hydrogen molecules and isotopically substituted hydrogen molecules, *J. Phys. Chem.*, 83, 1079, 1979; Sana, M., Reckinger, G., and Leroy, G., An internal coordinate invariant reaction pathway, *Theor. Chim. Acta (Berlin)*, 58, 145, 1981.
64. **Fukui, K.,** A formulation of the reaction coordinate, *J. Phys. Chem.*, 74, 4161, 1970.
65. **Fukui, K.,** The path of chemical reactions — the IRC approach, *Acc. Chem. Res.*, 14, 363, 1981.
66. **Truhlar, D. G. and Kuppermann, A.,** Exact tunneling calculations, *J. Am. Chem. Soc.*, 93, 1840, 1971.
67. **Pople, J. A., Krishnan, R., Schlegel, H. B., and Binkley, J. S.,** Derivative studies in Hartree-Fock and Moller-Plesset theories, *Int. J. Quantum Chem.*, S13, 225, 1979; Pulay, P., Direct use of the gradient for investigating molecular energy surfaces, in *Applications of Electronic Structure Theory*, Schaefer, H. F., III, Ed., Plenum Press, N.Y., 1977; Dupuis, M., Energy derivatives for configuration interaction wave functions, *J. Chem. Phys.*, 74, 5758, 1981.
68. **Evans, M. G. and Polanyi, M.,** Inertia and driving force of chemical reactions, *Trans. Faraday Soc.*, 34, 11, 1938.
69. **Hammond, G. S.,** A correlation of reaction rates, *J. Am. Chem. Soc.*, 77, 334, 1955.
70. **Johnston, H. S. and Parr, C.,** Activation energies from bond energies. I. Hydrogen transfer reactions, *J. Am. Chem. Soc.*, 85, 2544, 1963.
71. **Marcus, R. A.,** Theoretical relations among rate constants, barriers, and bronsted slopes of chemical reactions, *J. Phys. Chem.*, 72, 891, 1968.
72. **Mok, M. H. and Polanyi, J. C.,** Location of energy barriers. II. Correlation with barrier height, *J. Chem. Phys.*, 51, 1451, 1969.
73. **Agmon, N. and Levine, R. D.,** Energy, entropy, and the reaction coordinate: thermodynamic-like relations in chemical kinetics, *Chem. Phys. Lett.*, 52, 197, 1977.
74. **Agmon, N.,** Quantitative Hammond postulate, *J. Chem. Soc. Faraday Trans. II*, 74, 388, 1978.
75. **Dunning, T. H., Jr.,** Theoretical studies of the energetics of the abstraction and exchange reactions in H + HX, with X = F−I, *J. Phys. Chem.*, to be published.
76. **Bender, C. F., O'Neil, S. V. Pearson, P. K., and Schaefer, H. F., III,** Potential energy surface including electron correlation for $F + H_2 \rightarrow FH + H$: refined linear surface, *Science*, 176, 1412, 1972.
77. **Bender, C. F., Garrison, B. J., and Schaefer, H. F., III,** A critical test of semiempirical FH_2 potential energy surfaces: the barrier height for H + FH → HF + H, *J. Chem. Phys.*, 62, 1188, 1975.
78. **Ungemach, S. R., Schaefer, H. F., III, and Liu, B.,** Potential energy surfaces, *Faraday Disc. Chem. Soc.*, 62, 330, 1977.
79. **Botschwina, P. and Meyer, W.,** PNO-CEPA calculation of collinear potential energy barriers for thermoneutral exchange reactions, *Chem. Phys.*, 20, 43, 1977.
80. **Wadt, W. R. and Winter, N. W.,** Accurate characterization of the transition state geometry for the HF + H′→ H + H′F reaction, *J. Chem. Phys.*, 67, 3068, 1977.
81. **Dunning, T. H., Jar.,** The barriers for abstraction and exchange in H + HCl, *J. Chem. Phys.*, 66, 2752, 1977.
82. **Botschwina, P. and Meyer, W.,** A PNO-CEPA calculation of the barrier height for the collinear atom exchange reaction H′ + BrH → H′Br + H, *J. Chem. Phys.*, 67, 2390, 1977.
83. **Levy, M. R.,** Dynamics of reactive collisions, *Prog. React. Kinet.*, 10, 1, 1979.
84. **Bartoszek, F. E., Manos, D. M., and Polanyi, J. C.,** Effect of changing reagent energy. X. Vibrational threshold energies for alternative reaction paths HF(v) + D → F + HD and → H + DF, *J. Chem. Phys.*, 69, 933, 1978.
85. **Miller, J. C. and Gordon, R. J.,** Kinetics of the $Cl\text{-}H_2$ system. II. Abstraction vs. exchange in D + HCl, *J. Chem. Phys.*, 78, 3713, 1983.
86. **Balint-Kurti, G. G. and Yardley, R. N.,** Potential energy surfaces for simple chemical reactions: application of valence-bond techniques to the Li + HF → LiF + H reaction, *Faraday Disc. Chem. Soc.*, 62, 77, 1977.
87. **Chen, M. M. L. and Schaefer, H. F., III,** Potential energy surface for the Li + HF → LiF + H reaction, *J. Chem. Phys.*, 72, 4376, 1980.
88. **Huber, K. P. and Herzberg, G.,** *Molecular Spectra and Molecule Structure*, Vol. IV, Van Nostrand Reinhold, N.Y., 1979.
89. **Becker, C. H., Casavecchia, P., Tiedmann, P. W., Valentini, J. J., and Lee, Y. T.,** Study of the reaction dynamics of Li + HF, HCl by the crossed molecular beams method, *J. Chem. Phys.*, 73, 2833, 1980.

90. **Harding, L. B.,** Theoretical studies of the potential surface for the reaction $C(^3P) + H \rightarrow CH_2\ (^3B_1)$, *J. Phys. Chem.*, 87, 441, 1983.
91. **Goddard, J. D. and Schaefer, H. F., III,** The photodissociation of formaldehyde: potential energy surface features, *J. Chem. Phys.*, 70, 5117, 1979.
92. **Harding, L. B., Schlegel, H. B., Krishnan, R., and Pople, J. A.,** Moller-Plesset study of the H_3CO potential energy surface, *J. Phys. Chem.*, 84, 3394, 1980.
93. **Dunning, T. H., Jr.,** Theoretical characterization of the potential energy surface of the ground state of HCO system, *J. Chem. Phys.*, 73, 2304, 1980.
94. **Bruna, P., Buenker, R. J., and Peyerimhoff, S. D.,** *Ab initio* study of the structure, isomers, and vertical electronic spectrum of the formyl radical HCO, *J. Mol. Struct.*, 32, 217, 1976.
95. **Tanaka, K. and Davidson, E. R.,** A theoretical study on the potential energy surfaces of the lower electronic states of HCO, *J. Chem. Phys.*, 70, 2904, 1979.
96. **Adams, G. F., Bent, G. D., Purvis, G. D., and Bartlett, R. J.,** The electronic structure of the formyl radical HCO, *J. Chem. Phys.*, 71, 3697, 1979.
97. **Harding, L. B.,** unpublished.
98. **Warneck, P.,** Heat of formation of the HCO radical, *Z. Naturforsch., Teil,* A 29, 350, 1974.
99. **Wang, H. Y., Eyre, J. A., and Dorfman, L. M.,** Activation energy for the gas phase reaction of hydrogen atoms with carbon monoxide, *J. Chem. Phys.*, 59, 5199, 1973.
100. **Harding, L. B., Wagner, A. F., Bowman, J. M., Schatz, G. C., and Christoffel, K.,** *Ab initio* calculation of the transition-state properties and addition rate constants for $H + C_2H_2$ and selected isotopic analogues, *J. Phys. Chem.*, 86, 4312, 1982.
101. **Nagase, S. and Kern, C. W.,** *Ab initio* mechanistic study of radical reactions. Transition states and reaction barriers for the reaction of atomic hydrogen with acetylene, *J. Am. Chem. Soc.*, 101, 2544, 1979.
102. **Nagase, S. and Kern, C. W.,** *Ab initio* mechanistic study of radical reactions. Relative reactivity of olefinic and acetylenic bonds in addition reactions, *J. Am. Chem. Soc.*, 102, 4513, 1980.
103. **Hase, W. L., Mrowka, G., Brudzynski, R. J., and Sloane, C. S.,** An analytic function describing the $H + C_2H_4 \rightarrow C_2H_5$ potential energy surface, *J. Chem. Phys.*, 69, 3548, 1978.
104. **Bair, R. A. and Dunning, T. H., Jr.,** *J. Chem. Phys.*, to be published.
105. **Blint, R. J. and Newton, M. D.,** *Ab initio* potential energy surfaces for the reaction of atomic carbon with molecular hydrogen, *Chem. Phys. Lett.*, 32, 178, 1975.
106. **Casida, M. E., Chen, M. M. L., MacGregor, R. D., and Schaefer, H. F., III,** Walsh's rules and the small bond angle states of triatomic dihydride molecules, *Isr. J. Chem.*, 19, 127, 1980.
107. **Knowles, P., Handy, N. C., and Carter, S.,** A potential energy surface for the ground state of CH_2, *Mol. Phys.*, 49, 681, 1983.
108. **Walch, S. P. and Dunning, T. H., Jr.,** A theoretical study of the potential energy surface for $OH + H_2$, *J. Chem. Phys.*, 72, 1303, 1980.
109. **Dunning, T. H., Jr., Walch, S. P., and Wagner, A. F.,** Theoretical studies of selected reactions in the hydrogen-oxygen system, in *Potential Energy Surfaces and Dynamics Calculations*, Truhlar, D. G., Ed., Plenum Press, N.Y., 1981.
110. **Schatz, G. C. and Walch, S. P.,** An *ab initio* calculation of the rate constant for the $OH + H_2 \rightarrow H_2O + H$ reaction, *J. Chem. Phys.*, 72, 776, 1980; Isaacson, A. D. and Truhlar, D. G., Polyatomic canonical variational theory for chemical reaction rates. Separable mode formalism with application to $OH + H_2 \rightarrow H_2O + H$, *J. Chem. Phys.*, 76, 1380, 1982.
110a. **Truhlar, D. G. and Isaacson, A. D.,** Statistical-diabatic model for state-selected reaction rates. Theory and application of vibrational mode correlation analysis to $OH\ (n_{OH}) + H_2\ (n_{H2}) \rightarrow H_2O + H$, *J. Chem. Phys.*, 77, 3516, 1982.
111. **Herzberg, G.,** *Molecular Spectra and Structure, Vol. III,* Van Nostrand Reinhold N.Y., 1966.
112. **Bauer, S. H. and Ossa, E.,** Isotope exchange rates. III. The homogeneous four-center reaction $H_2 + D_2$, *J. Chem. Phys.*, 45, 434, 1966.
113. **Burcat, A. and Lifschitz, A.,** Further studies on the homogeneous exchange reaction $H_2 + D_2$, *J. Chem. Phys.*, 47, 3079, 1967.
114. **Wilson, C. W., Jr. and Goddard, W. A., III,** *Ab initio* calculations on the $H_2 + D_2 = 2HD$ four-center exchange reaction. I. Elements of the reaction surface, *J. Chem. Phys.*, 51, 716, 1969.
115. **Rubinstein, M. and Shavitt, I.,** Theoretical study of the potential surface for the H_4 system by double-zeta configuration-interaction calculations, *J. Chem. Phys.*, 51, 2014, 1969.
116. **Bender, C. F. and Schaefer, H. F., III,** Linear symmetric H_4, *J. Chem. Phys.*, 57, 217, 1972.
117. **Silver, D. M. and Stevens, R. M.,** Reaction paths on the H_4 potential energy surface, *J. Chem. Phys.*, 59, 3378, 1973.
118. **Hoffmann, R.,** Transition state for the hydrogen-iodine and the hydrogen exchange reactions, *J. Chem. Phys.*, 49, 3739, 1968.

119. **Brooks, B. R. and Schaefer, H. F., III,** Reactions of carbynes. Potential energy surfaces for the doublet and quartet methylidene (CH) reactions with molecular hydrogen, *J. Chem. Phys.*, 67, 5146, 1977.
120. **Dunning, T. H., Jr.,** Annual Report of the Theoretical Chemistry Group, Chemistry Division, Argonne National Laboratory, Illinois, October 1980—September 1981.
121. **Herman, M. R. and Lin, M. C.,** unpublished.

Chapter 2

SEMIEMPIRICAL POTENTIAL ENERGY SURFACES

P. J. Kuntz

TABLE OF CONTENTS

I. TYPES OF METHODS AND GENERAL REQUIREMENTS

Semiempirical methods are quite popular in the field of chemical dynamics, especially for processes involving medium-sized molecules, complicated electronic structures, excited electronic states, or nonadiabatic effects. They are used to compute potential energy surfaces which are essential to almost all dynamical calculations. Several books and reviews of this topic are available in the literature.[1-5] This article is more in the nature of an extended comment than a review — an account colored by my personal view of the subject.

Semiempirical methods are often viewed as a sort of poor second choice to *ab initio* methods, or as a supplement to the latter, filling in gaps so that dynamical calculations are feasible. I feel that this view, while not completely incorrect, is somewhat distorted. Except in a very few cases (e.g., H_3 in the ground state), it is quite unrealistic to expect *ab initio* calculations of "chemical accuracy" to be available in sufficient quantity for a dynamical calculation. Supplementary empirical data is almost always necessary to correct deficiencies in the computed potential surfaces and thus to effect a meaningful comparison between theory and experiment. As soon as such empirical data is introduced, even the most ambitious *ab initio* calculation becomes semiempirical. Such use of semiempiricism should not detract from the worth of a potential energy surface, for it allows simultaneous exploitation of the rigor of theory and the accuracy of experiment (which in most cases exceeds that of theory). In practice, it hardly ever comes to a choice between *ab initio* and semiempirical methods but rather it is a question of how much empirical data is to be introduced and how this is to be done effectively.

There are some who would argue that, although the introduction of empirical data into the description of the potential surface may still be essential to successful dynamical calculations today, it must surely be our long-term goal to reduce the empirical component of such calculations as much as possible. Militant reductionists would, of course, insist on this goal; I shall not, for I believe that we are now at a stage where potential surfaces obtained from a mixture of *ab initio* and empirical sources can have sufficient predictive power to suggest new experiments. Of course, there can be no objection to progress in substantially reducing the empirical component of potential surfaces, but I fear we are still a long way from realizing this goal.

Applications of semiempirical methods to molecular collision dynamics are legion. They range from small adjustments of *ab initio* surfaces to almost purely empirical formulations such as the BEBO (bond-energy bond-order) method.[6] This chapter will concentrate on only a few methods. Those based on SCF calculations (CNDO, MINDO, etc.) are mentioned only briefly, since they are not easily applicable to dynamics. The very highly developed information-theoretic approach is also not treated, although it has proved very useful and is widely applicable; it does not, however, lead to potential energy surfaces. The theory is rather specialized and has been adequately summarized in other articles.[7-8] I shall also say very little about fitting potential energy surfaces, except to point out when and how certain semiempirical methods may be suitable for this task. This topic is not trivial and would, I think, overstep the bounds of this chapter.

The basis for my discussion is the well-established route from the Schrödinger equation (SE) for all particles to the electronic SE (via the Born-Oppenheimer (BO) approximation), whose solution yields the potential energy surfaces which serve as the foundation for dynamical calculations. The various approximations to the potential surfaces serve as the objects of discussion in this chapter. It is convenient to divide the approximations into three types: (1) *ab initio* calculations plus empirical adjustments, (2) methods involving diagonalization of an approximate Hamiltonian matrix, and (3) largely empirical methods calibrated to reproduce known theoretical or experimental results. Clearly, type 1 follows the rigorous, well-established route most faithfully, whereas type 2 departs from it to a relatively small degree, type 3 to a large degree.

Before going on to discuss the approximations in detail, I should like to make some comments which apply to all three groups. In devising any semiempirical method for use in dynamical calculations, it is wise to consider the following checklist, part of which is repeated from earlier work:[3]

1. If a collision process is to be described properly, it is essential that the asymptotic regions of all open channels (i.e., energetically accessible products) are correctly described. Failure to do this makes comparison with experiment very difficult, if not impossible.
2. The approximation to the potential surface in the interaction regions should agree as well as possible with available theoretical calculations and with experimental information (e.g., spectroscopic information).
3. In regions where no information about the potential surface is available, the approximation should behave in a physically reasonable manner.
4. The asymptotic and interaction regions should be connected smoothly and in a physically reasonable way.
5. The approximation should reflect the spatial symmetry of the real potential. For example, in the reaction $H + CH_4 \rightarrow H_2 + CH_3$, all of the H-atoms should be equivalent: the potential ought to be invariant under exchange of any two of the H-atoms.
6. The approximation should reflect the mathematical structure of the real potential. This is especially important when two potential surfaces have regions of avoided crossings or when a conical intersection occurs in the region of interest.

If the semiempirical method is used directly in dynamical calculations or as a fit to *ab initio* points, then the following properties are desirable:[9]

7. The potential function and its derivatives should have as simple a form as possible, consistent with the desired goodness of fit.
8. The method should require as small a number of data points as possible to achieve an accurate fit.
9. The fit should converge to the true surface as more data become available.
10. The method should suggest where new points should best be computed.

The above requirements are not easy to satisfy and, indeed, there have been many successful dynamical calculations where one or more of these points has not been satisfied. This need not be a catastrophe but it should be borne in mind that failure to observe these guidelines can place rather severe limitations on the validity of calculated cross-sections and rate constants. Methods of type 1 and 2 usually satisfy points 2 to 6 with no great trouble, but they usually have difficulty with point 1; *ab initio* methods alone have the greatest difficulty with point 4. Methods of type 3 often fail badly with respect to points 5 and 6, and empirical fits to *ab initio* points often violate point 3.

II. TYPE 1 METHODS

A. Empirically Adjusted *Ab Initio* Surfaces

Let us now assume that we have a fairly large number of good quality *ab initio* points for a single potential surface and we are faced with the problem of using these in a dynamical calculation. If the points are very accurate and sufficient in number, points 1, 2, 4, 5, and 6 are automatically satisfied and it remains only to find a suitable interpolation scheme to satisfy point 3. Truhlar and Horowitz[10] show how a skillful blend of semiempiricism and curve-fitting can achieve an excellent fit to the H_3 potential surface. It is, however, very rare that such extensive calculations are available, so I shall not pursue this case further; the interested reader should refer to the literature cited.

It is more often the case that the *ab initio* points are not sufficiently accurate and that there are not enough of them. A typical example is illustrated in a paper by Carter and Murrell,[11] who attempt to find an analytic fit to the 214 *ab initio* points of Chen and Schaeffer[12] for the reaction Li + HF → LiF + H. A catalogue of woes and their remedies appears in the paper. They include:

1. There were not enough points available in the asymptotic region LiF + H to define the fitting function adequately. A *graphical* extrapolation in this region was necessary.
2. The positions of the potential energy curves for the diatomic fragments LiF and HF did not agree with experiment. Carter and Murrell found that reasonable agreement could be obtained by subtracting small amounts ΔR_{LiF} and ΔR_{HF} from R_{LiF} and R_{HF}, respectively, for all tabulated *ab initio* points; i.e., the *ab initio* surface was *empirically shifted*.
3. The computed barrier height was too high to agree with the experimental activation energy. A more agreeable barrier was obtained by subtracting from the *ab initio* points a Gaussian function of amplitude 10 kJ/mol centered at the position of the *ab initio* barrier.
4. The exothermicity was incorrect. This was adjusted by scaling the data in the LiF + H asymptotic region using an exponential factor.

The resulting points were then fitted to an analytical function which was subsequently used in trajectory calculations.[13] The approximate potential surface so obtained is a very good example of a type I semiempirical method. It attempts to incorporate as much as possible of the available *ab initio* information, which nonetheless needs to be empirically shifted, scaled, adjusted, and extrapolated in order to be suitable for dynamical calculations.

The reactions of $O(^3P)$ and $O(^1D)$ with H_2 have also been extensively investigated[14-16] using *ab initio* calculations[17] and functions fitted to them. Again, these must be adjusted to agree with experimental exothermicity in order to compute OH vibrational and rotational distributions[18] in a meaningful way.

B. General Approach to Representing Potential Surfaces

The problem of representing potential energy surfaces in a consistent way has been investigated in a general way by Murrell and co-workers.[19-20] Their ambitious program is aimed at expressing the potential energy of a polyatomic molecule as a many-body expansion involving atomic terms plus diatomic interaction terms plus triatomic interaction terms and so on.

$$E_{ABC...} = V_A + V_B + V_C + \ldots + V_{AB} + V_{BC} + V_{CA} + \ldots + V_{ABC} + \ldots \quad (1)$$

The term V_{ABC}, for example, is nonzero only when atoms A, B, and C are close together. The advantage of such a procedure is that the potential E_{ABC} automatically describes all of the asymptotic channels correctly; e.g., for the triatomic molecule ABC,

$$E(A + B + C) = V_A + V_B + V_C \quad (2)$$

$$E(A + BC) = V_A + V_B + V_C + V_{BC} \quad (3)$$

There are some practical difficulties with this approach, however. One problem arises from the fact that, for an N-atom molecule, the sum of the terms up to order (N-1) is not, in general, a good approximation to the full potential, so that the Nth-order term may have to cancel out the effect of the smaller order terms, which ultimately could lead to numerical instabilities for large molecules. As an example, consider the H_3 molecule in the D_{3h} configuration with $R_{HH} = r_e$. If we choose the configuration H + H + H to correspond to the zero of energy, then $V_A = V_B = V_C = 0$. The sum of $V_{AB} + V_{BC} + V_{AC}$ is then equal to $-3De$, where D_e is the well-depth of the H_2 molecule (4.75 eV). Now the binding energy of H_3 (relative to H + H + H) in the D_{3h} configuration lies above the value $-D_e/2$. This means that the term V_{ABC} must be of the order of 12 eV for the expansion to approximate the H_3 potential adequately; i.e., V_{ABC} is 5 times larger than the binding energy.

Another problem is that the N-body term may require discontinuities in the derivatives (cusps) in order to cancel out discontinuities present in the (N-1)-body terms. For example, the three-body terms for H_2O need to cancel the cusps arising from curve crossings in the two-body terms; the four-body terms for H_2F_2 need to cancel out conical intersections which (in general) occur in the three-body terms for H_2F.

Murrell and Carter found that the H_2F surface could only be obtained by considering the potential surface to be the lowest eigenvalue of a 2 × 2 matrix. This is because there appears to be a conical intersection in C_{2v} configurations. To investigate this property of the potential surface more closely, we make use of the approximate London formula, which describes the potential surfaces expected upon coupling together three atoms in doublet spin states:

$$E = Q \pm J \tag{4}$$

$$Q = Q_{AB} + Q_{BC} + Q_{AC} \tag{5}$$

$$J = [J_{AB}^2 + J_{BC}^2 + J_{AC}^2 - J_{AB}J_{BC} - J_{BC}J_{AC} - J_{AB}J_{AC}]^{1/2} \tag{6}$$

In the diatomics-in-molecules (DIM) and LEPS formulations, the diatomic terms Q_{ij} and J_{ij} are expressed in terms of the singlet and triplet interaction energies:

$$Q_{ij} = 0.5\,(S_{ij} + T_{ij}) \tag{7}$$

$$J_{ij} = 0.5\,(S_{ij} - T_{ij}) \tag{8}$$

Now in C_{2v} configurations (H − F − H) the AB and AC interactions are equal (taking A ≡ F). The potential energy (Equation 4) reduces to:[3]

$$E = 2\,Q_{HF} + Q_{HH} \pm |J_{HF} - J_{HH}| \tag{9}$$

In general, for a fixed value of R_{HH}, there may be some finite value of R_{HF} at which $J_{HF} = J_{HH}$, so that the two surfaces meet each other. In such regions, the three-body terms in Equation 1 must have a conical intersection.

The LEPS method appears to be inadequate for describing H_2F in $D_{\infty h}$ symmetry despite its having the correct mathematical structure. The superiority of the Murrell method (*with* diagonalization!) in this case is doubtless due to the much greater flexibility which is built into the 2 × 2 matrix elements.

We have seen that simple treatments such as the LEPS method can elucidate the nature of the potential energy surface, thereby suggesting improvements to more general methods. Carter and Murrell mention that LiFH is also a molecule which dissociates into atoms in doublet spin states, yet they find no evidence of a conical intersection in this molecule. The explanation may lie in the ionic nature of LiF: it is more apt to consider LiFH as arising from Li^+ (^1S_g) + F^- (^1S_g) + $H(^2S_g)$. The simplest DIM treatment of this system leads to the analytical potential surface:

$$E = V_{Li^+F^-} + V_{Li^+H} + V_{HF^-} \tag{10}$$

This function is, of course, very crude, but it indicates that in regions where LiFH is ionic, no conical intersection is expected. Probably much more important is the coupling between ionic and covalent structures.

C. Minimum Requirements for *Ab Initio* Surfaces

Type 1 semiempirical methods are determined largely by the quality and quantity of *ab initio* points available. Those doing *ab initio* calculations should bear in mind the problems of those who are later going to use these calculations. Of utmost importance to the kineticist is a knowledge of the asymptotic regions: he needs to know the energy of the separated atoms and the energies of the separated fragments corresponding to all possible reaction channels. Unfortunately, such information is often not reported. Several points at intermediate and long-range fragment-fragment distances are also very helpful. A common failing is to report hundreds of points at special configurations (e.g., $C_{\infty v}$, C_{2v}, etc.), but no points for lower spatial symmetries — this makes interpolation difficult. Where several potential surfaces are needed it is necessary to report all fragment states and separated-atom states which correlate with the surfaces of interest. The extra effort needed to acquire this information is small in comparison to the rest of the *ab initio* calculations and is of enormous help. Even if it is not so easy to obtain such information, the person doing the *ab initio* calculation is in a much better position to do this than the kineticist, who must otherwise rely on educated guesses.

III. TYPE 2 METHODS: DIAGONALIZATION OF APPROXIMATE HAMILTONIAN MATRIXES

We now consider methods which are akin to *ab initio* methods in that the electronic state eigenvalues result from the diagonalization of a Hamiltonian matrix (**H**) in a basis $\boldsymbol{\phi}$:

$$\mathbf{HC} = \mathbf{SCE} \tag{11}$$

where $\mathbf{H} = \langle\boldsymbol{\phi}|\mathcal{H}|\boldsymbol{\phi}\rangle$ and $\mathbf{S} = \langle\boldsymbol{\phi}|\boldsymbol{\phi}\rangle$. **C** is the eigenvector matrix and **E** the diagonal eigenvalue matrix. Matrix **C** simultaneously diagonalizes **H** and **S**:

$$\mathbf{C\ H\ C} = \mathbf{E} \tag{12}$$

$$\mathbf{C\ S\ C} = \mathbf{1} \tag{13}$$

In these methods, a semiempirical *model* is defined by the basis set $\boldsymbol{\phi}$, a row-vector. The elements of **H** and **S** are then approximated, usually with the help of empirical data, and Equation 11 is then solved for the eigenvalues **E** and eigenvectors **C**.

A. Molecular Orbital (MO) Methods

Examples of MO semiempirical methods are pseudopotential methods,[4] Hückel theory,[21] and methods involving neglect of differential overlap (CNDO, INDO, MINDO).[22] These methods have been largely restricted to ground-state, near-equilibrium problems, and as such do not lend themselves readily to dynamics. Nevertheless, MINDO/3 and MINDO/2 have been successfully applied to the study of organic reaction mechanisms, particularly with regard to the location of the transition state and the computation of molecular vibrational frequencies.[23]

CNDO/2 calculations[24] have aided the interpretation of experimental chemiluminescence studies of the reaction $F + HN_3 \rightarrow HF\ (v \leq 5) + N_3$. This type of method, which is an approximation to the SCF-MO procedure, is unsuited to dynamical calculations because of convergence difficulties in many geometrical configurations: the calculations on $H - N_3 - F$ converged only if $1.15\ \text{Å} \leq F - N \leq 1.7\ \text{Å}$. On the other hand, the calculations identify the configuration $H-N-N-N-F$ as being more strongly bound than $F-H-N-N-N$, thus providing valuable information about the strong interaction region.

A combination of *ab initio* and type 2 semiempirical models for different regions of the same molecule has been used to estimate a potential surface for the reaction $C_2H_5 \rightleftharpoons H + C_2H_4$.[25] An *ab initio* SCF calculation was used for the C_2H_5 region of the surface and semiempirical information in the form of force constants, dissociation energies, and supporting calculations was applied to C_2H_4. Again, for intermediate ranges $H - - - C_2H_4$, the SCF method proved unreliable so that an empirical interpolation was necessary.

B. Valence Bond (VB) Methods

Approximations based on the VB method are free from the convergence difficulties one encounters in the SCF methods. The best known of these are the London formula together with the LEPS modifications (see Equations 4 to 8) and the Porter-Karplus[26] formula for H_3 which introduces three-center terms. The latter method has been generalized to H_4,[27,28] and a similar approach has been developed for H_4[29] and H_n.[30] Zembekov[31] has used a semiempirical VB method to study the $^1A'$ states of H_2O and H_2F^+. He uses the atoms-in-molecules procedure (AIM) to correct the atomic energies and treats three-center integrals by the Mulliken approximation. Zeiri and Shapiro[32-33] have developed a VB method for which the matrix elements are obtained from ground state diatomic potentials only; it is formulated for alkali-hydrogen halide reactions. So far, the VB approximate methods have been specialized to a few types of molecules but they could be applied to a wider variety.

C. Diatomics-in-Molecules (DIM)

1. Outline

This is the most general of the type 2 methods: it is applicable to ground and excited states and to all regions of configuration space. It also describes diatomic and atomic fragments exactly, and for many processes, can be used in dynamical calculations directly. The main drawbacks are that (1) the polyatomic basis set $\boldsymbol{\phi}$ increases rapidly as the number of atoms increases and as the L-quantum number of the atomic states defining the basis increases and (2) a knowledge of the potential curves and wave functions of many diatomic states is needed. The DIM method is sometimes criticized because it does not explicitly construct a polyatomic wave function; however, quantities which would normally require a polyatomic wave function, such as nonadiabatic coupling coefficients or dipole transition probabilities can often be computed in terms of the corresponding diatomic quantities.[34] This, I feel, is one of the advantages of DIM and not a drawback.

The DIM method[35] has been adequately reviewed and formulated in the literature,[3-4,36-39] so that a detailed statement of the method is not needed here; instead, I shall concentrate more on how to define a DIM model suitable for a particular problem. Readers who are interested in learning more about the method should refer first to the orginal formulation of Ellison,[35] for all of the *essential* aspects of the method are treated in his paper. The paper by Kuntz and Roach[40] and the review articles of Tully[36-37] present the method in a compact notation and show how to treat atomic states in angular momentum states other than S-states. For a nice overview of the method, together with a critique of various aspects of the method, I recommend reading Tully.[37] The formulation by Steiner et al.[41] and extended by

Kuntz,[38] is a somewhat more systematic approach adaptable to programming by computer.* A more recent formulation by Faist and Muckerman[42-44] translates the previous formulations into the language of projection operators. All of the reformulations provide useful points of view but add essentially nothing new to Ellison's original idea.

To briefly outline the DIM method, we begin by writing the polyatomic Hamiltonian operator as a sum of diatomic and atomic fragments:

$$\mathcal{H}_{ABC} = \sum_{P=1}^{N-1} \sum_{Q=P+1}^{N} \mathcal{H}_{PQ} - (N - 2) \sum_{P=1}^{N} \mathcal{H}_{P} \tag{14}$$

$\mathcal{H}_{PQ}$ and $\mathcal{H}_{P}$ are diatomic and atomic Hamiltonian operators and N is the number of atomic centers in the polyatomic molecule. The polyatomic basis functions, $\mathbf{\Phi}$ are antisymmetrized products of N atomic eigenstates, $\phi(P)$:

$$\Phi_M = \mathcal{A} \prod_P \phi_M (P) \tag{15}$$

Diatomic basis functions are also written in terms of products of these same atomic eigenstates:

$$\chi_D = \mathcal{A}\ \phi_D(P_1)\ \phi_D(P_2) \tag{16}$$

Hence, by using the identity Equation 14 together with Equations 15 and 16, it is possible to express the polyatomic matrix, $\mathbf{S}^{-1}\mathbf{H} \equiv \mathbf{B}$ (defined in the basis $\mathbf{\Phi}$) in terms of a sum of diatomic matrixes ($\mathbf{B}_{PQ}$) and atomic matrixes ($\mathbf{B}_P$):

$$\mathbf{B} \approx \sum_{P=1}^{N-1} \sum_{Q=P+1}^{N} \mathbf{B}_{PQ} - (N - 2) \sum_{P=1}^{N} \mathbf{B}_P \tag{17}$$

$\mathbf{B}_P$ is a diagonal matrix with elements consisting of atomic state energies. $\mathbf{B}_{PQ}$ has elements from the diatomic fragment matrixes $\mathbf{b}_m \equiv \mathbf{s}^{-1}\mathbf{h}$ where $\mathbf{s}$ is the overlap matrix and $\mathbf{h}$ the Hamiltonian matrix defined in the basis χ_m, and m labels the appropriate diatomic species. Inherent in Equation 17 is the assumption that for the diatomic species m,

$$h_m\ \chi_m = \chi_m\ \mathbf{b}_m \tag{18}$$

which is only true if the basis χ_m is complete; this is the main assumption in DIM. In practice, completeness is assumed and the matrixes ($\mathbf{b}_m$) are defined in terms of the eigenvalues ($\mathbf{e}_m$) and eigenvectors ($\mathbf{c}_m$) of the matrixes ($\mathbf{h}_m$):

$$\mathbf{b} = \mathbf{c}\ \mathbf{e}\ \mathbf{c}^{-1} \tag{19}$$

If the diatomic fragment information is to be extracted from the eigenvectors ($|\mathbf{E}>$) and eigenvalues ($\mathbf{E}$) of extended basis VB or MO calculations, then $\mathbf{b}$ must be defined as the matrix of an effective Hamiltonian ($\mathbf{h}_{eff}$,) over the Löwdin orthogonalized functions $|\chi'_m>$:

$$\mathbf{b} = <\chi'_m|h_{eff}|\chi'_m> \tag{20}$$

$$|\chi'_m> = |\chi_m> s^{-1/2} \tag{21}$$

$$h_{eff} = \mathcal{P}'_M\ |\mathbf{E}>\mathbf{E}<\mathbf{E}|\ \mathcal{P}'_M \tag{22}$$

* On page 85 of Reference 38, the middle column of the matrix in Equation 9 should be multiplied by i. Also, in the paragraph 2.3.7 preceding Equation 49, m should be 3, $n_{B\beta} = 0$, and $\Delta e = 0$, resulting in a phase factor +1. $B_{BC}^{7,2}$ is then $-J_1c$, as in the equation.

where P'_M is a renormalized projector onto the DIM basis.[45] Diatomic properties needed to calculate polyatomic properties via DIM can likewise be extracted from *ab initio* calculations by replacing **E** in Equation 22 by the appropriate property matrix, $<\mathbf{E}|\mathcal{T}|\mathbf{E}>$. The DIM procedure can be applied in an *ab initio* manner in that the matrixes **c** and **e** in Equation 19 are calculated by a valence bond (VB) or molecular orbit (MO) method (in the latter case, using projectors to obtain the appropriate matrix **b**). An example is the work of Schubert[46] on the CH_2^+ molecule. His DIM model comprised 15 structures for the $^2A'$ and 12 structures for the $^2A''$ surfaces; the matrixes **b** were estimated from VB calculations in the DIM basis. Schubert found that a minimum basis set (formed by a direct product of atomic ground states) was not adequate, ^4P_g states from the sp^2 electronic configuration of C^+ and the $C(^1D)$ and $C(^1S)$ states being important. This fact was evident from the VB calculations on the CH^+ diatomic fragment states.

More often, DIM is applied in a semiempirical fashion in such a way as to make use of accurate potential curves for the diatomic fragments. The usual procedure is to replace the diagonal matrix (**e**) in Equation 19 with a matrix ($\mathbf{e}_s$) having some or all of the diagonal elements replaced by accurate diatomic energies obtained either from experiment or from large *ab initio* calculations:

$$\mathbf{b}_1 = \mathbf{c}\ \mathbf{e}_s\ \mathbf{c}^{-1} \tag{23}$$

This procedure allows the DIM method to treat asymptotic regions of the potential surface leading to diatomic products essentially correctly — it is probably the single most powerful aspect of the method. But great care must be exercised, for subtle errors can be introduced which could invalidate the DIM calculations. The common errors which occur will emerge from the following discussion on defining a DIM model.

2. Selecting the DIM Basis

The polyatomic basis $\boldsymbol{\phi}$ defines the overall model and determines the number and size of the diatomic matrixes $\mathbf{b}_m$. It is difficult to say in advance just how large the basis $\boldsymbol{\phi}$ should be — it is dictated by the nature of the bonding in the polyatomic as well as by the requirement of completeness of the various diatomic bases. Obviously, the latter requirement is impossible to fulfill with a finite basis, so a compromise has to be made. No general recipe for defining the basis exists, but the following procedure[47] has proven useful for triatomic and tetratomic molecules:

1. Guess a polyatomic basis using chemical horse sense. This choice implies a basis for a set of diatomic molecules in specific states.
2. For those diatomic states which correlate with the polyatomic potential surfaces of interest, carry out good quality *ab initio* calculations at all internuclear distances (R) of interest. Calculate the projection of the *ab initio* wave functions onto the DIM diatomic basis. If this projection is reasonably large (e.g., > 0.85) over the whole range of R, then the basis is probably adequate.

A more detailed list of considerations is given in Reference 45.

Let us now consider some concrete examples. Application of DIM to H_3 using a minimum basis set leads to the London formula (Equations 4 to 8). In this case, Equation 23 is trivial, since the matrixes (**b**) for the H_2 $^1\Sigma_g^+$ and $^3\Sigma_u^+$ states are both of dimension 1×1. Also, use of the exact *ab initio* singlet and triplet H_2 potential curves can be justified over the range $R_e \leq R < \infty$ for the singlet state and over the range 2 bohr $\leq R < \infty$ for the triplet state ($R_e = 1.4$ bohr).[48] Hence, the London formula for H_3 should be qualitatively correct, as it is. Note, however, that the corresponding DIM for H_4 fails.[49] It is not yet known whether an extended basis will help.

Turning now to H_2F, we could first attempt a description of the $^2A'$ states in terms of a minimum basis set obtained from $H(^2S_g) + H(^2S_g) + F(^2P_u)$. This leads to a 4 × 4 DIM secular equation but again, all the diatomic matrixes (**b**) are 1 × 1. The new diatomic states required are the $^1\Sigma^+$, $^3\Sigma^+$, $^1\Pi$, and $^3\Pi$ states of HF. A projection analysis[45] of the $X^1\Sigma^+$ state of HF indicates that the minimum DIM basis is adequate only for internuclear distances greater than 3.3 bohr ($R_e \sim$ 1.8 bohr) hence, a minimum basis DIM model is not expected to yield a good description of H_2F. Diatomic basis functions corresponding to ionic binding H^+F^- are necessary; these would result from a polyatomic basis enlarged by the addition of two functions corresponding to $H^+(^1S_g) + H(^2S_g) + F^-(^1S_g)$ and $H(^2S_g) + H^+(^1S_g) + F^-(^1S_g)$.

It is instructive now to consider the molecule Na_2F. The same arguments apply to this as to H_2F, except that at small NaF distances, the bonding in NaF could be considered almost 100% ionic. In such regions, an adequate DIM model might be obtained by considering *only* the two ionic structures in the polyatomic basis. The corresponding potential surface is[50]

$$E = Q - J \tag{24}$$

$$Q = Q_{AB} + Q_{BC} + Q_{AC} \tag{25}$$

$$J = [J_{AB}^2 + (J_{AC} - J_{BC})^2]^{1/2} \tag{26}$$

where A ≡ Na, B ≡ Na, and C ≡ F. The diatomic terms are defined here in terms of the $X^2\Sigma_g^+$ state of Na_2^+ (≡ G), the $^2\Sigma_u^+$ state of Na_2^+ (≡ U), the $X^2\Sigma^+$ state of NaF^- (≡ D), and the ground state, $X^1\Sigma_g^+$, of Na^+F^- (≡ S), which is, of course, assumed to be purely ionic. The **Q** and J terms are

$$\mathbf{Q}_{AB} = (G + U)/2 \tag{27}$$

$$J_{AB} = (G - U)/2 \tag{28}$$

$$\mathbf{Q}_{NaF} = (S + D)/2 \tag{29}$$

$$J_{NaF} = (S - D)/2 \tag{30}$$

Such a surface has a completely different mathematical structure from the covalent surface. In particular, J can be zero only as $R_{NaNa} \rightarrow \infty$, so that no conical intersection occurs for finite configurations. The covalent surface, however, has a conical interaction in C_{2v} configurations, as discussed earlier. It is clear, therefore, that use of the covalent model in regions where NaF is ionic would lead to a *qualitatively* incorrect potential surface. The effect of changes in the DIM basis is even greater for BeFH, where a low-lying excited state of Be as well as ionic BeF structures are important.[45,47]

DIM calculations with small basis sets often produce good results in collinear configurations, but are worse in C_{2v} symmetries. I think this often occurs because in the higher symmetries the basis functions (or linear combinations of them) must be distributed among several irreducible representations, thus cutting down the flexibility of the method there in comparison to other regions. Conical intersections arise because of this. Use of larger basis sets relaxes the demands of symmetry somewhat. A recent example is the comparison of a DIM surface for HF_2 (minimal basis + ionic structures H^+F^-) with an LEPS surface.[51]

3. Incorporation of Empirical Data

Assuming that an adequate polyatomic basis has been ascertained, there remains the problem of incorporating empirical data into the diatomic matrixes (**b**). In general, it is necessary to correct the matrix **c** and the matrix **e**, for the matrix **b** is usually obtained from an approximate *ab initio* calculation, which may not yield the correct atomic energies in the limit of large internuclear distances. It is in any case desirable that the elements of **c** not be too sensitive to the quality of the *ab initio* calculations.

Empirical atomic term values can be used to correct the matrix **c**. Since the DIM basis is defined in terms of products of atomic eigenfunctions, it is clear that as $R \rightarrow \infty$, **b** becomes diagonal with elements approaching the atomic term values. Let us denote the matrix **b** obtained from an approximate *ab initio* calculation as $\mathbf{b}'$ (R). The limit of this at large R is the diagonal matrix $\mathbf{e}'_A$:

$$\mathbf{e}'_A = \lim_{R \rightarrow \infty} \mathbf{b}'_A(R) \tag{31}$$

Now let $\mathbf{e}_A$ be the matrix obtained on replacing the approximate elements in $\mathbf{e}'_A$ by their empirical values. The corrections to $\mathbf{e}'_A$ can be written in the form of the matrix $\delta\mathbf{e}_A$:

$$\delta\mathbf{e}_A \equiv \mathbf{e}_A - e'_A \tag{32}$$

At each internuclear distance, a new matrix $\mathbf{b}''(R)$ can then be defined as

$$\mathbf{b}''(R) = \mathbf{b}'(R) + \delta\mathbf{e}_A \tag{33}$$

$\mathbf{b}''$ (R) then approaches the correct limiting value ($\mathbf{e}_A$) at large R. Diagonalization of $\mathbf{b}''$ (R) then yields the corrected eigenvector matrix, $\mathbf{c}_S$ (R), and a new eigenvalue matrix e'' (R).

This method of obtaining a semiempirical eigenvector matrix ($\mathbf{c}_s$) was tested for the two lowest OH^+ ($^3\Sigma^-$) states, for which the separated atom limits are $O(^3P_g) + H^+$ and $O^+(^4S_u) + H(^2S_g)$.[52] Since the ionization potentials of O (13.614 eV) and H (13.595) are nearly equal, the atomic limits of each state have almost the same energy. Accordingly, at intermediate and long-range values of R, matrix $\mathbf{c}'(R)$, obtained by diagonalizing $\mathbf{b}'(R)$, and matrix $\mathbf{c}_s(R)$ may differ greatly; i.e., $c'(R)$ is particularly sensitive to the quality of the *ab initio* calculation. In contrast to this, the semiempirical matrix $\mathbf{c}_s(R)$ was insensitive to variations in the *ab initio* basis set. It turns out that a DIM calculation of the $^4A''$ surface of OH_2^+ is very sensitive to the input matrix (**c**) for the OH^+ fragment.[53]

A further advantage of the above correction procedure is that avoided crossings in the diatomic eigenvalues ($\mathbf{e}'(R)$) present no difficulties: $\mathbf{e}'(R)$ is replaced by $\mathbf{e}''(R)$ in such a way that the avoided crossings are merely shifted to new values of R.

The final empirical manipulation comes in the adjustment of some of the eigenvalues in $\mathbf{e}''$ (R). It is important here to adjust *only* those eigenstates for which the DIM basis has a sufficiently large projection onto the *ab initio* wave function. Obviously, special care must be taken in regions of avoided crossings. The final semiempirical input matrix ($\mathbf{b}_s$) is defined in Equation 19 by replacing **c** by $\mathbf{c}_s$ and **e** by $\mathbf{e}_s$, the latter being the adjusted $\mathbf{e}''(R)$:

$$\mathbf{b}_s = \mathbf{c}_s\, \mathbf{e}_s\, \mathbf{c}_s^{-1} \tag{34}$$

In concluding this section, I should emphasize that blunders in DIM calculations (very easy to make!) occur chiefly for one of two reasons: (1) the DIM basis is inadequate or even irrelevant to the problem at hand or (2) empirical information is introduced incorrectly in the definition of $\mathbf{b}_s$. The second error occurs very easily when no *ab initio* calculations are needed for $\mathbf{c}_s$ or are simply not available. As an example, I cite some of our own calculations on FH_2^+.[54] We were mainly concerned with calculating the $^3A'$ and $^3A''$ surfaces and accordingly carried out the necessary *ab initio* calculations for the HF^+ and HF molecule. As an aside, we wished to estimate one of the *singlet* surfaces of H_2F^+ for which we needed a $^2\Pi$ state of HF^+. Since we had no data on this state, we used a potential curve for the isoelectronic OH molecule. This was incorrect because the state we needed correlated with $H^+ + F$, not with $H + F^+$, i.e., we used information from the wrong state of OH. This sort of error is easy to make and occurs in a number of papers in the literature.

In reporting a DIM calculation, it is important to recall that the DIM model is defined not only by the number of polyatomic basis functions and their genealogy[55] but also by the exact specification of the matrixes (**b**) for all of the diatomic fragments. Such information is essential because it allows others to reproduce the results and to compare them more easily with other models.

IV. TYPE 3 METHODS: MAINLY EMPIRICAL METHODS

A. Introduction

Somewhat more *ad hoc* in nature than methods of type 1 and 2, these procedures tend to view the polyatomic potential surface as an interpolation between the potential energy functions of the reagent and product. They tend to be used for ground state surfaces, usually for triatomic molecules, and, in some cases, are restricted to collinear configurations. They usually aim to obtain the interaction potential directly, exploiting electrostatic interactions as does the Rittner model for alkali-halide molecules,[56-57] or applying qualitative notions of chemical bonding such as the BEBO method.[6] Some of these methods appear to have a remarkable predictive power for their simplicity, and the considerable flexibility of many makes them attractive for fitting *ab initio* points.

B. Rotated Morse Potentials

At the one extreme are the methods which may be regarded as pure interpolation schemes. Best known of these is perhaps the rotated Morse method for triatomic molecules A – B – C.[2,9,58] This is particularly suitable for describing the reaction A + BC → AB + C. For an ABC angle (ϕ) the potential surface, a "fixed-angle surface" (FAS), can be represented as a series of contour lines in the $R_{AB} - R_{BC}$ plane. A reference point ($R^{\circ}_{AB}, R^{\circ}_{BC}$) is chosen somewhere in the region corresponding to separated atoms A + B + C with little interaction between them. It is then assumed that the potential along a line through ($R^{\circ}_{AB}, R^{\circ}_{BC}$) making an angle ($\theta$) with the R_{BC}-axis is a Morse function with parameters $D(\theta,\phi)$, $\beta(\theta,\phi)$, and $R_e(\theta,\phi)$. The Morse curves for $\phi = 0$ represent the molecule BC; those for $\phi = \pi/2$ represent AB. The functions $D(\theta,\phi)$, $\beta(\theta,\phi)$, and $R_e(\theta,\phi)$, which are often represented by spline functions, determine the features of the surface. The method is straightforward and flexible; however, potentials which have "sudden" character,[59] i.e., which change rapidly over a small range of θ cannot be represented well, if at all. Also, the method as outlined here is not suited to describing the reaction A + BC → AC + B.

C. The Hyperbolic Map Function

A method similar in spirit to the above is the hyperbolic map function. As originally formulated, it too could only treat A + BC → AB + C but Valencich[60] has reported a modification which allows a description of A + BC → AC + B as well. The rotated Morse and hyperbolic map methods are best suited for surface fitting or for investigating the dynamical effects of systematically varying features of the potential surface. Because they are not based on physical models, they have little predictive power.

D. Simple DIM Formulas

Simple DIM functions can also be used freely as interpolations. The most celebrated of these is the LEPS function, which is based on the London formula described above. There are, in fact, many similar DIM formulas (see Equations 24 to 30 above) which might be suitable for a wide variety of reactions. For example, triatomic potentials for doublet states obtained by coupling a ^2S atom with two atoms in ^nS states ($n = 2,3,4,5$) have the analytical form of Equation 24 but with J replaced by

$$J_n = [J_1^2 + J_2^2 + J_3^2 - 2\,J_1(J_2 + J_3)/n - 2(n^2 - 2)\,(J_2J_3)/n^2]^{1/2} \tag{35}$$

and with the diatomic terms

$$J_1 = 0.5(S - T) \tag{36}$$

$$J_k = 0.5\,(V_{n-1} - V_{n+1})_k, \quad k = 2,\,3 \tag{37}$$

The potential curves $V_{n\pm1}$ correspond to the $^{n\pm1}\Sigma^+$ states correlating with atoms ^2S + ^nS; S and T refer to the $^1\Sigma^+$ and $^3\Sigma^+$ states correlating with the atoms ^nS + ^nS. Equation 35 also describes the surfaces arising from an atom in a ^nS state coupled with two ^2S atoms. In this case, S and T refer to the $^1\Sigma^+$ and $^3\Sigma^+$ states arising from ^2S + ^2S. In C_{2v} configurations, $J_2 = J_3$, so that J_n^2 can be written as a perfect square:

$$J_n^2 = (J_1 - 2J_2/n)^2 \tag{38}$$

hence, all of these surface can have conical intersections.

There have also been reasonably successful attempts at defining self-consistent parameters which, in the framework of the London formalism, can be applied to 3- and 4-center reactions. In a study of the reactions $HBr + Cl_2 \rightarrow HCl + BrCl$ and $HBr + BrCl \rightarrow HCl + Br_2$, atomic parameters (used to construct the triplet curves in the London formula) were obtained from three-body reactions such as $Cl + HBr \rightarrow HCl + Br$.[61]

E. Parameterized DIM

Attempts to go beyond the simple DIM expressions empirically have been made in two different ways exemplified by the generalized DIM (GDIM) approach and the DIM-3C method.

1. Generalized DIM

The GDIM method, developed by Wu[62-65] and applied by him and others,[66] tries to increase the flexibility of DIM by introducing parameters into the partitioning of the Hamiltonian, Equation 14:

$$\mathcal{H}_N = \sum_{P,Q} G_{PQ}H_{PQ} + \sum_{P}\left(1 - \sum_{Q\neq P} G_{PQ}\right) H_P + \sum_{P,Q} (1 - G_{PQ})V_{PQ} \tag{39}$$

The functions G_{PQ} depend on the internuclear distances and bond angles of the polyatomic and contain adjustable parameters. If all $G_{PQ} = 1$, GDIM reduces to Ellison's DIM; if $G_{PQ} = 0$, GDIM reduces to Moffitt's AIM — the V_{PQ} are the interactions not included in the sum over the atomic Hamiltonian. The GDIM method has been used successfully to fit the Liu and Siegbahn *ab initio* calculations for H_3.[56-66]

2. DIM-3C

The DIM-3C method, developed by Last and Baer[67-68] makes the assumption that the true potential surface can be approximated by the addition of a term involving (parameterized) three-center terms (W) to a DIM minimum basis potential: $E = E_{DIM} + W$. The method appears to be useful in rationalizing the features of potential surfaces for a whole family of reactions ($X + H_2 \rightarrow HX + H$ and $HX + H \rightarrow H + XH$) in terms of a few parameters. The idea of supplementing E_{DIM} by a 3-center term came from a comparison of the 2×2 DIM Hamiltonian matrix, for H_3 ($\mathbf{H}_{DIM}$) with the VB matrix (H_{VB}) in the same basis. VB expressions for the diatomic singlet and triplet H_2 potential curves were used to effect this comparison, since the DIM matrix elements are expressed in terms of the energies of these diatomic states (see Equation 19). The terms neglected by DIM were three-center terms. When the DIM-3C method is applied, however, the empirical DIM method (Equation 23) is used so that the difference between $\mathbf{H}_{DIM}$ and $\mathbf{H}_{VB}$ can no longer consist only of three-center terms. This does not seem to be important in practice, however, for the parameters of the method are always calibrated and can therefore make up for such deficiencies.

One problem with the GDIM and DIM-3C methods is that they are difficult to extend to more complicated systems. For example, the application of DIM-3C to reactions involving the chalcogens (e.g., $Ca + F_2 \rightarrow CaF + F$) would be difficult, for a minimum basis DIM calculation is known to be completely inadequate,[47] and this inadequacy arises mainly from the neglect of ionic structures and of excited states of the metal atom. The three-center term W would, in this case, not correspond to the true deficiencies of DIM.

F. The Method of Agmon and Levine

A particularly intriguing new method is that of Agmon and Levine[69] using orthogonal bond-order coordinates. Consider the potential energy surface for collinear A – B – C defined in the r_1r_2 plane, where $r_i = R_i - R_i^{eq}$ and R_1 and R_2 are the BC and AB internuclear distances, respectively. The goal is to express the potential surface $E(r_1, r_2)$ in terms of another set of coordinates (m,c) which are more natural to the description of a reaction. To achieve this, new variables (n_1,n_2) are defined by:

$$n_i = e^{-r_i/a_i} \qquad (i = 1, 2) \tag{40}$$

The reaction coordinate is defined by the condition $n_1 + n_2 = 1$, its shape being determined by the constants a_1 and a_2. A family of curves parallel to the reaction coordinate are defined in terms of a parameter (m):

$$n_1 + n_2 = m \tag{41}$$

A family of curves orthogonal to these is defined by a parameter (c):

$$1/n_1 + 1/n_2 = c \tag{42}$$

Equations 40 to 42 define a transformation from (r_1, r_2) to the orthogonal coordinates (m,c).

At this point, Agmon and Levine decide to write the potential energy surface in terms of a linear interpolation between the two diatomic potential curves, $V_1(r_1)$ and $V_2(r_2)$, plus another term which allows the surface to have a barrier:

$$E(r_1,r_2) = \omega_1 V_1(\sigma) + \omega_2 V_2(\sigma) - \lambda V^*(\sigma)\, M(n) \tag{43}$$

Here, $\sigma = -a \log m$ and is a measure of the distance from the reaction coordinate (where $\sigma = 0$); n is the value of n_2 at the point (1,c) (i.e., the point on the reaction coordinate corresponding to the point (m,c) a distance σ away from the reaction coordinate). It is, therefore, determined by the condition:

$$\frac{1}{1 - n} - \frac{1}{n} = c(r_1, r_2) \tag{44}$$

$$n = 0.5 - 1/c + 0.5[c^2 + 4]^{1/2}/c \tag{45}$$

i.e., n = 0 for the separated reagents ($c \rightarrow -\infty$), reaches the value $^1/_2$ at the transition state (c = 0), and then increases further to 1 for the separated products ($c \rightarrow \infty$).

The functions V* and M are arbitrarily defined as:

$$V^*(\sigma) = \{\omega_1 V_1(\sigma) + \omega_2 V_2(\sigma)\}/(\omega_1 D_1 + \omega_2 D_2) \tag{46}$$

where D_i is the dissociation energy of V_i (r_i) and

$$\omega_i = n_i/(n_1 + n_2) \tag{47}$$

$$M(n) = -n \log n - (1 - n) \log (1 - n) \tag{48}$$

For symmetrical reactions ($a = a_1 = a_2$), the entire triatomic surface is specified by two parameters, a and λ.

Agmon and Levine[69] report that the 134 *ab initio* points for the H_3 collinear surface can be fitted with an r.m.s. deviation of 0.6 kcal/mol — an impressive achievement. They are also able to describe a whole series of reactions with the *same* set of parameters.[70] More recently, Garrett et al.[71] have computed kinetic isotope effects in the ClH_2 system using canonical variation TST applied to the Agmon-Levine potential extended to 3D configurations by the addition of an anti-Morse bend potential. Agreement with experiment is much better for this surface than for others (rotated Morse BEBO and extended LEPS) and this surface has "more predictive power" according to the authors. This suggests that the Agmon-Levine coordinate system may be more natural than others for fitting global functions to *ab initio* potential surfaces.

V. EXTENSIONS TO LARGER MOLECULES

It is slowly becoming necessary to address the problem of obtaining potential energy surfaces for medium-sized molecules. My personal opinion is that the methods of type 2 should be developed in this direction, since they have a firm theoretical foundation and can be more easily extended to cope with new problems. They are, however, somewhat inflexible (this may be an advantage, for a completely flexible method has little predictive ability): one cannot help but admire the idea of Murrell and co-workers to make the best use of n-body potentials in constructing (n + 1)-body potentials. I should now like to mention two ways of using DIM theory in this direction, the one a rather ambitious procedure and the other a more rough-and-ready procedure which may lend itself to dynamical calculations.

A. Molecules-in-Molecules (MIM)[37]

The first way is the method of MIM. The theory of organic chemistry, largely a theory of functional groups, is successful in near-equilibrium problems, which suggests that a semiempirical theory based on groups of triatomic fragments may be used for calculations of potential energy surfaces. MIM is just such a theory.

For molecules with more than three atoms, the Hamiltonian may be partitioned in terms of fragments with three, two, and single atoms. As an example, let us see how one might apply MIM to the H_4 molecule. By applying the partitioning Equation 14 to the Hamiltonian $\mathcal{H}_{ABCD}$ and to each of the triatomic fragment Hamiltonians and then eliminating the diatomic Hamiltonians, there results:

$$\mathcal{H}_{ABCD} = 0.5\,\{(\mathcal{H}_{ABC} + \mathcal{H}_{BCD} + \mathcal{H}_{CDA} + \mathcal{H}_{DAB}) - (\mathcal{H}_A + \mathcal{H}_B + \mathcal{H}_C + \mathcal{H}_D)\} \quad (49)$$

The minimum basic set, obtained by coupling four $H(^2S_g)$ states to get a resultant singlet state, contains two functions $\mathbf{\Phi}^{(s)} \equiv (\phi_1^{(s)}), \phi_2^{(s)}$ where the superscript (s) labels the coupling scheme. There are four such schemes corresponding to the four triatomic fragment Hamiltonians in the partitioning:

1. ((AB) + C) + D
2. ((BC) + D) + A
3. ((CD) + A) + B
4. ((DA) + B) + C

Function $\phi_1^{(1)}$ is obtained by coupling A to B to form a singlet AB, then C to form a doublet ABC, then D to form a singlet ABCD. The functions $\mathbf{\Phi}^{(s)}$ for s = 2,3,4 are defined similarly by permuting the letters A to D in the above definition) and are related to $\mathbf{\Phi}^{(1)}$ by unitary transformations:

$$\mathbf{\Phi}(s) = \mathbf{\Phi}(1)\,\mathbf{U}_s \qquad s = 2, 3, 4 \quad (50)$$

Now, for each of the triatomic fragments, there is also a 2 + 2 secular problem:

$$\mathbf{h\,c} = \mathbf{c\,e} \quad (51)$$

whose eigenvalues **e** are given by the London formula and whose eigenvectors (**c**) can also be written in terms of the diatomic H_2 singlet and triplet states. Hence,

$$\mathcal{H}_{ABC} \equiv \langle \mathbf{\Phi}^{(1)} | \mathcal{H}_{ABC} | \mathbf{\Phi}^{(1)} \rangle = \mathbf{c}_1\mathbf{e}_1\mathbf{c}_1^{-1} \equiv \mathbf{h}_1 \quad (52)$$

where the subscripts 1 on the RHS mean that $\mathbf{c}_1$ and $\mathbf{e}_1$ are evaluated for the triatomic fragment appropriate to the coupling scheme used in $\Phi^{(1)}$. Similar relations hold for the other triatomic fragments, so that the MIM Hamiltonian matrix may be written as (taking the zero of energy to be the ground state of the H-atom):

$$\mathbf{H}_{ABCD} = 0.5\,\{\mathbf{h}_1 + \sum_{s=2}^{4} (\mathbf{U}_s\,\mathbf{h}_s\,\mathbf{U}_s^{-1})\} \quad (53)$$

At this point, the exact H_3 potential surface can be introduced into the fragment matrixes **e**; however, this must be done in such a way as to preserve the mathematical structure. Eigenvalue e_1 can be replaced by the exact ground state potential surface e'_1 but e_2 should be replaced by:

$$e'_2 = e'_1 + 2\,J \quad (54)$$

where J is given by Equation 6, the diatomic terms being expressed in terms of exact H_2

singlet and triplet curves. Such a procedure would allow full use of the exact *ab initio* calculations on H_3 and H_2.

Eaker and Parr[49] have shown that such an MIM theory does not in fact do well for H_4 in highly symmetrical configurations. For example, in tetrahedral configurations, each H_3 fragment has D_{3h} symmetry, so that the term J in Equation 2 is zero. In this case, the four-atom Hamiltonian is diagonal, each element having the value 2 E_{H_3} — this is far too low compared with *ab initio* calculations. This appears to be a persistent drawback of DIM (at least for minimum basis calculations): configurations with high symmetry are incorrectly described. Hence, MIM in itself is not a cure for the troubles of DIM: as mentioned earlier, it may be necessary to consider larger basis sets in order to retain enough flexibility in the high symmetries. Nevertheless, a theory along the lines of MIM provides a practical way of approaching larger molecules.

B. Approximate DIM

For the second example, I should like to show how a simplified DIM model can be applied to a reaction involving six atoms in such a way that all degrees of freedom are represented, the symmetry of the molecule is taken into account, and known potential surfaces of the reactant and product fragments can be exploited. Consider the substitution reaction:

$$H + CH_4 \rightarrow H\text{---}CH_3\text{---}H \rightarrow CH_4 + H$$

and assume that the potential surface for CH_4 is known. We wish to describe the simultaneous breaking of the old bond and formation of the new bond with concomitant inversion of the CH_3 group. Following the example of Roach and Gimzewski,[72] we break the molecule into groups of atoms which will be treated as units in the DIM formulation: the C "atom", considered to be in a quintet valence state (which we shall treat as an S-state), the H-atom furthest away from C, and the H_4 group, considered also to be in a quintet state (no H-H bonds are possible among these 4 atoms; each H-atom must bond to the carbon atom). The overall state is a doublet, since the separated reactants (products) consist of a doublet atom and a methane molecule in the ground singlet state. The three groups (doublet, quintet, and quintet) can be coupled in two ways to form a doublet state of the CH_5 molecule: ^5C + $^5H_4$ couple to form $^1CH_4$ and then the ^2H yields doublet CH_5, or ^5C + $^5H_4$ couple to form $^3CH_4$ and then the ^2H yields doublet CH_5. The first basis function describes the reactants and products. The second function can describe the transition state, where only three of the H-atoms are bonded to the C-atom. The appropriate potential function is then given by Equations 24, 25, and 35 (with $n = 5$) minus the energies of the three atoms. The energies of H and C can be chosen as zero but the H_4 energy must be a sum of six H-H triplet interactions, since H_4 is in a quintet state (in the absence of any other directional forces, this alone would ensure that equilibrium CH_4 in the ground state were tetrahedral). The diatomic energies S and T in Equation 33 should depend on all the degrees of freedom in CH_4 and should describe, respectively, the ground state of CH_4 and the (repulsive) interaction between a CH_3 radical and an H-atom. The H-H_4 sextet interaction should be a sum of 10 H-H triplet repulsions (one for each H-H pair) and the quartet interaction should allow for the formation of H_2 + 3H, i.e., H + 4H $\rightarrow$ H_2 + 3H. The CH sextet curve must be repulsive and the quartet curve must be attractive but does not have to correspond to the ground state of CH.

Such a model would allow participation of all degrees of freedom in a dynamical calculation, it could be set up so that all H-atoms are equivalent, the leaving atom being determined by the dynamics, and it could make use of the best surfaces available for CH_3 and CH_4 in the spirit of Murrell et al. If the H_5 quartet state were carefully constructed, the product channel H_2 + CH_3 could also be approximately treated.

VI. CONCLUSION

Semiempirical methods for use in reaction dynamics are many and varied. They provide the key to carrying out dynamical calculations intelligently and are the means through which a meaningful contact between theory and experiment can be achieved. Considerable progress has been made with methods of type 2 and 3. In the near future, it is expected that methods of type 1 will receive more attention, especially since *ab initio* calculations are becoming available in greater numbers.

Improvements in type 1 methods require a more systematic approach to the introduction of empirical information. This goal can perhaps best be approached by using semiempirical methods of types 2 or 3 to estimate the corrections to the *ab initio* points. For example, for a triatomic molecule, let $\tilde{E}(\mathbf{R})$ be an *ab initio* surface depending on the coordinates $\mathbf{R}$. Let $\tilde{E}_s(\mathbf{R})$ be a semiempirical surface which equals $\tilde{E}(\mathbf{R})$ in the asymptotic regions and approximates it elsewhere. Let $E_s(\mathbf{R})$ be the surface obtained by adjusting the diatomic parameters in $E_s(\mathbf{R})$ such that $E_s(\mathbf{R})$ goes to the correct diatomic and atomic limits. Then, a corrected *ab initio* surface could be written as $E(R) = \tilde{E}(R) + \{E_s(R) - \tilde{E}_s(R)\}$. A treatment along these lines would help in the development of type 1 semiempirical methods.

ACKNOWLEDGMENT

I should like to thank Mrs. G. Snoei, Miss K. Charisius, and Mrs. K. Gfrörer for their assistance in preparing the manuscript.

REFERENCES

1. **Segal, G. A., Ed.,** *Modern Theoretical Chemistry. Semiempirical Methods of Electronic Structure Calculation,* Plenum Press, N.Y., 1979.
2. **Connor, J. N. L.,** Reactive molecular collision calculations, *Comput. Phys. Comm.,* 17, 117, 1979.
3. **Kuntz, P. J.,** Features of potential energy surfaces and their effect on collisions, in *Modern Theoretical Chemistry,* Vol. 2B, Miller, W. H., Ed., Plenum Press, N.Y., 1976, 53.
4. **Bottcher, C.,** Excited-state potential surfaces and their applications, *Adv. Chem. Phys.,* 42, 169, 1980.
5. **Truhlar, D. G., Ed.,** *Potential Energy Surfaces and Dynamics Calculations,* Plenum Press, N.Y., 1980.
6. **Kafri, O. and Berry, M. J.,** A new empirical potential surface for bimolecular reaction systems, *Faraday Disc. Chem. Soc.,* 62, 125, 1977.
7. **Levine, R. D. and Bernstein, R. B.,** Thermodynamic approach to collision processes, in *Modern Theoretical Chemistry,* Vol. 2B, Miller, W. H., Ed., Plenum Press, N.Y., 1976, 323.
8. **Levine, R. D. and Kinsey, J. L.,** Information-theoretic approach; application to molecular collisions, in *Atom-Molecule Collision Theory,* Bernstein, R. B., Ed., Plenum Press, N.Y., 1979, chap. 22, 693.
9. **Wright, J. S. and Gray, S. K.,** Rotated morse curve spline potential for A + BC reaction dynamics, *J. Chem. Phys.,* 69, 67, 1978.
10. **Truhlar, D. G. and Horowitz, C. J.,** Functional representation of Liu and Siegbahn's accurate *ab initio* potential energy calculations for $H + H_2$, *J. Chem. Phys.,* 68, 2466, 1978; 71, 1514, 1979.
11. **Carter, S. and Murrell, J. N.,** Analytic potentials for triatomic molecules. VII. Application to repulsive surface, *Mol. Phys.,* 41, 567, 1980.
12. **Chen, M. M. L. and Schaefer, H. F.,** Potential surface for the $Li + HF \rightarrow LiF + H$ reaction, *J. Chem. Phys.,* 72, 4376, 1980.
13. **Alvarino, J. M., Casavecchia, P., Gervasi, O., and Lagana, A.,** Trajectory study of Li + HF, *J. Chem. Phys.,* 77, 6341, 1982.
14. **Whitlock, P. A., Muckerman, J. T., and Kroger, P. M.,** Reactions of the ^4D state of oxygen and carbon, in *Potential Energy Surfaces and Dynamics Calculations,* Truhlar, D. G., Ed., Plenum Press, N.Y., 1980, 551.

15. **Schinke, R. and Lester, W. A., Jr.,** Trajectory study of O + H_2 reactions on fitted *ab initio* surfaces. I. Triplet case, *J. Chem. Phys.*, 70, 4893, 1979.
16. **Schinke, R. and Lester, W. A., Jr.,** Trajectory studies of O + H_2 reactions on fitted *ab initio* surfaces. II. Singlet cases, *J. Chem. Phys.*, 72, 3754, 1980.
17. **Howard, R. E., McLean, A. D., and Lester, W. A., Jr.,** Extended basis first-order CI study of $^1A'$, $^1A''$, $^3A''$, B^1A' potential surfaces of $O(^3P, ^1D)$ + $H_2(^1\Sigma_g^+)$, *J. Chem. Phys.*, 71, 2412, 1979.
18. **Luntz, A. C., Schinke, R., Lester, W. A., Jr., and Günthard, Hs. H.,** Product state distributions in the reactions $O(^1D)$ + $H_2 \rightarrow$ OH + H: comparison of experiment with theory, *J. Chem. Phys.*, 70, 5908, 1979.
19. **Varandas, A. J. C. and Murrell, J. N.,** A many-body expansion of polyatomic potential energy surfaces: application to H_n, *Faraday Disc. Chem. Soc.*, 62, 92, 1977.
20. **Carter, S., Mills, I. M., and Murrell, J. N.,** Analytical potentials for triatomic molecules from spectroscopic data. VI, *Mol. Phys.*, 37, 1885, 1979.
21. **Trinajstić, N.,** Hückel theory and topology, in *Theoretical Chemistry*, Vol. 7A, Segal, G. A., Ed., Plenum Press, N.Y., 1977, chap. 1.
22. **Klopman, G. and Evans, R. C.,** The neglect-of-differential overlap methods of molecular orbital theory, in *Modern Theoretical Chemistry*, Vol. 7A, Segal, G. A., Ed., Plenum Press, N.Y., 1977, chap. 2.
23. **Dewar, M. J. S.,** Studies of the mechanisms of organic reactions by semiempirical SCF-MO methods, *Faraday Disc. Chem. Soc.*, 62, 197, 1977.
24. **Sloan, J. J., Watson, D. G., and Wright, J. S.,** Strong interactions in gas phase reactions: F + HN_3, *J. Chem. Phys.*, 43, 1, 1979.
25. **Sloane, C. S. and Hase, W. L.,** Ethyl radical potential surface, *Faraday Disc. Chem. Soc.*, 62, 210, 1977.
26. **Porter, R. N. and Karplus, M.,** Potential energy surface for H_3, *J. Chem. Phys.*, 40, 1105, 1964.
27. **Silver, D. M. and Brown, N. J.,** Valence-bond model potential surface for H_4, *J. Chem. Phys.*, 72, 3859, 1980.
28. **Brown, N. J. and Silver, D. M.,** Reactive and inelastic scattering of H_2 + D_2 using four semiempirical potential surfaces, *J. Chem. Phys.*, 72, 3869, 1980.
29. **Tanaka, N. and Nomura, O.,** A semiempirical method for the calculation of potential energy surfaces of H_n systems, *J. Chem. Phys.*, 77, 1373, 1982.
30. **Tanaka, N.,** Semiempirical calculation of potential surfaces for polyatomic systems, *Chem. Phys. Lett.*, 54, 551, 1978.
31. **Zembekov, A. A.,** Semiempirical VB potential surfaces for $^1A'$ states of H_2O and H_2F^+, *Mol. Phys.*, 44, 1399, 1981.
32. **Zeiri, Y. and Shapiro, M.,** Semiempirical potential surfaces for electron transfer reactions: Li + FH $\rightarrow$ LiF + H and Li + $F_2 \rightarrow$ LiF + F, *J. Chem. Phys.*, 31, 217, 1978.
33. **Shapiro, M. and Zeiri, Y.,** Semiempirical potential surfaces for the alkali-hydrogen halide reactions, *J. Chem. Phys.*, 70, 5264, 1979.
34. **Tully, J. C.,** Calculation of molecular properties by the method of diatomics-in-molecules, *J. Chem. Phys.*, 64, 3182, 1976.
35. **Ellison, F. O.,** The method of diatomics in molecules, *J. Am. Chem. Soc.*, 85, 3540, 1963.
36. **Tully, J. C.,** Semiempirical DIM potential energy surfaces, *Adv. Chem. Phys.*, 42, 63, 1980.
37. **Tully, J. C.,** Diatomics-in-molecules, in *Modern Theoretical Chemistry*, Vol. 7A, Segal, G. A., Ed., Plenum Press, N.Y., 1977, chap. 6.
38. **Kuntz, P. J.,** Semiempirical-molecule potentials for collision theory, in *Atom-Molecule Collision Theory*, Bernstein, R. B., Ed., Plenum Press, N.Y., 1979, chap. 3.
39. **Kuntz, P. J.,** Diatomics-in-molecules: a present-day extension of the LEP formalism, *Ber. Bunsenges. Phys. Chem.*, 86, 367, 1982.
40. **Kuntz, P. J. and Roach, A. C.,** Ion-molecule reactions of the rare gases with hydrogen. I. Diatomics-in-molecules potential energy, *J. Chem. Soc. Faraday Trans. II*, 68, 259, 1972.
41. **Steiner, E., Certain, P. R., and Kuntz, P. J.,** Extended DIM calculations, *J. Chem. Phys.*, 59, 47, 1973.
42. **Faist, M. B. and Muckerman, J. T.,** On the valence bond DIM method. I. A projection operator formalism, *J. Chem. Phys.*, 71, 225, 1979.
43. **Faist, M. B. and Muckerman, J. T.,** On the valence bond DIM method. II. Application to the valence states of FH_2, *J. Chem. Phys.*, 71, 233, 1979.
44. **Isaacson, A. D. and Muckerman, J. T.,** VB-DIM collinear interactions of Ca with HCl, *J. Chem. Phys.*, 73, 1729, 1980.
45. **Kuntz, P. J. and Schreiber, J. L.,** A systematic procedure for extracting fragment matrices for the method of diatomics-in-molecules from *ab initio* calculations on diatomics, *J. Chem. Phys.*, 76, 4120, 1982.
46. **Schubert, J. G.,** DIM studies of CH_2^+, Report WIS-TCI-591 of the Theoretical Chemistry Institute of the University of Wisconsin, 1978.

47. **Schreiber, J. L. and Kuntz, P. J.,** A criterion for the applicability of the method of diatomics-in-molecules to potential surface calculations. I. Selection of the DIM basis, *J. Chem. Phys.*, 76, 1872, 1982.
48. **Kuntz, P. J. and Chang, C. C.,** Why does the diatomics-in-molecules method appear to fail for H_4?, *J. Phys. Chem.*, 86, 1212, 1982.
49. **Eaker, C. W. and Parr, C. A.,** Optimized DIM potential for H_3 and H_4, *J. Chem. Phys.*, 65, 5155, 1976.
50. **Kuntz, P. J.,** Use of DIM in fitting *ab initio* potential surfaces: HeH_2^+ *Chem. Phys. Lett.*, 16, 581, 1972.
51. **Duggan, J. J. and Grice, R.,** Topography of potential energy surfaces: DIM for H + F_2, *J. Chem. Phys.*, 78, 3842, 1983.
52. **Kuntz, P. J. and Chang, C. C.,** Semiempirical corrections to the mixing coefficients for the diatomics-in-molecules method, *Chem. Phys.*, 75, 79, 1983.
53. **Whitton, W. N. and Mahling, S.,** private communication.
54. **Kendrick, J., Kuntz, P. J., and Hillier, I. H.,** Theoretical study of reactive processes in the FH^+_2 system by *ab initio* MCSCF-CI and diatomics-in-molecules calculations, *J. Chem. Phys.*, 68, 2373, 1978.
55. **Pickup, B. T.,** The symmetric group and the method of diatomics-in-molecules, *Proc. R. Soc. (London)*, A, 333, 69, 1973.
56. **Rittner, E. S.,** Binding energy and dipole moment of alkalic halide molecules, *J. Chem. Phys.*, 19, 1030, 1951.
57. **Alexander, M. H.,** Semiempirical potential surfaces for Li + O_2 and Na + O_2 collisions, *J. Chem. Phys.*, 69, 3502, 1978.
58. **Connor, J. N. L., Jakubetz, W., and Manz, J.,** Quantum collinear reaction probabilities for vibrationally excited reactants: F + $H_2 \rightarrow$ FH + H, *Mol. Phys.*, 39, 799, 1980.
59. **Polanyi, J. C. and Sathyamurthy, N.,** Location of energy barriers. VII. Sudden and gradual late-energy barriers, *Chem. Phys.*, 33, 287, 1978.
60. **Valencich, T.,** The modified hyperbolic map function potential, *Faraday Disc. Chem. Soc.*, 62, 154, 1977.
61. **Brown, J. C., Bass, H. E., and Thompson, D. L.,** Trajectory study of four-center reactions, *J. Chem. Phys.*, 70, 2326, 1979.
62. **Wu, A. A.,** Generalized DIM I. Application to H_3, *Mol. Phys.*, 38, 843, 1979.
63. **Wu, A. A.,** Generalized DIM theory II. Zero differential overlap, *Mol. Phys.*, 39, 1287, 1980.
64. **Wu, A. A.,** GDIM theory. III. Accurate fit to *ab initio* H_3 surface, *Mol. Phys.*, 42, 379, 1981.
65. **Wu, A. A. J.,** GDIM theory. IV. Single parameter, near chemical accuracy three dimensional H_3 potential surface, *Mol. Phys.*, 43, 1459, 1981.
66. **Eaker, C. W. and Allard, L. R.,** Generalized DIM potential surfaces for H_3 and H_4, *J. Chem. Phys.*, 74, 1821, 1981.
67. **Last, I.,** Limitations of the DIM method with neglect of overlap, *Chem. Phys.*, 55, 237, 1981.
68. **Baer, M. and Last, I.,** Potential energy surfaces for the H + HX (X = F, Cl, Br, I) abstraction and exchange reaction channels calculated by the modified DIM method, in *Potential Energy Surfaces and Dynamics Calculations*, Truhlar, D. G., Ed., Plenum Press, N.Y., 1981, chap. 21.
69. **Agmon, N. and Levine, R. D.,** Empirical triatomic potential energy surfaces defined over orthogonal bond order coordinates, *J. Chem. Phys.*, 71, 3034, 1979.
70. **Agmon, N. and Levine, R. D.,** Structural considerations in chemical kinetics, *Isr. J. Chem.*, 19, 330, 1980.
71. **Garrett, B. C., Trukhlar, D. G., and Magnuson, A. W.,** A new semiempirical method of modeling potential energy surfaces, *J. Chem. Phys.*, 76, 2321, 1982.
72. **Roach, A. C. and Gimzewski, E.,** A semiempirical valence bond pi-electron theory related to the method of diatomics-in-molecules, *J. Chem. Phys.*, 73, 1294, 1980.

Chapter 3

THE GENERAL THEORY OF REACTIVE SCATTERING: THE DIFFERENTIAL EQUATION APPROACH

Michael Baer

TABLE OF CONTENTS

I. INTRODUCTION

In this chapter the application of the differential equation approach to the quantum mechanical treatment of three-atom exchange collisions is reviewed. The discussion is divided into two major sections; the first is concerned with the collinear system in which three particles are forced to move along a straight line during the reaction process and the second deals with the system in its full dimensionality. The two sections are arranged quite differently: in the first, the collinear Schrödinger equation (SE) is obtained and various techniques for solving it are described. A large portion of the section is devoted to the analysis of exact numerical results and physical models of the reaction process. In the second section we first derive the SE in three dimensions and then mainly discuss background information needed in order to be able to solve it. This information is particularly important for the approximate treatments as discussed by other authors in Volume II, Chapters 1 and 2.

In certain respects this review somewhat overlaps another recently published review.[1] However, many changes were made; the material is organized differently, details were added to make the information more comprehensive, and recent developments were included, but the major changes were with respect to the numerical aspects of the SE. This material is discussed in considerable detail because it is important for anyone who would like to enter the field.

Finally, this chapter is limited to the case where the reaction takes place on one single (adiabatic) potential energy surface. Those reactions which occur on more than one potential energy surface are treated in a separate chapter.

II. THE COLLINEAR SYSTEM

A. Background

In 1931, Eyring and Polanyi[2] analyzed that $H + H_2$ potential energy surface that they had calculated and found that the minimum energy path (MEP) lay in the linear configuration. Later, Pelzer and Wigner,[3] while using this surface to calculate reaction rates, confirmed the uniqueness and importance of the collinear arrangement, at least for the $H + H_2$ system. Most of the exact and approximate treatments in the 50 years that followed were related to the collinear arrangement. In the first part of this section we analyze the collinear system in some detail and as we go along, we discuss the various coordinates usually employed in treating reactive systems and describe the two arrangement channels involved in this (collinear) case. Special emphasis is placed on the numerical aspects. We refer to most of the methods that have been employed in recent years to solve the SE. The second part is devoted to simplified models. One-dimensional (mathematically speaking) models are discussed with the main emphasis on tunneling and hard-sphere models, which for several decades served as one of the main sources for understanding chemical reactions. In the third part, the (vibrational) inversion in exothermic reactions and the relevance of the curvature of the reaction coordinate, as well as the Franck-Condon-type models, are considered; in the fourth part the possibility of the formation of Feschbach-type resonances due to shallow dips along the vibronic curves is discussed.

B. The Schrödinger Equation

1. Cartesian Coordinates

The Hamiltonian for three particles A, B, and C, in a collinear configuration, is written in the form:

$$\mathcal{H} = \frac{1}{2m_A}\mathbf{P}_A^2 + \frac{1}{2m_B}\mathbf{P}_B^2 + \frac{1}{2m_C}\mathbf{P}_C^2 + V(\mathbf{r}_A, \mathbf{r}_B, \mathbf{r}_C) \tag{1}$$

where $\mathbf{r}_A$, $\mathbf{r}_B$, and $\mathbf{r}_C$ are the distances from some reference point and $\mathbf{P}_A$, $\mathbf{P}_B$, and $\mathbf{P}_C$ are the momenta of A, B, and C, respectively (see Figure 1A). Eliminating the center of mass coordinate and expressing the potential energy in terms of interatomic distances $\mathbf{r}_{CB} = (\mathbf{r}_C - \mathbf{r}_B)$ and $\mathbf{r}_{BA} = (\mathbf{r}_B - \mathbf{r}_A)$, one obtains:

$$\mathcal{H} = \frac{1}{2\mu_{AB}} \mathbf{P}_{AB}^2 + \frac{1}{2\mu_{BC}} \mathbf{P}_{BC}^2 - \frac{1}{m_B} \mathbf{P}_{AB} \cdot \mathbf{P}_{BC} + V(\mathbf{r}_{BA}, \mathbf{r}_{BC}) \tag{2}$$

where μ_{AB} and μ_{BC} are the reduced masses:

$$\mu_{AB} = \frac{m_A m_B}{m_A + m_B} \quad ; \quad \mu_{BC} = \frac{m_B m_C}{m_B + m_C} \tag{3}$$

Now writing $\mathcal{H}$ in an operator form, we get the corresponding SE:[4]

$$\mathcal{H} = -\frac{\hbar^2}{2\mu_{AB}} \frac{\partial^2}{\partial r_{BA}^2} - \frac{\hbar^2}{2\mu_{BC}} \frac{\partial^2}{\partial r_{CB}^2} + \frac{\hbar^2}{m_B} \frac{\partial}{\partial r_{BA}} \frac{\partial}{\partial r_{CB}} + V(\mathbf{r}_{BA}, \mathbf{r}_{BC}) \tag{4}$$

A more common set of coordinates is that where, as before, one coordinate is an interatomic distance, but the other is translational.[5] In the reagents channel (A,BC) we have the coordinates r_{CB} and R_A, the latter being defined as:

$$R_A = r_{BA} + \frac{m_C}{m_C + m_B} r_{CB} \tag{5}$$

and in the products channel (AB, C), we have r_{BA} and R_C defined as:

$$R_C = r_{CB} + \frac{m_A}{m_A + m_B} r_{BA} \tag{6}$$

Here and in what follows the vector signs are dropped since in the collinear system all vectors become scalars.

The corresponding Hamiltonian in the (A,BC) channel is

$$\mathcal{H} = -\frac{\hbar^2}{2\mu_{A,BC}} \frac{\partial^2}{\partial R_A^2} - \frac{\hbar^2}{2\mu_{BC}} \frac{\partial^2}{\partial r_{CB}^2} + V(R_A, r_{CB}) \tag{7}$$

where $\mu_{A,BC}$ is given as:

$$\mu_{A,BC} = \frac{m_A(m_B + m_C)}{m_A + m_B + m_C} \tag{8}$$

A similar equation exists for the (AB,C) channel.

The disadvantage in this formulation is that in each channel there are not only different coordinates but different masses as well. This yields a nonorthogonal transformation matrix from one arrangement channel to the other, and consequently, kinematic effects are not well presented. In 1959, Smith[6] and Delves[7] each independently suggested scaling the coordinates in both channels in the following way:

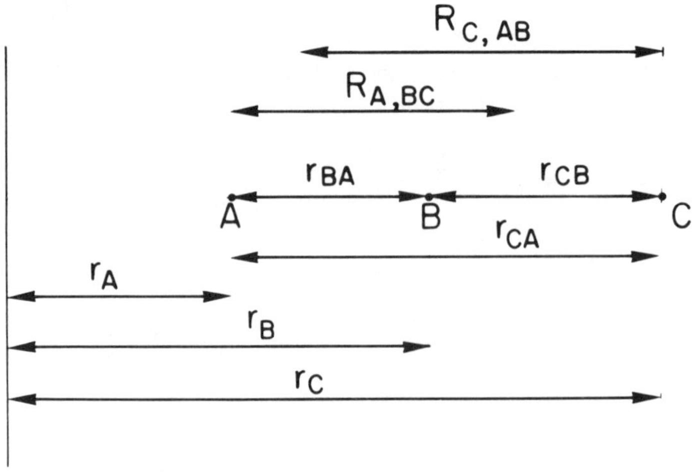

A

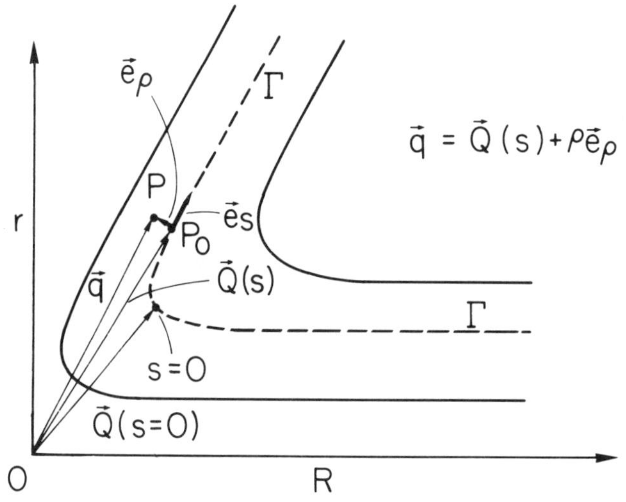

B

FIGURE 1. The three-body collinear reactive system. (A) Absolute, interatomic, and center of mass distances; (B) the Marcus system of coordinates (s stands for the reaction coordinate and ρ for the vibrational one); (C) polar coordinates with respect to an origin in the plateau region; (D) the division of the reaction coordinate Γ into five sectors: two asymptotic ones (sectors I and V), two quasiasymptotic ones (sectors II and IV), and the strong interaction region (sector III). The curvature k is equal to zero in sectors I, II, IV, and V but differs from zero in sector III.

$$\begin{aligned} r_\lambda &= a_\lambda r_{CB} \quad ; \quad R_\lambda = a_\lambda^{-1} R_A \quad ; \quad a_\lambda = \left(\frac{\mu_{BC}}{\mu_{A,BC}}\right)^{1/4} \\ r_\nu &= a_\nu r_{BA} \quad ; \quad R_\nu = a_\nu^{-1} R_C \quad ; \quad a_\nu = \left(\frac{\mu_{AB}}{\mu_{C,AB}}\right)^{1/4} \end{aligned} \tag{9}$$

where λ and ν stand for the reagent and the product channels, respectively. With these sets of coordinates, the Hamiltonian in the two channels looks the same, i.e.,

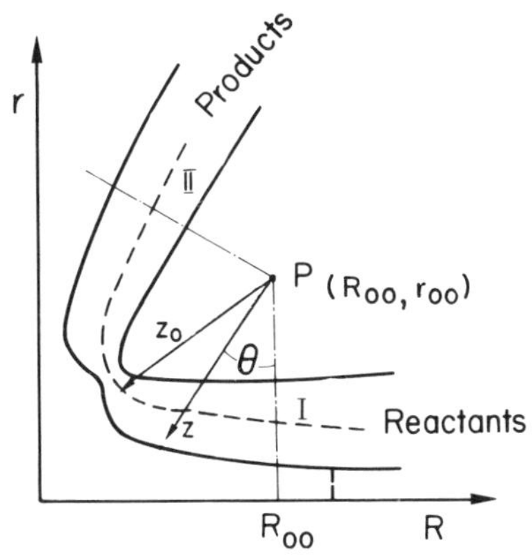

FIGURE 1C

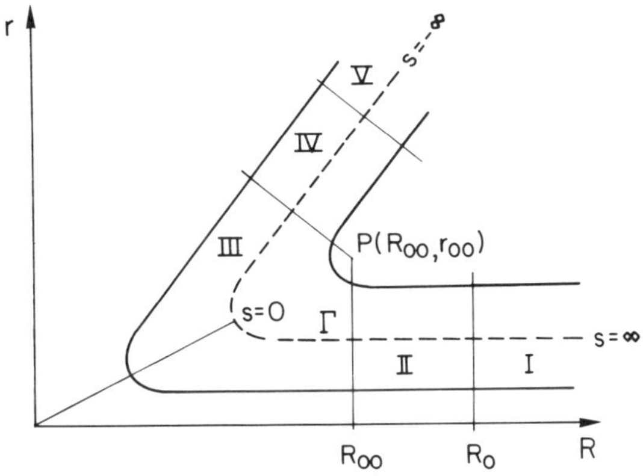

FIGURE 1D

$$\mathcal{H} = -\frac{\hbar^2}{2\mu}\left(\frac{\partial^2}{\partial R^2} + \frac{\partial^2}{\partial r^2}\right) + V(R,r) \tag{10}$$

where $R = R_\lambda, R_\nu$ and $r = r_\lambda, r_\nu$, and μ, which from now on will be considered as the characteristic mass of the system, takes the form:

$$\mu = \left(\frac{m_A m_B m_C}{m_A + m_B + m_C}\right)^{1/2} \tag{11}$$

With this scaling, the transformation matrix from the reagents' coordinates to those of the products' becomes orthogonal:

$$\begin{pmatrix} R_\nu \\ r_\nu \end{pmatrix} = \begin{pmatrix} \cos\beta & -\sin\beta \\ \sin\beta & \cos\beta \end{pmatrix} \begin{pmatrix} R_\lambda \\ r_\lambda \end{pmatrix} \tag{12}$$

where

$$\cos\beta = -\left(\frac{m_A m_C}{(m_A + m_B)(m_C + m_B)}\right)^{1/2} ; \qquad \sin\beta = \left(\frac{m_B(m_A + m_B + m_C)}{(m_A + m_B)(m_C + m_B)}\right)^{1/2} \tag{12'}$$

It can be seen that whenever m_B (the central atom) is much larger than m_A and m_C, β becomes close to $\pi/2$ and when m_B is much smaller than m_A and m_C, β approaches zero. As will be seen later, this kinematic effect has a very important influence on the reaction process.

Since there are three arrangement channels, we have different reactive processes, each of which is characterized by an angle β as described above. It can be shown that the sum of the three β angles is π.

2. The Reaction Coordinate

a. The Continuous Path Approach

The idea of a coordinate which continuously leads from reagents to products is at least as old as the concept of the potential energy hypersurface and may be even older. Langer used it in 1929,[8] without explicitly mentioning it, in his theory of chemical reactions. Most of the theories and processes that Wigner, Eyring, Polanyi, Hirschfelder, and others discussed[2,3,9-13] during the 1930s were based on this concept. However, no serious attempts were made to use it strictly as a coordinate until the 1960s when this possibility was first discussed and analyzed by Hofacker.[14] In 1966, Marcus[15,16] presented his version of the SE for a collinear reactive system in which the reaction coordinate appears explicitly. The basic idea involved using coordinates defined with respect to an origin which move along a continuous curve (Γ).[17] As can be seen from Figure 1B, the radius vector describing the position of an arbitrary point (P) can be written as:

$$\mathbf{q} = \mathbf{Q}(s) + \rho \mathbf{e}_\rho \tag{13}$$

where $\mathbf{Q}(s)$ is the radius vector of P with respect to a given origin, s is the distance between 0′ (the point for which s is defined as being zero), and P_0 is measured along the defined curve (Γ). The unit vector $\mathbf{e}_\rho$ is perpendicular to the curve ($\mathbf{\Gamma}$) at P_0, and ρ is a scalar. The Frenet formulas[17] for any smooth curve are given in the form:

$$\frac{d\mathbf{e}_\rho}{ds} = k\mathbf{e}_s \quad ; \quad \frac{d\mathbf{e}_s}{ds} = -k\mathbf{e}_\rho \tag{14}$$

where k is the curvature of Γ (see Section II.D.2). By defining the conjugate momenta P_s and P_ρ, it can now be shown that the classical Hamiltonian takes the form:[15]

$$\mathcal{H} = \frac{1}{2\mu}\frac{P_s^2}{\eta^2} + P_\rho^2 + V(s,\rho) \tag{15}$$

and the corresponding quantum mechanical representation is[16,18]

$$\mathcal{H} = -\frac{\hbar^2}{2\mu}\left(\frac{1}{\eta}\frac{\partial}{\partial s}\frac{1}{\eta}\frac{\partial}{\partial s} + \frac{1}{\eta}\frac{\partial}{\partial \rho}\eta\frac{\partial}{\partial \rho}\right) + V(s,\rho) \tag{16}$$

where η is defined as:

$$\eta = 1 + k\rho \tag{17}$$

The convenient part of this presentation is that there is no need to consider each arrangement channel independently because there is a smooth transition from the reagent to the product channel. In this sense, the equation describing the reactive process becomes very similar to that describing the inelastic process. The similarity becomes even stronger when the curvature (k) along the reaction path is small. If this is the case, η can then be assumed to be one, and consequently, the Hamiltonian in Equation 16 takes the form:

$$\mathcal{H} = -\frac{\hbar^2}{2\mu}\left(\frac{\partial^2}{\partial s^2} + \frac{\partial^2}{\partial \rho^2}\right) + V(s,\rho) \tag{16'}$$

In general, Equation 16 can be applied for most cases where the curvature is not too large. In case the curvature is large, e.g., in the heavy-light-heavy (HLH) systems, the application of the Marcus equation becomes more complicated, because difficulties such as branch cuts and multivalueness of the wavefunction[19] are encountered. Equation 16 can be simplified as follows:

1. The potential can be written in the form:

$$V(s,\rho) = V_1(s) + V_2(s,\rho) \tag{18}$$

   where $V_2(s,\rho)$ is a "Morse-type potential" which fulfills the condition:

$$V_2(s,\ \rho=0) = 0 \tag{19}$$

   for any s. Consequently, one obtains:

$$V(s,\ \rho=0) = V_1(s) \tag{20}$$

   namely, $V_1(s)$ is the potential along a path which leads continuously from reagents to products. If the route is chosen such that:

$$V_2(s,\ \rho=0) = \min(V(s,\rho)) \tag{21}$$

   for any s, then the route $\rho = \rho(s) = 0$ is the MEP.
2. If $\psi(s,\rho)$ is the total wavefunction, then substitution[20] of

$$\psi(s,\rho) = \eta^p\chi(s,\rho) \tag{22}$$

   leads to

$$\eta_p\left\{-\frac{\hbar^2}{2\mu}\left(\frac{1}{\eta^2}\frac{\partial^2 X}{\partial s^2} + \frac{\partial^2 X}{\partial \rho^2} + \frac{1}{\eta^2}M(s,\rho)\right)\chi + (V(s,\rho) - E)\chi\right\} = 0 \tag{23}$$

   where $M(s,\rho)$ is

$$M(s,\rho) = (2p-1)\frac{\rho}{\eta}\frac{dk}{ds}\frac{\partial}{\partial s} + p(p-2)\left(\frac{\rho}{\eta}\right)^2\left(\frac{dk}{ds}\right)^2 + p\frac{\rho}{\eta}\frac{d^2k}{ds^2} + (2p+1)\,\eta k\frac{\partial}{\partial \rho} + p^2k^2 \tag{24}$$

   Equations 23 and 24 are valid for every p as long as the path $\rho = \rho(s) = 0$ is chosen in such a way that $\eta \neq 0$ in the region of interest. If, however, η becomes zero for certain s values (namely, when k is large), then the choice of p is limited. In most treatments, p was taken to be either $-1/2$,[15,16,21,22] 4,[23-25] or $+1/2$.[26-28] Different reasonings were suggested to justify the various choices. We shall refer to this later.

3. Our next step is to expand $\chi(s,\rho)$ in terms of a local basis set, (Hirschfelder and Wigner[13] were the first to suggest using a local basis set), namely,

$$\chi(s,\rho) = \Sigma\xi_n(s)\phi_n(s,\rho) = \boldsymbol{\xi}^*\boldsymbol{\phi} \tag{25}$$

where $\phi_n(s,\rho)$: n = 0,1,2 . . . , are the eigenfunctions of the equation:

$$\left(-\frac{\hbar^2}{2\mu}\frac{d^2}{d\rho^2} + V_2(s,\rho) - \epsilon_n(s)\right)\phi_n(s,\rho) = 0 \tag{26}$$

and $\xi_n(s)$: n = 0,1,2 . . . , are the corresponding translational functions. Substituting Equation 25 in Equation 23 and employing Equation 26, one obtains:

$$-\frac{\hbar^2}{2\mu}\left(\frac{\partial^2}{\partial s^2} + 2\boldsymbol{\tau}^{(1)}\frac{\partial}{\partial s} + \boldsymbol{\tau}^{(2)} + \mathbf{M}\right)\boldsymbol{\xi} + \boldsymbol{\eta}^{(2)}(\mathbf{U} - \mathbf{E}\mathbf{I})\boldsymbol{\xi} = 0 \tag{27}$$

where

$$\begin{aligned}
M_{nm} &= \langle\phi_n|M|\phi_m\rangle \\
\eta^{(2)}_{nm} &= \langle\phi_n|\eta^2|\phi_m\rangle \\
U_{nm} &= (\epsilon_n(s) + V_1(s))\delta_{nm} \\
\tau^{(1)}_{nm} &= \langle\phi\left|\frac{\partial}{\partial s}\right.\phi_m\rangle + \left(p - \frac{1}{2}\right)\frac{dk}{ds}\langle\phi_n\left|\frac{\rho}{\eta}\right|\phi_m\rangle \\
\tau^{(2)}_{nm} &= \langle\phi_n\left|\frac{\partial^2}{\partial s^2}\right.\phi_m\rangle
\end{aligned} \tag{28}$$

Further simplification can be achieved by introducing the adiabatic-diabatic transformation[29-32] (see also Volume II, Chapter 4).

$$\boldsymbol{\xi}(s) = \mathbf{A}(s)\boldsymbol{\zeta}(s) \tag{29}$$

and choosing **A**(s) to satisfy the equation:

$$\frac{d\mathbf{A}}{ds} + \boldsymbol{\tau}^{(1)}\mathbf{A} = 0 \tag{30}$$

Once this is done it can be shown that **A** is orthogonal and that the corresponding equation for $\boldsymbol{\zeta}(s)$ is

$$\left(-\frac{\hbar^2}{2\mu}\frac{d^2}{ds^2} + \mathbf{M}_A\right)\boldsymbol{\zeta} + \boldsymbol{\eta}^{(2)}_A(\mathbf{W} - E)\boldsymbol{\zeta} = 0 \tag{31}$$

where

$$\mathbf{M}_A = \mathbf{A}^*\mathbf{M}\mathbf{A};\ \boldsymbol{\eta}^{(2)}_A = \mathbf{A}^*\boldsymbol{\eta}^{(2)}\mathbf{A};\ \mathbf{W} = \mathbf{A}^*\mathbf{U}\mathbf{A} \tag{32}$$

The matrix **W** is called the adiabatic-diabatic potential matrix. For cases where the curvature is small enough, one may ignore $\mathbf{M}_A$ and replace $\boldsymbol{\eta}_A^{(2)}$ by a unity matrix so that the SE becomes:

$$\left(-\frac{\hbar^2}{2\mu}\frac{d^2}{ds^2} + (\mathbf{W} - E)\right)\zeta = 0 \tag{33}$$

Equation 27 has been used several times to obtain exact reactive transition probabilities and cross-sections.[19,23,25,26,33-35]

b. The Broken Path Approach

The broken path approach (BPA) differs from the continuous path approach (CPA) in that the reaction path is divided into segments, each characterized by a constant curvature. This idea was suggested and implemented by Johnson[36] in his studies of the H + H_2 reaction. Therefore, for a given segment, Equation 16 becomes:

$$\mathcal{H} = -\frac{\hbar^2}{2\mu}\left(\frac{\partial^2}{\partial\rho^2} + \frac{k}{\eta}\frac{\partial}{\partial\rho} + \frac{1}{\eta^2}\frac{\partial^2}{\partial s^2}\right) + V(s,\rho) \tag{34}$$

The main two regions with constant curvatures are the entrance and the exit channels which are characterized by zero curvature. Substituting k = 0 in Equation 34 yields Equation 16′. Thus, s and ρ are identified as the ordinary Cartesian coordinates R and r, respectively, and consequently, Equation 34 becomes:

$$\mathcal{H} = -\frac{\hbar^2}{2\mu}\left(\frac{\partial^2}{\partial R^2} + \frac{\partial^2}{\partial r^2}\right) + V(R,r) \tag{35}$$

Here $R = R_\lambda, R_\nu$ and $r = r_\lambda, r_\nu$ where λ and ν refer to the entrance and exit channels, respectively. The interaction region where $k \neq 0$ is divided into a number of sectors, each characterized by an arc of a given radius and therefore of a constant curvature. If we define P as the center of a circle with a radius z_0 (see Figure 1C), the curvature of a corresponding arc is

$$k = z_0^{-1} \tag{36}$$

For a given arc (and z_0) two polar coordinates are introduced; a distance z and an angle θ:

$$z = \rho + z_0 \tag{37}$$

$$\theta = s/z_0 \tag{38}$$

From Equations 36 and 37, η takes the form:

$$\eta = 1 + k\rho = z/z_0 \tag{39}$$

By substituting Equations 36 to 39 in Equation 29, it can be seen that $\mathcal{H}$ becomes:

$$\mathcal{H} = -\frac{\hbar^2}{2\mu}\left(\frac{\partial^2}{\partial z^2} + \frac{1}{z}\frac{\partial}{\partial z} + \frac{1}{z^2}\frac{\partial^2}{\partial\theta^2}\right) + V(z,\theta) \tag{40}$$

It is somewhat unexpected that Equation 40 does not depend on z_0. In what follows it is assumed that the reaction coordinate is made up of arcs characterized by radii of different lengths having one single center (see Figures 1C,D). In any subsequent treatment the Hamiltonian will be either of the form given in Equation 35 ($k = 0$) or in Equation 40 ($k \neq 0$).

The asymptotic region (sector I or IV in Figure 1D) is characterized by the fact that the potential V(R,r) becomes independent of R and, up to a constant V_0 equals the diatomic potential v(r), i.e.,

$$\lim_{R \to \infty} V(R,r) = v(r) + v_0 \tag{41}$$

Usually, v_0 is equal to zero in the reactant channel but may differ from zero in the product channel. Substituting Equation 41 in Equation 35, one can write the general solution of Equation 35 as

$$\psi(R,r) = \sum_n \chi_n(R)\phi_n(r) \tag{42}$$

where $\phi_n(r)$; $n = 0,1,\ldots$. are the vibrational basis set functions of the diatomic molecule, i.e.,

$$-\frac{\hbar^2}{2\mu}\left(\frac{\partial^2}{\partial r^2} + v(r) + v_0 - \epsilon_n\right)\phi_n(r) = 0 \tag{43}$$

Here ϵ_n; $n = 0,1,\ldots$ are the eigenvalues of the diatomic molecule shifted by an amount v_0. The translational functions, $\chi_n(R)$, may consequently be written as:

$$\chi_n(R) = A_n e^{-ik_m R} + B_n e^{ik_n R} \tag{44}$$

where

$$k_n = \sqrt{\frac{2\mu}{\hbar^2}(E - \epsilon_n)} \tag{45}$$

and A_n and B_n are constants. Equation 44 describes a superposition of two waves — one incoming and one outgoing. Consequently, Equation 42 describes a situation of many incoming and outgoing waves. The reactant channel is characterized by the fact that there are both incoming and outgoing waves, but the incoming one belongs to a single, well-defined state (n_o). Therefore, the general solution for this channel is of the form:

$$\psi^R_{n_o}(R,r) = \phi^R_{n_o}(R)e^{-ik^R_{n_o}R} + \sum_n \left(\frac{k^R_{n_o}}{k^R_n}\right)^{1/2} T^R_{nn_o}\,\phi^R_n(r)e^{ik^R_n R} \tag{46}$$

The index "R" stands for "Reactants" and $T^R_{nn_o}$; $n = 0,1,\ldots$ are identified as the *inelastic* **T** matrix elements. The product channel contains only outgoing waves and therefore, the general solution for this channel is of the form:

$$\psi^P_{n_o}(R,r) = \sum_\ell \left(\frac{k^R_{n_o}}{k^P_\ell}\right)^{1/2} T^P_{\ell n_o}\,\phi^P_\ell(r)e^{ik^P_\ell R} \tag{47}$$

where the index "P" stands for "Products" and $T^P_{\ell n_o}$ are the *reactive* **T** matrix elements. The summation limits in Equations 46 and 47 are equal to the number of open states in each channel, namely N_R and N_P, respectively. The $\mathbf{T}^R_{nn_o}$ and the $\mathbf{T}^P_{\ell n_o}$ matrix elements are closely related to the inelastic and reactive transition probabilities. Thus,

$$P^{R}_{nn_0} = |T^{R}_{nn_0}|^2 \qquad (48)$$
$$P^{P}_{\ell n_0} = |T^{P}_{\ell n_0}|^2$$

so that

$$\sum_n P^{R}_{nn_0} + \sum_\ell P^{P}_{\ell n_0} = 1 \qquad (49)$$

The strong interaction region ($k \neq 0$) is further divided into several sectors, each characterized by a given radius. For each sector (or region) the SE is solved independently. The task is to match any two solutions defined in two adjacent sectors (or regions). Therefore, we consider two regions; region 1 and region 2 in which two solutions $\psi_1(x_1,y_1)$ and $\psi_2(x_2,y_2)$, are defined, respectively. Here x_i; $i = 1,2$ are the translational coordinates (not necessarily identical) and y_i; $i = 1,2$ are the vibrational coordinates [(x_1y_1) and (x_2,y_2) represent the same point in configuration space]. The borderline between the two regions is defined in terms of $x_1 = C_1$ and $x_2 = C_2$ where C_i; $i = 1,2$ are given constants. In order to have continuous and physically well-behaved solutions everywhere, the following requirements have to be fulfilled on each borderline:

$$\psi_1(x_1,y_1)\Big|_{x_1=C_1} = \psi_2(x_2,y_2)\Big|_{x_2=C_2} \qquad (50)$$
$$\frac{\partial\psi_1(x_1,y_1)}{\partial n_1}\Big|_{x_1=C_1} = \frac{\partial\psi_2(x_2,y_2)}{\partial n_2}\Big|_{x_2=C_2}$$

where n_i are the normals to the lines $x_i = C_i$. Thus, the continuity requirements for solutions of two nearby sectors in the interaction region are:

$$\psi_1(\theta,z)\Big|_{\theta=\theta_0} = \psi_2(\theta,z)\Big|_{\theta=\theta_0} \qquad (51)$$
$$\frac{\partial\psi_1(\theta,z)}{\partial\theta}\Big|_{\theta=\theta_0} = \frac{\partial\psi_2(\theta,z)}{\partial\theta}\Big|_{\theta=\theta_0}$$

Similar relations should hold between solutions defined in the asymptotic and the quasi-asymptotic regions.

The requirements on the borderline between the quasiasymptotic region and the strong interaction region (see Figure 1D) are somewhat different, namely:

$$\psi_\theta(\theta,z)\Big|_{\theta=0} = \psi_R(R,r)\Big|_{R=R_{00}} \qquad (52)$$
$$\frac{1}{z}\frac{\partial\psi_\theta(\theta,z)}{\partial\theta}\Big|_{\theta=0} = -\frac{\partial\psi_R(R,r)}{\partial R}\Big|_{R=R_{00}}$$

It should be noted that the relation between (θ,z) and (R,r) is

$$z\sin\theta = R_{00} - R \qquad (53)$$
$$z\cos\theta = r_{00} - r$$

where (R_{00},r_{00}) are the coordinates of P (see Figure 1C).

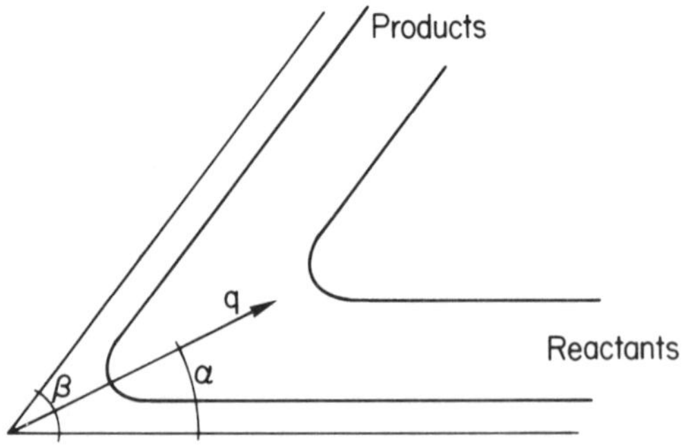

FIGURE 2. Delves coordinates; q and α are ordinary polar coordinates measured from the origin. The range of α is $0 \leq \alpha \leq \beta$.

3. Hyperspherical Coordinates

The Marcus equation (ME) in its original or modified form serves as the main tool for studying (quantum mechanically) simple exchange collisions. Still, it was found that in certain cases this approach becomes very inefficient. The skewing angle β, which is the angle between the translational axes of the two arrangement channels, i.e., R_λ and R_ν, was introduced in Equation 12. It was realized in 1975 by Baer[37] that for systems for which β is close to zero (i.e., the case when the central atom, atom B in Equation 12′ is very light compared with the other two atoms) the ordinary reaction path approach (RPA) might fail. He found that due to the large skew, the interaction region becomes very broad which, in turn, forces in the close-coupling expansion (see Equation 25 and also next section), including a large number of vibrational states in order to span the energy range needed for convergence. In another group of cases the limitation of this approach is even more severe. Recently, due to extensive experimental investigation, the interest in breakup collision has increased appreciably. Whereas those processes with a single-bond arrangement channel were treated in a rather efficient way, hardly any method is available to treat systems with two or more arrangement channels. As mentioned, the ME seems to be inadequate for such cases.

In 1959, Delves[7] published a set of new coordinates which were found to be most suitable for treating the various kinds of exchange collisions, including the ones mentioned above. These are now known as the hyperspherical coordinates. Delves' paper deals with three-dimensional systems but we will consider only the collinear case.[38-42]

In this particular case the Delves coordinates become simply the ordinary polar coordinates q and α (Figure 2), i. e.,

$$\begin{aligned} q &= \sqrt{R^2 + r^2} \\ \alpha &= \tan^{-1}(r/R) \end{aligned} \tag{54}$$

This set of coordinates is found to be suitable for the following reasons:

1. Except for the asymptotic regions where the **T** matrix analysis is best performed employing the relevant arrangement channel coordinates, these coordinates are simple and most convenient over the entire configuration space.
2. In this representation the coordinate α is the bound coordinate irrespective of whether breakup processes are ignored or included. The range of α is

$$0 \leq \alpha \leq \beta$$

where β is the skew angle. The coordinate q is unbound.

3. It was established that in realistic cases the convergence towards the exact results is achieved much faster employing these kinds of coordinates.[42]

Starting with Equation 10 and performing the transformation (Equation 54), one obtains for $\mathcal{H}$:

$$\mathcal{H} = -\frac{\hbar^2}{2\mu}\left(\frac{\partial^2}{\partial q^2} + \frac{1}{q}\frac{\partial}{\partial q} + \frac{1}{q^2}\frac{\partial^2}{\partial \alpha^2}\right) + V(q,\alpha) \tag{55}$$

Once the equation

$$\mathcal{H}\psi(q,\alpha) = E\psi(q,\alpha) \tag{56}$$

is solved (see next section), the value of $\psi(q = q_{as},\alpha)$ and its first derivative in terms of q and α at some asymptotic q value ($q = q_{as}$) are known. These functions are then matched with their corresponding asymptotic expressions (see Equations 46 and 47) so the **T**-matrix elements can be extracted. The method becomes somewhat more involved in case the energy is high enough for dissociation to occur. Still, no principal complications were encountered and cases of this type were treated.[40-41] This method was tested on several occasions and its reliability and efficiency have been clearly demonstrated.[38-42] This method was found to be particularly convenient for the HLH mass combination. As an example, the Cl-H-Br system is considered. It was studied twice: once, 8 years ago by Baer,[37] employing the ordinary propagative method (see previous and also next section), and just recently by Kay and Kuppermann[42] employing the Delves coordinates. The results were the same, but in the first treatment, 40 vibrational states were required, whereas in the more recent one only 12 states were included in the expansion.

An interesting feature, which is related to both the HLH system and the Delves coordinates, is the possibility of a new chemical bonding. This idea was first suggested by Pollak,[43] and then confirmed employing the Delves coordinates by Manz et al.[44]

4. Numerical Treatment

a. The Close-Coupling Technique

The equations that will be considered are either of the form:

$$\left[-\frac{\hbar^2}{2\mu}\left(\frac{\partial^2}{\partial R^2} + \frac{\partial^2}{\partial r^2}\right) + V(R,r) - E\right]\psi(R,r) = 0 \tag{57}$$

or

$$\left[-\frac{\hbar^2}{2\mu}\left(\frac{\partial^2}{\partial z^2} + \frac{1}{z}\frac{\partial}{\partial z} + \frac{1}{z^2}\frac{\partial^2}{\partial \theta^2}\right) + V(z,\theta) - E\right]\psi(z,\theta) = 0 \tag{58}$$

Equation 58 looks somewhat different than Equation 57, but by making the following substitutions:

$$\psi(z,\theta) = \frac{1}{\sqrt{z}}\chi(z,\theta) \tag{59}$$

it becomes similar:

$$\left[-\frac{\hbar^2}{2\mu}\left(\frac{\partial^2}{\partial z^2}+\frac{1}{z^2}\frac{\partial^2}{\partial\theta^2}\right)+\widetilde{V}(z,\theta)-E\right]\chi(z,\theta)=0 \tag{60}$$

where

$$V(z,\theta) = V(z,\theta) - \frac{\hbar^2}{8\mu z^2} \tag{61}$$

In each case, one of the coordinates (say y) is bound and the other (x) is not bound. The method most extensively used is to expand $\psi(x,y)$ in terms of a basis set with respect to y and to propagate the x-dependent coefficients from one x value to the other. This is accomplished by dividing the whole x range into a certain number of intervals. The ith interval is bound by the point x_{i-1} at the beginning of this interval and x_i at its end. We define $\bar{x}_i$ as:

$$\bar{x}_i = (x_{i-1} + x_i)/2 \tag{62}$$

The aim is to calculate $\psi(x_i,y)$ once $\psi(x_{i-1},y)$ is given. This is done in several steps. First, the adiabatic close-coupling expansion is employed for the i-th interval:

$$\psi(x,y) = \sum_{n=1}^{N} \zeta_n^{(i)}(x)\phi_n(\bar{x}_i,y) \tag{63}$$

where $\phi_n(\bar{x}_i,y)$ are eigenfunctions obtained from:

$$-\frac{\hbar^2}{2\mu}\left(\frac{\partial^2}{\partial y^2} + V(\bar{x}_i,y) - \epsilon_n(\bar{x}_i)\right)\phi_n(\bar{x}_i,y) = 0 \tag{64}$$

Here $\epsilon_n(\bar{x}_i)$ are the corresponding eigenvalues. Next, Equations 63 and 64 are substituted in Equation 57 (or Equation 60), and the result is multiplied by $\phi_\ell(\bar{x}_i,y)$ (or by $y^2\phi_\ell(\bar{x}_i,y)$ in the second case) and integrated over y. Thus the equation:

$$-\frac{\hbar^2}{2\mu}\frac{\partial^2}{\partial x^2}\boldsymbol{\zeta}^{(i)} + (\mathbf{W} - E\mathbf{I})\boldsymbol{\zeta}^{i} = 0 \tag{65}$$

is obtained where $\boldsymbol{\zeta}^{(i)}$ is a column vector with the functions $\zeta_n^{(i)}(x)$; $n = 1,2 \ldots N$, $\mathbf{I}$ is the diagonal unity matrix, and $\mathbf{W}(x)$ is a square matrix of the order $N \times N$ with the following elements:

$$W_{\ell n}(\bar{x}_i,x) = \langle \phi_\ell(\bar{x}_i,y)|V(x,y) - V(\bar{x}_i,y)|\,\phi_n(\bar{x}_i,y)\rangle + \epsilon_\ell(\bar{x}_i)\delta_{\ell n} \tag{66}$$

Once $\mathbf{W}(x)$ is known, the next step is to propagate the vector $\zeta^{(i)}$ from its initial value at $x = x_{i-1}$ to its value at $x = x_i$. (This procedure is described to a certain extent in Section II.4.c). Assuming this procedure has been completed, the function $\psi(x_i,y)$ is in fact derived, as can be seen from Equation 63. In order to proceed in the next interval we have to obtain the value of $\boldsymbol{\zeta}^{(i+1)}$ at $x = x_i$:

$$\psi(x_i,y) = \sum_\ell \zeta_\ell^{(i+1)}(x)\psi_\ell(\bar{x}_{i+1},y) = \sum_n \zeta_n^{(i)}(x)\phi_n(\bar{x}_i,y) \tag{67}$$

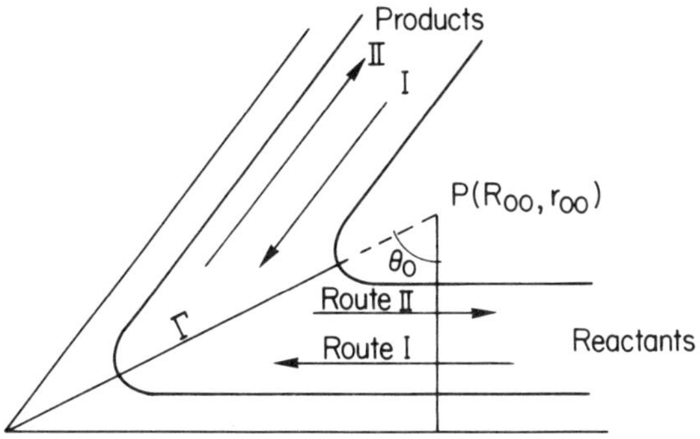

FIGURE 3. Routes for integrating the Schrödinger equation.

It can be seen that

$$\zeta^{(i+1)} = S^{(i+1,i)}\zeta^{(i)} \tag{68}$$

where $S^{(i+1,i)}$ is an overlap matrix with elements:

$$S_{\ell n}^{(i+1,i)} = \langle \phi_\ell(\bar{x}_{i+1},y) | \phi_n(\bar{x}_i,y) \rangle \tag{69}$$

This process is repeated from the first interval, which may be either in the asymptotic region or deep in the interaction region or at any other place.

Equation 65 is of the second order (like Equation 57) and consequently has 2N-independent solutions. In contrast to the situation encountered in nonreactive systems, here all 2N-independent solutions are needed to obtain meaningful physical results. Since the solution matrix ζ contains N columns (each column represents one solution), solving Equation 65 leads to N solutions only. Usually, such a set of solutions is obtained either by propagating from the asymptotic region towards the strong interaction region (route I in Figure 3) or, vice versa, by propagating from the deep interaction region towards the asymptotic region (route II in Figure 3). It turns out that both procedures yield N solutions (each one arranged in one matrix) which are independent of each other. To distinguish between the two the $(\pm)$ indexes are added and we obtain $\zeta^{(\pm)}$. In the collinear case, the two arrangement channels together with the interaction region form one single channel, and consequently, ζ^+ can be obtained by propagating from the reagent channel to the product channel and ζ^- by propagating backwards from the product channel to the reagent channel. In case the system contains more than two arrangement channels, this procedure can no longer be used. Instead, the propagation inside is stopped at some location in the interaction (line Γ in Figure 3). At the same location, the propagation outside is started and continued up to the asymptotic region (see Figure 3). This procedure is repeated for all channels so that at each channel 2N-independent solutions are now available. In order to get one single set of 2N physical solutions, the solutions and their first derivatives between any two channels have to be matched at the proper location (line Γ in Figure 3). As an example, let us consider the collinear case again. The solution is obtained by solving Equation 57 in the quasiasymptotic region and Equation 58 in the interaction region. The solutions of the two channels are matched in the interaction region. Since the propagating coordinate is in this region θ, a convenient location to perform the matching is $\theta = \theta_0$ (see Figure 3). Recalling Equation 63, one can verify that the matching conditions for the functions and their first derivatives with respect to θ are:

$$\begin{pmatrix} \zeta_\lambda^+(\theta_0) & \zeta_\lambda^-(\theta_0) & -\zeta_\nu^+(\theta_0) & -\zeta_\nu^-(\theta_0) \\ (\zeta_\lambda^+(\theta_0))' & (\zeta_\lambda^-(\theta_0))' & (\zeta_\nu^+(\theta_0))' & (\zeta_\nu(\theta_0))' \end{pmatrix} \begin{pmatrix} \mathbf{A}_\lambda^+ \\ \mathbf{A}_\lambda^- \\ \mathbf{A}_\nu^+ \\ \mathbf{A}_\nu^- \end{pmatrix} = \begin{pmatrix} \mathbf{0} \\ \mathbf{0} \end{pmatrix} \tag{70}$$

Here the matrixes $\mathbf{0}$ are zero matrixes and $(\zeta_\alpha(\theta_0))'$; $\alpha = \lambda,\nu$ are the first derivatives of the corresponding $\zeta_\alpha(\theta_0)$; $\alpha = \lambda,\nu$ matrixes. The unknowns to be solved are the 4-column vectors $\mathbf{A}_\alpha^{(\pm)}$; $\alpha = \lambda,\nu$. However, it can be seen that there are 2N equations and 4N unknowns. This is so because out of the 2N unmatched solutions given in each channel, we are interested in forming 2N smoothly matched solutions. Therefore, by assigning numerical values, say to $\mathbf{A}_\lambda^+$ and $\mathbf{A}_\nu^+$, the other two matrixes $\mathbf{A}_\alpha^-$; $\alpha = \lambda,\nu$ are uniquely determined and 2N smoothly matched solutions are formed. In the case of three channels or more, a similar procedure is performed but the extension is not straightforward due to the bifurcation which has to be taken into account.[45]

Two more points are to be considered. The first is concerned with matching at the boundary between the Cartesian and the polar regions; the matching conditions are given in Equation 52, or by employing Equation 59, they become:

$$\begin{aligned} \chi(\theta = 0,x) &= z^{1/2}\,\psi_R(R = R_{00},r) \\ \frac{d\chi(\theta = 0,z)}{d\theta} &= -z^{3/2}\,\frac{d\psi_R(R = R_{00},r)}{dR} \end{aligned} \tag{71}$$

where

$$z = r_{00} - r \tag{72}$$

Introducing the expansion of each of the functions (see Equation 63), we can verify that Equation 71 takes the form:

$$\begin{aligned} \zeta_\theta^\pm(\theta = 0) &= \mathbf{S}^{(1/2)}\,\zeta_R^\pm(R = R_{00}) \\ (\zeta_\theta^\pm(\theta = 0))' &= -\mathbf{S}^{(3/2)}(\zeta_R^\pm(R = R_{00}))' \end{aligned} \tag{73}$$

where

$$S_{nm}^{(k)} = \langle \phi_n(\theta = 0,z)\,|z^k|\phi_m(\theta = 0,z)\rangle \quad ; \quad k = {}^1\!/_2,\ {}^3\!/_2 \tag{74}$$

The second point is related to the matching at the asymptotic region which is accomplished by applying Equations 46, 47, and 63. Integrating over the basis set functions (assuming the same basis is employed in both expansions) one obtains relations between the various **T**-matrix elements and the values of the functions and their derivatives at the corresponding asymptotic $R = R_{\alpha o}$ values. To shorten the notation, matrix equations are employed:

$$\begin{pmatrix} \zeta_\lambda^+ & \zeta_\lambda^- & \mathbf{E}_\lambda^+ & \mathbf{0} \\ (\zeta_\lambda^+)' & (\zeta_\lambda^-)' & (\mathbf{E}_\lambda^+)' & \mathbf{0} \\ \zeta_\nu^+ & \zeta_\nu^- & \mathbf{0} & \mathbf{E}_\nu^+ \\ (\zeta_\nu^+)' & (\zeta_\nu^-)' & \mathbf{0} & (\mathbf{E}_\nu^+)' \end{pmatrix} \begin{pmatrix} \mathbf{B}_\lambda^+ & \mathbf{B}_\nu^+ \\ \mathbf{B}_\lambda^- & \mathbf{B}_\nu^- \\ \mathbf{T}_{\lambda\lambda}^R & \mathbf{T}_{\lambda\nu}^P \\ \mathbf{T}_{\nu\lambda}^P & \mathbf{T}_{\nu\nu}^R \end{pmatrix} = \begin{pmatrix} \mathbf{E}_\lambda^- & \mathbf{0} \\ (\mathbf{E}_\lambda^-)' & \mathbf{0} \\ \mathbf{0} & \mathbf{E}_\nu^- \\ \mathbf{0} & (\mathbf{E}_\nu^-)' \end{pmatrix} \tag{75}$$

where

$$\begin{aligned}
\zeta^{\pm}_{\alpha n\ell} &= \zeta^{\pm}_{n\ell}\,(R_\alpha = R_{\alpha 0}) \quad ; \quad \alpha = \lambda,\nu \\
\left(\zeta^{\pm}_{\alpha n\ell}\right)' &= (\zeta^{\pm}_{n\ell}(R_\alpha = R_{\alpha 0}))' \quad ; \quad \alpha = \lambda,\nu \\
E^{\pm}_{\alpha n\ell} &= \delta_{\ell n}\frac{1}{\sqrt{k_{n\alpha}}}\,e^{\pm ik_{n\alpha}R_{\alpha 0}} \quad ; \quad \text{for open states} \\
\left.\begin{aligned} E^{+}_{\alpha n\ell} &= \delta_{\ell n}\frac{1}{\sqrt{|k_{n\alpha}|}}\,e^{-|k_{n\alpha}|R_{\alpha 0}} \\ E^{-}_{\alpha n\ell} &= 0 \end{aligned}\right\} &\quad \text{for closed states} \\
(E^{\pm}_{\alpha n\ell})' &= \pm\delta_{\ell n}\, i\sqrt{k_{n\alpha}}\,e^{\pm ik_{n\alpha}R_{\alpha 0}} \quad ; \quad \text{for open states} \\
\left.\begin{aligned} (E^{+}_{\alpha n\ell})' &= -\delta_{\ell n}\sqrt{|k_{n\alpha}|}\,e^{-|k_{n\alpha}|R_{\alpha 0}} \\ (E^{-}_{\alpha n\ell})' &= 0 \end{aligned}\right\} &\quad \text{for closed states}
\end{aligned} \tag{76}$$

The solution of Equation 75 yields the entire **T**-matrix where each open state, whether in the λ- or the ν-arrangement channel, serves exactly once as an initial state.

The functions $\mathbf{E}^{\pm}_{\alpha}$; $\alpha = \lambda,\nu$ are complex and therefore, the algebra has to be carried out using complex numbers. If the exponential functions given in Equation 76 are replaced by sine and cosine functions, then the following matrix equation is obtained:

$$\begin{pmatrix} \boldsymbol{\zeta}^{+}_{\lambda} & \boldsymbol{\zeta}^{-}_{\lambda} & \mathbf{C}_{\lambda} & \mathbf{0} \\ (\boldsymbol{\zeta}^{+}_{\lambda})' & (\boldsymbol{\zeta}^{-}_{\lambda})' & \mathbf{C}'_{\lambda} & \mathbf{0} \\ \boldsymbol{\zeta}^{+}_{\gamma} & \boldsymbol{\zeta}^{-}_{\gamma} & \mathbf{0} & \mathbf{C}_{\nu} \\ (\boldsymbol{\zeta}^{+}_{\gamma})' & (\boldsymbol{\zeta}^{-}_{\nu})' & 0 & \mathbf{C}'_{\nu} \end{pmatrix} \begin{pmatrix} \overline{\mathbf{B}}^{+}_{\lambda} & \overline{\mathbf{B}}^{+}_{\nu} \\ \overline{\mathbf{B}}^{-}_{\lambda} & \overline{\mathbf{B}}^{-}_{\nu} \\ \mathbf{K}^{R}_{\lambda\lambda} & \mathbf{K}^{P}_{\lambda\nu} \\ \mathbf{K}^{P}_{\nu\lambda} & \mathbf{K}^{R}_{\nu\nu} \end{pmatrix} = \begin{pmatrix} \mathbf{S}_{\lambda} & \mathbf{0} \\ \mathbf{S}'_{\lambda} & \mathbf{0} \\ \mathbf{0}_{\chi} & \mathbf{S}_{\nu} \\ \mathbf{0} & \mathbf{S}'_{\nu} \end{pmatrix} \tag{77}$$

where $\boldsymbol{\zeta}^{\pm}_{\alpha}$; $\alpha = \lambda,\nu$ retain their previous presentations but $\mathbf{C}_{\alpha}$ and $\mathbf{S}_{\alpha}$; $\alpha = \lambda,\nu$ are defined as:

$$C_{\alpha n\ell} = \begin{cases} \delta_{n\ell}\dfrac{1}{\sqrt{k_{\alpha n}}}\cos(k_{\alpha n}R_{\alpha 0}) & \text{for open states} \\[2ex] \delta_{n\ell}\dfrac{1}{\sqrt{|k_{\alpha n}|}}\,e^{-|k_{\alpha n}|R_{\alpha 0}} & \text{for closed states} \end{cases}$$

$$S_{\alpha n\ell} = \begin{cases} \delta_{n\ell}\dfrac{1}{\sqrt{k_{\alpha n}}}\sin k_{\alpha n}R_{\alpha 0} & \text{for open states} \\[2ex] 0 & \text{for closed states} \end{cases} \tag{78}$$

The matrixes of the derivatives, $\mathbf{C}'_{\alpha}$ and $\mathbf{S}'_{\alpha}$ ($\alpha = \lambda,\nu$) are defined accordingly. The **T** and **K** matrixes are related to each other through the Heitler's damping equation,[46] namely:

$$\mathbf{T} = \frac{1 + i\mathbf{K}}{1 - i\mathbf{K}} \tag{79}$$

Equations 70 and 77 do not have to be solved separately. In fact, they can be combined to form one single set of equations which have to be solved explicitly only for the various **K** matrixes:

$$\begin{pmatrix} \zeta^+_\lambda & \zeta^-_\lambda & \mathbf{0} & \mathbf{0} & \mathbf{C}_\lambda & \mathbf{0} \\ \zeta^{+\prime}_\lambda & \zeta^{-\prime}_\lambda & \mathbf{0} & \mathbf{0} & \mathbf{C}'_\lambda & \mathbf{0} \\ \mathbf{0} & \mathbf{0} & \zeta^+_\nu & \zeta^-_\nu & \mathbf{0} & \mathbf{C}_\nu \\ \mathbf{0} & \mathbf{0} & \zeta^{+\prime}_\nu & \zeta^{-\prime}_\nu & \mathbf{0} & \mathbf{C}'_\nu \\ \zeta^+_{\lambda\theta} & \zeta^-_{\lambda\theta} & -\zeta^+_{\nu\theta} & -\zeta^-_{\nu\theta} & \mathbf{0} & \mathbf{0} \\ \zeta^{+\prime}_{\lambda\theta} & \zeta^{-\prime}_{\lambda\theta} & \zeta^{+\prime}_{\nu\theta} & \zeta^{-\prime}_{\nu\theta} & \mathbf{0} & \mathbf{0} \end{pmatrix} \begin{pmatrix} \mathbf{D}^+_{\lambda\lambda} & \mathbf{D}^+_{\lambda\nu} \\ \mathbf{D}^-_{\lambda\lambda} & \mathbf{D}^-_{\lambda\nu} \\ \mathbf{D}^+_{\nu\lambda} & \mathbf{D}^+_{\nu\nu} \\ \mathbf{D}^-_{\nu\lambda} & \mathbf{D}^-_{\nu\nu} \\ \mathbf{K}^R_{\lambda\lambda} & \mathbf{K}^P_{\lambda\nu} \\ \mathbf{K}^P_{\nu\lambda} & \mathbf{K}^R_{\nu\nu} \end{pmatrix} = \begin{pmatrix} \mathbf{S}_\lambda & \mathbf{0} \\ \mathbf{S}'_\lambda & \mathbf{0} \\ \mathbf{0} & \mathbf{S}_\nu \\ \mathbf{0} & \mathbf{S}'_\nu \\ \mathbf{0} & \mathbf{0} \\ \mathbf{0} & \mathbf{0} \end{pmatrix} \tag{80}$$

In the actual calculation the square matrix on the l.h.s. is inverted and then multiplied by the matrix on the r.h.s. to form the **K**-matrix elements for the open states only. Thus, the various **D** matrixes are, in fact, never calculated.

b. The Vibrational Basis Set

In order to carry out the close-coupling treatment described in the previous section, one needs a method to calculate the vibrational basis sets. In general, two approaches exist. The first is based on employing vibrational basis sets of a given analytic form. Such basis sets can be obtained for the harmonic oscillator (HO) potential and the Morse potential. Thus, the idea is, while moving along the reaction coordinate, to fix the HO or the Morse parameters in such a way that the reference potential will fit the actual potential as closely as possible. Consequently the potential difference (see Equation 66)

$$W^{(i)}(x,y) = V(x,y) - V(\bar{x}_i,y) \tag{81}$$

is expanded, for a given x and x_i, in a Taylor series in y, i.e.,

$$W^{(i)}(x,y) = \sum_{\ell=3} a^{(i)}_\ell(x)\,(y - y_{0i})^\ell \tag{82}$$

so that the corresponding **W** matrix elements can be calculated analytically. This approach was undertaken by Walker and Light[47] while developing their **R**-matrix approach. A possible disadvantage in employing this method is that one might need more vibrational states to obtain convergence than, for instance, when an adiabatic basis set is used. (It should be pointed out that in the most recent computer programs the (energy *independent*) potential matrix element W are calculated only *once*.)

The most efficient basis sets are the local adiabatic basis sets. In what follows a method to calculate them is described.

Let us consider the equation:

$$\left(-\frac{\hbar^2}{2\mu}\frac{\partial^2}{\partial y^2} + V(y) - \epsilon\right)\phi(y) = 0 \tag{83}$$

where V(y) is an arbitrary bound potential of a given depth. We are interested in the discrete spectrum ϵ only, and consequently, the solutions are subject to the following boundary conditions:

$$\phi(y = 0) = \phi(y = M) = 0 \tag{84}$$

Here M is a given large number (for which all eigenfunctions of interest are expected to be negligibly small).

To shorten the notation the following expressions are introduced:

$$U(y) = \frac{2\mu}{\hbar^2} V(y)$$
$$\lambda = \frac{2\mu}{\hbar} \epsilon \tag{85}$$

so that Equation 83 becomes:

$$\left(\frac{\partial^2}{\partial y^2} - U(y) + \lambda\right) \phi(y) = 0 \tag{86}$$

The finite differences method, which assumes that the range of interest $0 \leq y \leq M$ is divided by $N - 1$ grid points into N-intervals of equal length h, is employed. The second-order derivative of the function at some grid point y_i is then given in the form:[48]

$$\phi''(y_i) = \frac{1}{h^2}\left(\delta^2 - \frac{1}{12}\delta^4 + \frac{1}{90}\delta^6 - \frac{1}{560}\delta^8 \ldots\right)\phi(y_i) \tag{87}$$

where

$$\delta^{2n}\phi_i = \sum_{\ell=n}^{n} \binom{2n}{n-\ell} (-1)^{n-\ell}\phi_{i+\ell} \tag{88}$$

and

$$\phi_j = \phi(y_j) \tag{89}$$

Thus, for $n = 1$:

$$\delta^2\phi_i = \phi_{i-1} - 2\phi_i + \phi_{i+1} \tag{90}$$

and for $n = 2$, we have

$$\delta^4\phi_i = \phi_{i-2} - 4\phi_{i-1} + 6\phi_i - 4\phi_{i+1} + \phi_{i+2} \tag{91}$$

Sustitution of Equation 90 in Equation 87 yields:

$$\phi''(y_i) = \frac{1}{h^2}(\phi_{i-1} - 2\phi_i + \phi_{i+1}) \tag{90'}$$

and substitution of Equation 91 in Equation 87 leads to:

$$\phi''(y_i) = \frac{1}{12h^2}(-\phi_{i-2} + 16\phi_{i-1} - 30\phi_i + 16\phi_{i-1} - \phi_{i+2}) \tag{91'}$$

Substituting Equations 87 and 88 in Equation 86, one can see that Equation 86 becomes an algebraic eigenvalue problem:

$$(\mathbf{G} + \lambda)\,\boldsymbol{\phi} = 0 \tag{92}$$

where **G** is a square matrix which contains the various coefficients calculated from Equations 87 and 88 as well as the values of the potential at the various grid points. In general, the matrix is symmetric with nonzero elements around the diagonal. Thus, in case only the first term of Equation 86 is taken, the matrix is a triple diagonal matrix with the elements:

$$G_{ij} = \begin{cases} -(2h^{-2} + U_i) & i = j \\ h^{-2} & i = j \pm 1 \\ 0 & \text{all other cases} \end{cases} \tag{93}$$

and in case the second term is also included, the elements of **G** are:

$$G_{ij} = \begin{cases} -((5/2)h^{-2} + U_i) & i = j \\ (4/3)h^{-2} & i = j \pm 1 \\ (1/12)h^{-2} & i = j \pm 2 \\ 0 & \text{all other cases} \end{cases} \tag{94}$$

For many practical cases, Equation 93 with 300 to 400 division points yields quite good results (3 to 4 significant digits), whereas Equation 94 is certainly enough for all cases of interest (6 to 7 significant digits). It should be noted that for these kind of matrixes, very efficient computer routines are available to calculate the eigenvalues and the eigenvectors.

The case in which the second derivative $\phi''(y)$ is approximated by the three grid-point formula (Equation 90) is very attractive due to its simplicity. However, although the results are not always accurate enough, they can still be improved within the framework of the three grid-point formula by a method known as the Richardson h^2 extrapolation. The method was suggested and carefully tested by Truhlar.[49] The idea is the following. We consider a magnitude A(h) which was calculated with the three grid-point formula employing a step size h. For this purpose A(h) can be either an eigenvalue or an eigenfunction or even more complicated magnitudes like a potential matrix element of the kind:

$$v_{n\ell} = h \sum_{j}^{N} \phi_n(j)\phi_\ell(j)\, U_j \tag{95}$$

The same magnitude can be calculated again with the same method but applying other step sizes. Once J calculations are performed with step sizes h_i; $i = 0 \ldots J - 1$, the Richardson h^2 extrapolation method assumes that any of the $A(h_i)$ can be represented as a power series with fixed coefficients:

$$A(h_i) = \sum_{\ell=0}^{J-1} a_\ell h_i^{2\ell} \qquad i = 0, \ldots J - 1 \tag{96}$$

The above expressions are in fact a system of J equations with J unknown a_ℓ; $\ell = 0 \ldots J - 1$. The most interesting unknown is a_0 which is the value of A calculated with an *infinitely small step size.*

c. The Translational Wavefunction

Once the vibrational eigenvalues and eigenfunctions are calculated and the corresponding matrix elements in Equation 66 computed, the equation to be solved is usually of the form (see Equation 65):

$$\frac{d^2\boldsymbol{\zeta}}{dx^2} - \mathbf{M}(x)\boldsymbol{\zeta} = 0 \tag{97}$$

The matrix $\mathbf{M}(x)$ is not always symmetric (for instance, in the case where x and y are polar coordinates) but is guaranteed to have only real eigenvalues. To integrate this equation in a given direction, the x range is divided into a certain number of (not necessarily equal) intervals. For the sake of simplicity we concentrate on one of these intervals, say the jth interval, with the end points (x_{j-1},x_j). We wish to calculate $\boldsymbol{\zeta}(x)$ at $x = x_j$, assuming its value and that of its first derivative are known at $x = x_{j-1}$. The most straightforward approach is to transform the set of N second-order differential equations into a set of 2N first-order differential equations and then to apply one of the various standard methods like the Runge Kutta method or more efficient methods which were recently developed and are characterized by the fact that the propagation is performed through large step sizes. To carry out the transformation, the vector $\boldsymbol{\xi}(x)$, defined as

$$\boldsymbol{\xi}(x) = \boldsymbol{\zeta}'(x) \tag{98}$$

is introduced. Combining Equations 98 and 97 yields 2N first-order differential equations:

$$\begin{aligned} \frac{d\boldsymbol{\zeta}}{dx} &= \boldsymbol{\xi}(x) \\ \frac{d\boldsymbol{\xi}}{dx} &= \mathbf{M}(x)\boldsymbol{\zeta} \end{aligned} \tag{99}$$

If we now define the two new matrixes $\mathbf{g}$ and $\mathbf{G}$ as:

$$\mathbf{g} = \begin{pmatrix} \boldsymbol{\zeta} \\ \boldsymbol{\xi} \end{pmatrix} \quad : \quad \mathbf{G} = \begin{pmatrix} \mathbf{0} & \mathbf{I} \\ \mathbf{M} & \mathbf{0} \end{pmatrix} \tag{100}$$

Equation 97 becomes:

$$\frac{d\mathbf{g}}{dx} = \mathbf{G}(x)\mathbf{g} \tag{101}$$

The solution of this equation is given in the form:

$$\mathbf{g}(x) = \mathbf{U}(x_{j-1},x)\ \mathbf{g}(x_{j-1}) \tag{102}$$

where $\mathbf{U}(x_{j-1},x)$ has different forms according to the method employed. The simplest version is due to Nielsen and Gordon[50] and is based on approximating G(x) by

$$\mathbf{G}(x) \sim \mathbf{G}(\bar{x}_j) \tag{103}$$

where

$$\bar{x}_j = (x_{j-1} + x_j)/2$$

For this case, $\mathbf{U}(x_{j-1},x)$ takes the form:

$$\mathbf{U}(x_{j-1},x) = \mathbf{A}_j^{-1}\,\mathbf{E}_j(x)\mathbf{A}_j \tag{104}$$

where $\mathbf{E}_j(x)$ is a diagonal matrix with the elements:

$$E_{j\ell n}(x) = \delta_{\ell n}\exp[\lambda_\ell^{(j)}(x_{j-1} - x)]\ ; \tag{105}$$

$\lambda_\ell^{(j)}$ are the eigenvalues of $\mathbf{G}(\bar{x}_j)$, and $\mathbf{A}_j$ is the matrix of the corresponding eigenvectors. The method is simple, but since it is based on the first Taylor term of the potential, the step size $\Delta x_j = x_j - x_{j-1}$ is relatively small. A more efficient procedure was suggested by Miller and Light,[51] who employ the Magnus approximation. Here two terms of the Taylor expansion were taken, i.e.,

$$\mathbf{G}(x) = \mathbf{G}(\bar{x}_j) + (x - x_j)G'(\bar{x}_j) \tag{106}$$

According to this method, $\mathbf{U}(x_{j-1},x)$ takes the form:

$$\mathbf{U}(x_{j-1},x) = \mathbf{I} + \mathbf{\Omega} + {}^1\!/_2\mathbf{\Omega}^2 \tag{107}$$

where

$$\mathbf{\Omega}(x_{j-1},x) = \begin{pmatrix} \mathbf{0} & \mathbf{I} \\ \mathbf{M}(x_j) & \mathbf{0} \end{pmatrix}(x - x_{j-1}) + \frac{1}{12}\begin{pmatrix} \mathbf{M}'(\bar{x}_j) & \mathbf{0} \\ \mathbf{0} & \mathbf{M}'(\bar{x}_j) \end{pmatrix}(x - x_{j-1})^3 \tag{108}$$

A third method which takes advantage of the fact that Equation 101 is of first order is due to Billing and Baer.[52] Here the matrix $\mathbf{G}(x)$ is approximated by

$$\mathbf{G}(x) = \mathbf{A}_j\left(\mathbf{\Lambda}(\bar{x}_j) + \mathbf{\Omega}(\bar{x}_j)(x - x_{j-1}) + \frac{1}{2}\mathbf{\Gamma}(\bar{x}_j)(x - x_{j-1})^2\right)\mathbf{A}_j^{-1} \tag{109}$$

where $\mathbf{\Lambda}(\bar{x}_j)$ is a diagonal matrix which contains the eigenvalues of $G(\bar{x}_j)$, $\mathbf{A}_j$ is the corresponding transformation matrix and $\mathbf{\Omega}(\bar{x}_j)$ and $\mathbf{\Gamma}(\bar{x}_i)$ are accordingly defined as:

$$\mathbf{\Omega}(\bar{x}_j) = \mathbf{A}_j^{-1}\frac{d\mathbf{G}}{dx}\mathbf{A}_j \tag{110}$$

$$\mathbf{\Gamma}(\bar{x}_j) = \mathbf{A}_j^{-1}\frac{d^2\mathbf{G}}{dx}\mathbf{A}_j \tag{111}$$

Due to this transformation, Equation 101 becomes:

$$\frac{d\mathbf{f}(x)}{dx} = \left(\mathbf{\Lambda} + \mathbf{\Omega}(x - \bar{x}_j) + \frac{1}{2}\mathbf{\Gamma}(x - \bar{x}_j)^2\right)\mathbf{f} \tag{112}$$

where $\mathbf{f}(x)$ is related to $\mathbf{g}(x)$ in the form:

$$\mathbf{g}(x) = \mathbf{A}\,\mathbf{f}(x) \tag{113}$$

Here and in what follows the index j is deleted.

The next step is to write $\mathbf{f}(x)$ as a product of two matrixes:

$$\mathbf{f}(x) = \mathbf{F}(x)\mathbf{f}_0(x) \tag{114}$$

where $\mathbf{f}_0(x)$ is the fast varying part of the solution and $\mathbf{F}(x)$ is the slowly varying part. It was shown that $\mathbf{F}(x)$ can be *approximated* by:

$$\mathbf{F}(x) = \mathbf{I} + \frac{1}{2}\,\mathbf{\Omega}_0(x - \bar{x}_j) + \frac{1}{6}\,\mathbf{\Gamma}_0(x - \bar{x}_j)^2 \tag{115}$$

where $\mathbf{\Omega}_0$ and $\mathbf{\Gamma}_0$ are defined as:

$$\mathbf{\Omega}_0 = \mathbf{\Omega} - \mathbf{\Omega}_D \quad ; \quad \mathbf{\Gamma}_0 = \mathbf{\Gamma} - \mathbf{\Gamma}_D \tag{116}$$

Here $\mathbf{\Omega}_D$ and $\mathbf{\Gamma}_D$ are the diagonal parts of $\mathbf{\Omega}$ and $\mathbf{\Gamma}$, respectively. The fast varying function $\mathbf{f}_0(x)$ was shown to satisfy the equation:

$$\frac{d\mathbf{f}_0}{dx} = \left(\mathbf{\Lambda} + \mathbf{\Omega}_D(x - \bar{x}_j) + \frac{1}{2}\,\mathbf{\Gamma}_D\,(x - \bar{x}_j)^2\right)\mathbf{f}_0(x) \tag{117}$$

where $\mathbf{\Lambda}$ is a diagonal matrix of the eigenvalues of $\mathbf{G}(\bar{x}_j)$. The solution of this equation will be written in the form:

$$\mathbf{f}_0(x) = \mathbf{E}(x)\mathbf{f}_0(x_{n-1}) \tag{118}$$

where $\mathbf{f}_0(x_{n-1})$ is calculated employing Equations 113 to 115 and using the values of $\mathbf{g}(x)$ at $x = x_{j-1}$ as obtained in the previous step. The matrix $\mathbf{E}(x)$ is a diagonal matrix with the elements:

$$E_{\ell k}(x) = \delta_{\ell k}\exp\left[\lambda_\ell\left(\bar{x} + \frac{1}{2}\,\Delta x_j\right) + \frac{1}{2}\,\omega_\ell\left(\bar{x}^2 - \frac{1}{4}\,\Delta x_j^2\right) + \frac{1}{3}\,\gamma_\ell\left(\bar{x}^3 + \frac{1}{8}\,\Delta x_j^3\right)\right] \tag{119}$$

where $\bar{x} = x - \bar{x}_j$; $\Delta x_j = x_j - x_{j-1}$ and λ, ω_ℓ and γ_ℓ are the diagonal elements of $\mathbf{\Lambda}$, $\mathbf{\Omega}$, and $\mathbf{\Gamma}$, respectively. Combining all the transformations and solutions, we obtain for $U(x_{j-1},x_j)$ (see Equation 102) the expression:

$$\mathbf{U}(x_{j-1},x_j) = \mathbf{A}_j\,\mathbf{F}(x_j)\mathbf{E}(x_j)\mathbf{F}^{-1}(x_j)\mathbf{A}_j^{-1} \tag{120}$$

The most widely used method is the Gordon method[53] which deals directly with Equation 97. We consider the j-th sector with the end points (x_{j-1},x_j) and the center $\bar{x}_j$. Expanding the potential in a Taylor series around $x = \bar{x}_j$ and employing the transformation:

$$\boldsymbol{\zeta} = \mathbf{A}\,\boldsymbol{\chi} \tag{121}$$

yields:

$$\frac{d^2\boldsymbol{\chi}}{dx^2} = \left[\mathbf{\Lambda} + \mathbf{\Omega}(x - \bar{x}_j) + \frac{1}{2}\,\mathbf{\Gamma}(x - \bar{x}_j)^2\right]\boldsymbol{\chi} \tag{122}$$

where $\mathbf{\Lambda}$, $\mathbf{\Omega}$, and $\mathbf{\Gamma}$ are defined in a very similar way as before (see Equations 110, 111, and 117 but with respect to $\mathbf{M}(\tilde{x}_j)$ instead of $\mathbf{G}(\tilde{x}_j)$). Next, the two independent solutions of the (uncoupled) equation are considered:

$$\frac{d^2\chi^{\pm}_{0\ell}}{dx^2} = (\lambda_\ell + \omega_\ell(x - \bar{x}_j))\chi^{\pm}_{0\ell} \tag{123}$$

where λ_ℓ and ω_ℓ are the diagonal elements of $\mathbf{\Lambda}$ and $\mathbf{\Omega}$. The $\chi^{\pm}_{0\ell}$; $\ell = 1, \ldots N$ are the well-known Airy functions used to construct two diagonal matrixes:

$$Z^{\pm}_{\ell k} = \delta_{\ell k}\, \chi^{\pm}_{0\ell} \tag{124}$$

The solution of Equation 121 is now written as:

$$\chi = \mathbf{Z}^+\mathbf{P}^+ + \mathbf{Z}^-\mathbf{P}^- \tag{125}$$

where $\mathbf{P}^{\pm}$ are two vector matrixes which can be shown to change slowly over a relatively large interval. Employing this feature, the derivative matrix χ' takes the form:

$$\chi' = (\mathbf{Z}^+)'\mathbf{P}^+ + (\mathbf{Z}^-)'\mathbf{P}^- \tag{126}$$

Solving Equations 125 and 126 for $\mathbf{P}^+$ and $\mathbf{P}^-$ we obtain (recalling that $\mathbf{Z}^{\pm}$ and $(\mathbf{Z}^{\pm})'$ are diagonal matrixes),

$$\mathbf{P}^{\pm} = \mathbf{W}^{-1}(\mathbf{Z}^{\mp}\chi' - (\mathbf{Z}^{\mp})'\chi) \tag{127}$$

where $\mathbf{W}$ is the Wronskian matrix, i.e.,

$$\mathbf{W} = \mathbf{Z}^+(\mathbf{Z}^-)' - (\mathbf{Z}^+)'\mathbf{Z}^- \tag{128}$$

and can be proven to be a constant. Differentiating Equation 127 leads to:

$$\frac{d\mathbf{P}^{\pm}}{dx} = \mathbf{W}^{-1}\left(\mathbf{Z}^{\mp}\frac{d^2\chi}{dx^2} - \frac{d^2\mathbf{Z}^{\mp}}{dx^2}\chi\right) \tag{129}$$

Substituting Equations 122 and 123 for the second derivatives of χ and $\mathbf{Z}^{\pm}$ leads to the expression:

$$\frac{d\mathbf{P}^{\pm}}{dx} = \mathbf{W}^{-1}\mathbf{Z}\left[\mathbf{\Omega}_0(x - \bar{x}_j) + \frac{1}{2}\mathbf{\Gamma}(x - x_j)^2\right]\chi \tag{130}$$

where $\mathbf{\Omega}_0$ was defined in Equation 116 or:

$$\frac{d\mathbf{P}^{\pm}}{dx} = \mathbf{W}^{-1}\mathbf{Z}^{\pm}\left[\mathbf{\Omega}_0(x - \bar{x}_j) + \frac{1}{2}\mathbf{\Gamma}(x - \bar{x}_j)^2)\right]\left[\mathbf{Z}^+(x)\mathbf{P}^+(x) + \mathbf{Z}^-(x)\mathbf{P}^-(x)\right] \tag{131}$$

To solve these equations we take advantage of the fact that $\mathbf{P}^{\pm}$ are slowly varying functions and consequently replace them on the r.h.s. by their values at $x = x_{j-1}$ which are known from the calculations carried out at the previous $(j - 1)$ sector. Consequently, Equation 129 becomes an integral which can be integrated numerically:

$$\mathbf{P}^{\pm}(x) = \mathbf{P}^{\pm}(x_j) + \mathbf{W}^{-1}\int_{x_{j-1}}^{x} dx\left\{\mathbf{Z}^{\pm}(x)\left[\mathbf{\Omega}_0(x - \bar{x}_j) + \frac{1}{2}\mathbf{\Gamma}(x - \bar{x}_j)^2\right]\right. \quad (132)$$

$$\left.\left[\mathbf{Z}^{+}(x)\mathbf{P}^{+}(x_{j-1}) + \mathbf{Z}^{-}(x)\mathbf{P}^{-}(x_{j-1})\right]\right\}$$

Very efficient and highly accurate algorithms to carry out these integrals were given by Gordon.[53]

Once this is done all the necessary information is available to perform the propagation:

1. Diagonalization of $\mathbf{M}(x)$ at $x = x_{j-1}$ and calculations of A_j, $\mathbf{\Lambda}_j$, $\mathbf{\Omega}_j$, and $\mathbf{\Gamma}_j$
2. Derivation of $\mathbf{Z}^{\pm}$ by solving Equation 123.
3. Calculation of $\mathbf{P}^{\pm}(x_j)$ by performing the integrals in Equation 132.
4. Calculations of $\boldsymbol{\chi}$ and $\boldsymbol{\chi}'$ applying Equations 125 and 126.
5. By using the back transformation given in Equation 121, the value of $\boldsymbol{\zeta}(x)$ at $x = x_j$ is obtained.

Finally, we consider stabilization of the solution matrix while propagation takes place. The need for the stabilization is a direct consequence of the inclusion of closed states. It turns out that during the propagation the columns of the $\boldsymbol{\zeta}$ (solution) matrix, which correspond to the closed states, start to grow exponentially, taking on large numerical values. Since part of the matrix now contains large numbers and the other part relatively small numbers, the accuracy of the small ones is destroyed. To prevent this, we want to form new columns from the given columns of the solution matrix (each column corresponds to one solution) by taking linear combinations of the old ones. The new matrix formed in this way is also a solution matrix if the same transformation is applied for the derivative matrix $\boldsymbol{\zeta}'(x)$ as well. Also, since the *initial* (starting) values of the corresponding solutions, i.e., $\boldsymbol{\zeta}(x_0)$ and $\boldsymbol{\zeta}'(x_0)$ are also required, the same transformation has to be applied to them. Thus, if $\mathbf{Q}$ is such a transformation (regular) matrix, the following transformations will keep all the information, gathered so far during the propagation process, unchanged:

$$\begin{aligned} \bar{\boldsymbol{\zeta}}(\bar{x}) &= \boldsymbol{\zeta}(\bar{x})\mathbf{Q} \quad ; \quad \bar{\boldsymbol{\zeta}}'(x) = \bar{\boldsymbol{\zeta}}'(\bar{x})\mathbf{Q} \\ \bar{\boldsymbol{\zeta}}(x_0) &= \bar{\boldsymbol{\zeta}}(x_0)Q \quad ; \quad \boldsymbol{\zeta}'(x_0) = \bar{\boldsymbol{\zeta}}'(x_0)\mathbf{Q} \end{aligned} \quad (133)$$

It is found that the most efficient choice for Q is $(\boldsymbol{\zeta}(\bar{x}))^{-1}$ and consequently we have:

$$\begin{aligned} \bar{\boldsymbol{\zeta}}(\bar{x}) &= I \quad ; \quad \bar{\boldsymbol{\zeta}}'(\bar{x}) = \boldsymbol{\zeta}'(\bar{x})\,\boldsymbol{\zeta}^{-1}(\bar{x}) \\ \bar{\boldsymbol{\zeta}}(x_0) &= \boldsymbol{\zeta}(x_0)\,\boldsymbol{\zeta}^{-1}(\bar{x}) \quad ; \quad \bar{\boldsymbol{\zeta}}'(x_0) = \boldsymbol{\zeta}'(x_0)\boldsymbol{\zeta}^{-1}(\bar{x}) \end{aligned} \quad (134)$$

This procedure was found to be most reliable.

The list of methods for solving the translational part of the SE is by far not complete. There are still other propagative methods which were recently introduced, like the Noumerov method or the methods which propagate the **S** or the **R** matrix[47,54] (these last ones are very similar to the Gordon method[53] except that the Airy functions are replaced by sine and cosine functions). Among the nonpropagative methods to be mentioned are the finite difference method introduced by Diestler and McKoy[55] and the finite elements method of Askar et al.[56]

C. Simplified Models

1. Single Curve Models

The models relating to a single curve which extends from $s = -\infty$ to $s = \infty$ are based on the equation:

$$\left(-\frac{\hbar^2}{2\mu}\frac{d^2}{ds^2} + W(s) - E\right)\zeta = 0 \tag{135}$$

Here Equation 135 follows from Equation 27 provided that the nonadiabatic coupling terms $\tau_{nm}^{(1)}$ and $\tau_{nm}^{(2)}$ and the curvature of the reaction coordinate are negligibly small. From Equation 28 it can be seen that:

$$U_{00}(s) = W(s) = \epsilon_0(s) + V_1(s) \tag{136}$$

Equation 135 was used in different studies on transmission probabilities and tunneling. In one of the earliest, Langer[8] considered a potential curve as given in Figure 4A. He assumed that every chemical reaction proceeds only due to what was later called "tunneling", and that a necessary and sufficient condition for a chemical reaction to occur is that the initial energy level of the "somewhat distorted" original AB molecule coincides with the "somewhat distorted" final energy level of the product molecule BC. Although the Langer theory is incorrect in general, two basic concepts were nevertheless introduced: (1) the progress of a chemical reaction is due to tunneling and (2) the behavior of a system moving from reagents to products is nearly (vibrational) adiabatic. Langer's two main errors resided in that, as was established 2 years later, (1) the barrier in the interaction region is much lower than the dissociation energy[2] and therefore many reactions could occur classically without having to tunnel and (2) the motion from reagents to products is described by two coordinates of which one is not bound. Since the two modes of motion interacted with each other, the system could always arrange itself so as to be in any given open eigenstate. It is interesting to note that, in 1954, Bauer and Wu[57] considered a very similar potential form but assumed a different mechanism. Contrary to Langer, they ignored tunneling (although the existence of a barrier was essential for their model) and assumed the reaction to be accompanied by a vibrational nonadiabatic process. The importance of the Bauer-Wu treatment lies not in the mechanism, which by an exact treatment was later found to be inappropriate, but in the fact that these authors were the first to apply the Distorted Wave Born Approximation (DWBA) to reactive scattering.

The first treatment of Equation 135 was given by Eckart,[58] who did not refer to a chemical reaction. (He studied the penetration of a potential barrier by electrons.) However, because of the form of the potential W(s) that he used, Bell[59] and Wigner[60] recognized in 1933 that Eckart's treatment could apply to chemical reactions as well.

Eckart assumed a "reasonable" potential for which he was able to obtain an analytic expression for the reflection (and the transmission) coefficient. The function he considered is of the form:

$$W(s) = -Ax/(1 - x) - Bx/(1 - x)^2 \tag{137}$$

where

$$x = x(s) = -\exp(2\pi s/\ell) \tag{138}$$

and A, B, and ℓ are constants. Here, A (>0) is the asymptotic value of the potential at $s = \infty$ (the potential becomes zero for $s = -\infty$), ℓ is half the width of the transition region,

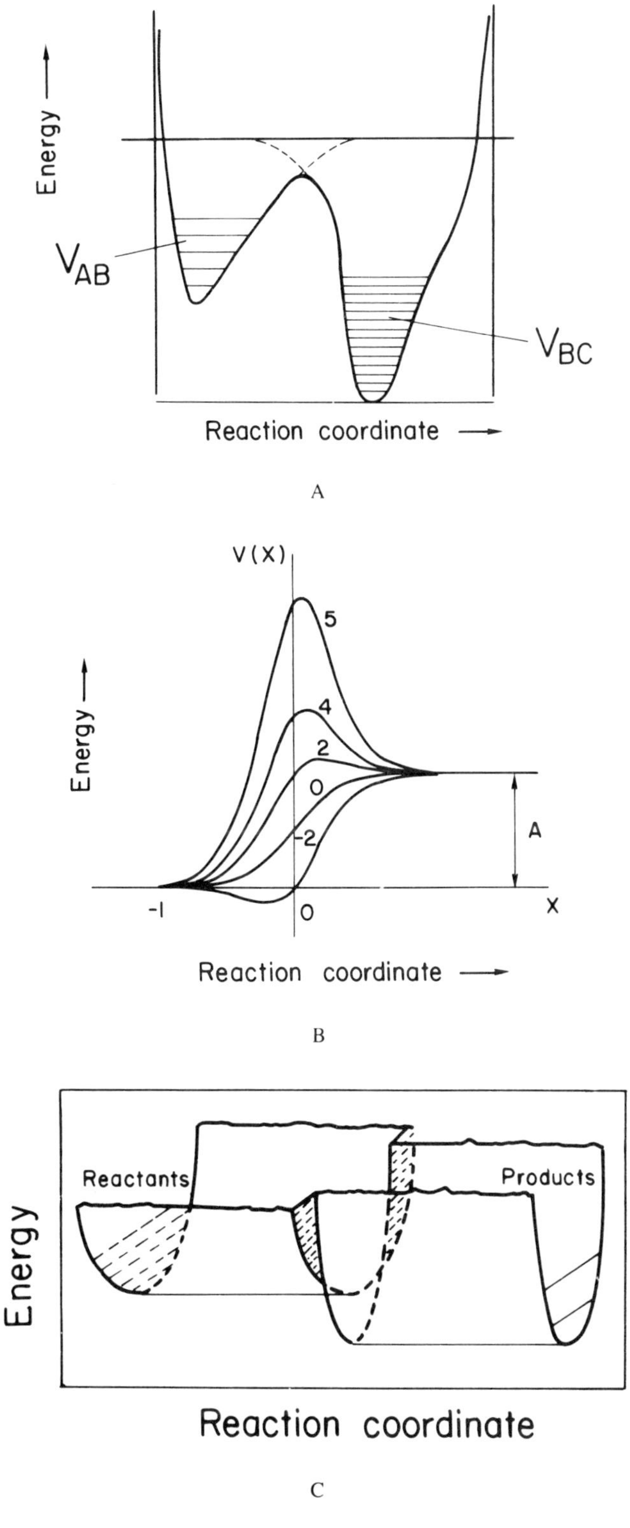

FIGURE 4. Simplified reaction coordinate models. (A) The Langer model [8] showing the Morse potentials of the reagent and the product molecules; (B) the Eckart model.[58] The numbers on the curves are the values of B/A; (C) the Hirschfelder-Wigner model.[13]

and B is responsible for the actual form of the potential in that region (see Figure 4B). For example, when B > 0, W(x) has a maximum given by:

$$W_m = \frac{(A + B)^2}{4B} \tag{139}$$

Substituting Equation 137 for W(s) in Equation 135 and replacing s by x as given in Equation 138, one obtains a hypergeometric-type equation:

$$x^2 \frac{d^2\zeta}{dx^2} + x \frac{d\zeta}{dx} + \frac{2\mu\ell^2}{\hbar^2}(Ax/(1 - x) + Bx/(1 - x)^2 + E)\zeta = 0 \tag{140}$$

Considering the asymptotic form of $\zeta(x)$ for $x \to -\infty$ (i.e., $s \to \infty$), one can show it to be of the form:

$$\zeta(s) = a_1 e^{2\pi i x/\lambda} + a_2 e^{-2\pi i x/\lambda} \tag{141}$$

where

$$\lambda = \hbar/(2\mu E)^{1/2} \tag{142}$$

Consequently, the reflection coefficient ρ, defined as:

$$\rho = |a_2/a_1|^2 \tag{143}$$

becomes:

$$\rho = \left[\frac{\Gamma(^1/_2 + i(\delta - \beta - \alpha))\Gamma(^1/_2 + i(-\delta - \beta - \alpha))}{\Gamma(^1/_2 + i(\delta - \beta + \alpha))\Gamma(^1/_2 + i(-\delta - \beta + \alpha))}\right]^2 \tag{144}$$

Here

$$\alpha = \frac{\ell}{\lambda} = {}^1/_2\left(\frac{E}{C}\right)^{1/2} \;;\; \beta = \frac{\ell}{\lambda} = {}^1/_2\left(\frac{E - A}{C}\right)^{1/2} \;;\; \delta = {}^1/_2\left(\frac{B - C}{C}\right)^{1/2} \tag{145}$$

where:

$$\lambda' = \hbar/(2\mu(E - A))^{1/2} \;:\; C = \hbar^2/(8\ell^2\mu) \tag{146}$$

In analyzing this expression one should distinguish between two cases, i.e., when δ is real and when it is imaginary. When δ is real, it can be shown that ρ becomes:

$$\rho = \frac{\cosh[2\pi(\alpha - \beta)] + \cos(2\pi\delta)}{\cosh[2\pi(\alpha + \beta)] + \cosh(2\pi\delta)} \tag{147}$$

while for an imaginary δ:

$$\rho = \frac{\cosh[2\pi(\alpha - \beta)] + \cos(2\pi|\delta|)}{\cosh[2\pi(\alpha + \beta)] + \cos(2\pi|\delta|)} \tag{148}$$

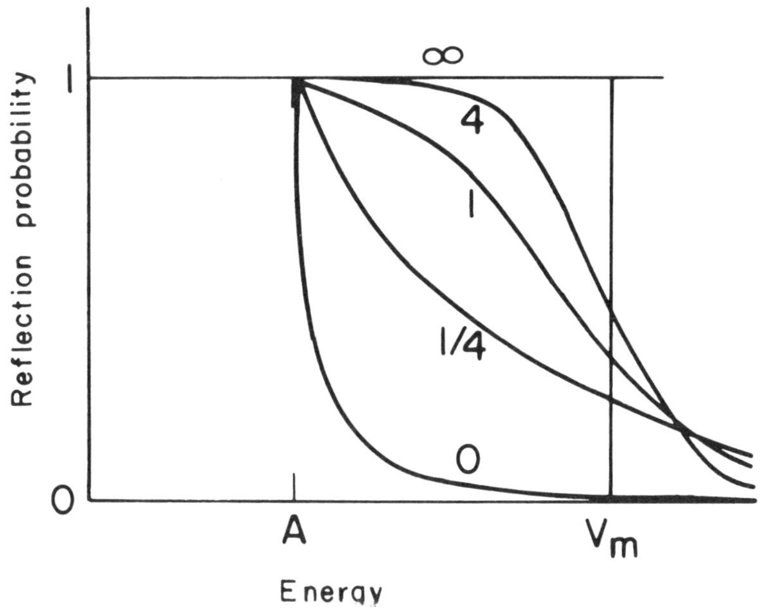

FIGURE 5. The reflection coefficients (nonreactive transition probabilities) as a function of energy for the Eckart potential. The numbers are the values of A/C. V_m stands for the maximum of the potential.

The values of ρ as a function E for A = 8B and various ℓ values are shown in Figure 5. Eckart also showed that the expression for the reflection coefficient based on Jeffrey's[61] approximate solution for the SE:

$$\rho = 1 - \exp\left\{-\frac{4\pi}{\hbar}\int_{x_1}^{x_2} (2\mu(W - E))^{1/2}dx\right\} \tag{149}$$

fits reasonably well with his exact results as long as $E < W_m$ (here x_i; i = 1,2, are the points for which W(x) = E).

Eckart's approach and results were adopted by Wigner[60] and Bell[59,62-63] in order to determine the importance of quantum effects on (heavy particle) exchange reactions. However, whereas Wigner was only looking for first-order quantum corrections to the classical results, Bell extended Eckart's approach and applied it to the study of the temperature dependence of rate constants. Applying Eckart's potential and integrating analytically, Bell found large deviations between the classical and the quantum mechanical rate constants for hydrogenic systems. For a parabolic barrier:

$$W(x) = W_m\left(1 - \frac{x^2}{a^2}\right) \tag{150}$$

and using Jeffrey's formula,[51] Bell was able to show that the transmission coefficient (τ) is given by:

$$\tau = \exp\left[-\frac{2\pi^2 a}{h}\left(\frac{2\mu}{W_m}\right)^{1/2}(W_m - E)\right] \tag{151}$$

where 2a is the width of the parabola. Applying this expression, Bell derived an analytic form for the rate q of particles passing the barrier (i.e., reacting) by extending the validity of the above expression for the whole energy range of interest:

$$\tau(E) = \begin{cases} \exp\left[-\frac{2\pi^2 a}{\hbar}\left(\frac{2\mu}{W_m}\right)(W_m - E)\right] & ; \text{ for } W_m \leqslant E \\ 1 & ; \text{ for } W_m \geqslant E \end{cases} \tag{152}$$

The total reaction yield (q) is defined as:

$$q = \frac{1}{kT}\int_0^\infty \tau(E)e^{-E/kT}\,dE \tag{153}$$

and consequently, the corresponding quantum mechanical reacting yield (q_Q) is

$$q_Q = (\beta - \alpha)^{-1}(\beta e^{-\alpha} - \alpha e^{-\beta}) \tag{154}$$

where

$$\alpha = \frac{W_m}{kT} \quad ; \quad \beta = \frac{2\pi^2 a(2\mu W_m)^{1/2}}{h} \tag{155}$$

The corresponding classical expression is given as:

$$q_c = e^{-\alpha} \tag{156}$$

Figure 6 shows a comparison between the classical and the quantum mechanical results for different potential parameters. The deviations are largest for low temperature T, and become smaller as T increases. It can also be seen that the higher or the narrower the potential, the larger are the deviations.

Having both q_c and q_Q, we may define the tunneling correction factor (Q) as:

$$Q = q_Q/q_c \tag{157}$$

From Equations 155 to 157 we obtain:

$$Q = \frac{\beta}{\beta - \alpha}\left(1 - \frac{\alpha}{\beta}e^{-(\beta-\alpha)}\right) \tag{158}$$

As mentioned, Wigner was able to derive a formula for Q in the high energy (temperature) limit:

$$Q = 1 + \frac{1}{2\pi}\left(\frac{h\nu}{kT}\right)^2 \tag{159}$$

where ν stands for:

$$\nu = \left(-\frac{1}{\mu}\frac{d^2W(s)}{ds^2}\right)^{1/2} \tag{160}$$

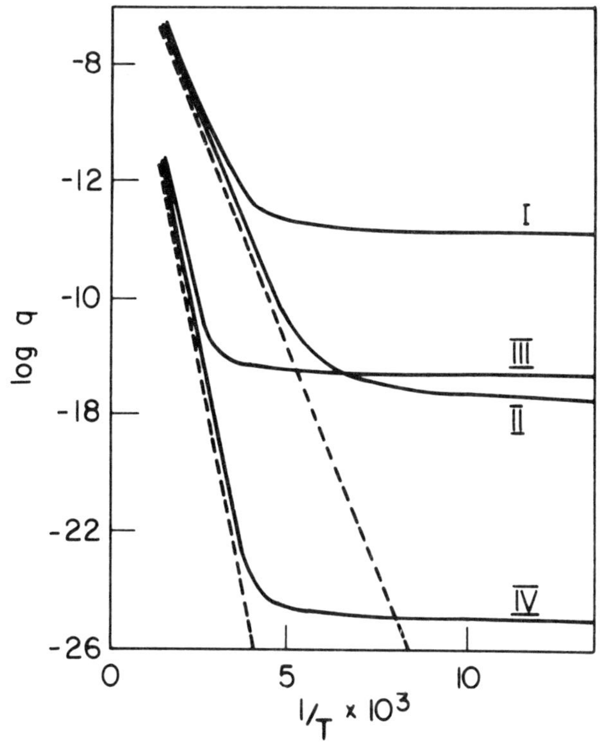

FIGURE 6. The reaction flux (rate) according to Bell[59] as a function of T^{-1} for different barrier heights and barrier widths (———, quantum mechanical treatment;-----, classical treatment).

which in this case becomes:

$$\nu = \left(\frac{2W_m}{\mu a^2}\right)^{1/2} \tag{161}$$

On the other hand, in the high-temperature region, Bell's result takes the form:

$$Q = 1 + \frac{1}{24}\frac{h\nu}{kT}e^{-\beta} \tag{162}$$

and it is apparent that the expressions of Wigner and Bell differ. The reason for this is that Bell's approximate transition probability function ($\tau(E)$) is not accurate enough in the corresponding energy region.

About 20 years later, Shavitt[64] suggested a reconsideration of the Eckart potential. He assumed the potential to be symmetric and of the form:

$$W(s) = W_m \operatorname{sech}(\pi s/\ell) \tag{163}$$

for which the transition coefficient is

$$\tau = (\cosh(4\pi\alpha) - 1)/(\cosh(4\pi\alpha) + \cos(2\pi\delta)) \tag{164}$$

where

$$\alpha = \frac{\ell}{\hbar}(2\mu E)^{1/2} \quad : \quad \delta = \frac{\ell}{\hbar}\left[8\mu\left(W_m - \frac{h^2}{4\ell^2}\right)\right]^{1/2} \tag{165}$$

Applying the approximate form for τ, Shavitt[64] was able to show that his result coincides with Wigner's formula for a potential with a nonzero fourth derivative:

$$Q = 1 + \frac{1}{2\pi}\left(\frac{h\nu}{kT}\right)^2\left(1 - \frac{kT}{4}\frac{W_4}{W_2}\right) \tag{166}$$

where

$$W_\ell = \frac{d^\ell W}{ds^\ell} \quad : \quad \ell = 2,4 \tag{167}$$

Shavitt also suggested obtaining reactive rate constants by fitting the potential of the H + H_2 system along the MEP to an Eckart potential. The first attempt was made with respect to an H_3 potential obtained by Boys and Shavitt.[65] Treatments of a similar kind were proposed by Weston,[66] who fitted a parabolic potential to a LEPS potential, and again by Shavitt,[67] who fitted Eckart's curve to the Shavitt-Stevens-Min-Karplus (SSMK) potential.[68] More work along these lines was done by Truhlar and Kuppermann,[69] who numerically solved the one-dimensional SE for the potential along the MEP using the SSMK potential. Recent interest in this subject has been aroused by the interesting work of Pollak and co-workers[70-73] employing semiclassical methods to determine the existence of dynamic barriers and their effect on the reaction rate. A detailed discussion on this subject is given in Volume III, Chapter 2.

Several attempts were also made to extend the ordinary one-dimensional treatments to cases where the collinear system as a whole was allowed to rotate; see for instance, Johnson and Rapp,[74] Child,[75] and the most recent, Walker and Hayes.[76]

2. Hard-Sphere Models

Although the main concepts of the reactive process were available from the beginning of the 1930s, it was only in 1943 that Hulburt and Hirschfelder (HH)[77] carried out the first numerical treatment with respect to a reactive hypersurface. Earlier exact numerical studies consisted mainly of treating single-coordinate systems (see previous section) or two-coordinate-type models which, however, were linear in the sense that a Cartesian coordinate changes from $-\infty$ to $+\infty$ (with zero curvature)[12] (see Figure 4C). The model potential considered by HH was simple enough to permit obtaining numerical results and, moreover, close enough to a realistic potential hypersurface governing the motion of three particles constrained to move along a line (the collinear case). The importance of the HH study is not only historical; in many of the present day methods use is still made of some of the techniques already applied in this very first treatment. HH considered a case which, in present day terminology, would be called the "Light-Heavy-Light" (LHL) case, where the "heavy" mass is infinitely heavy. As can be seen from Figure 7A, the potential channel is divided into three regions: a reagent (I), a product (II), and the interaction region (III). The wavefunction in region I is presented in terms of incoming and outgoing waves, i.e.,

$$\psi_I(x,y) = \frac{1}{\sqrt{2\pi}}\left(e^{-ik^I_{n_o}x}\,\phi^I_{n_o}(y) + \sum_{n=0}^{\infty}\left(\frac{k^I_{n_o}}{k^I_n}\right)^{1/2} R_{nn_o}\,e^{ik^I_n x}\,\phi^I_n(y)\right) \tag{168}$$

The wavefunction in region II is presented only in terms of outgoing waves, i.e.,

$$\psi_{II}(x,y) = \frac{1}{\sqrt{2\pi}}\left(\sum_{m=0}^{\infty}\left(\frac{k^I_{n_o}}{k^{II}_m}\right)^{1/2} T_{mn_o}\,e^{ik^{II}_m y}\,\phi^{II}_m(x)\right) \tag{169}$$

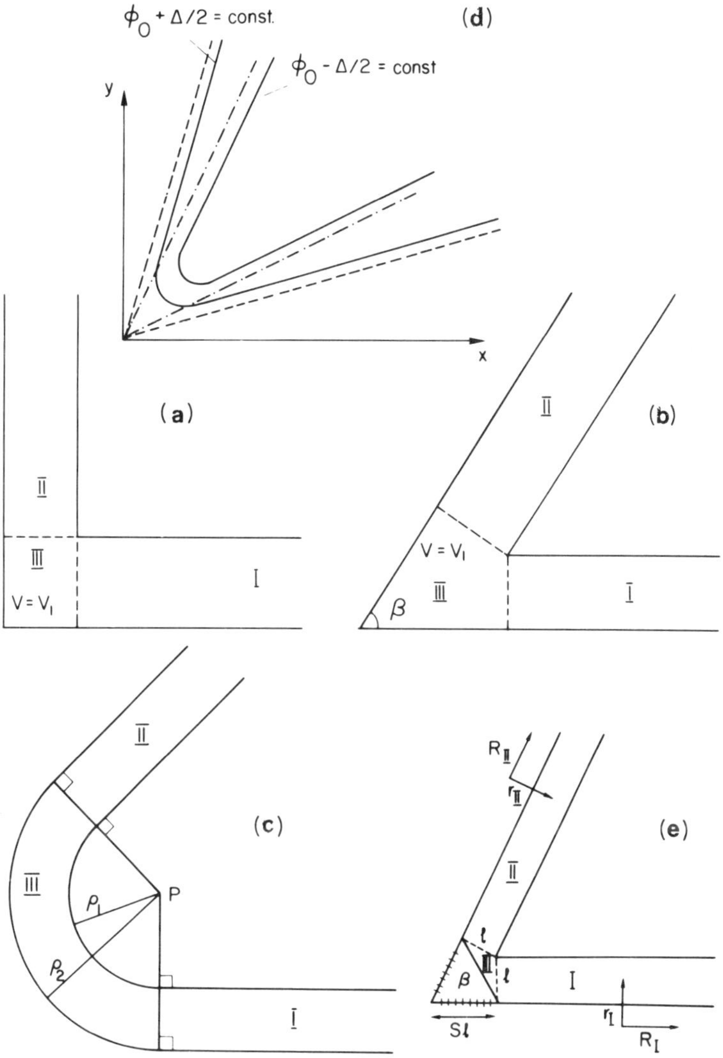

FIGURE 7. Simplified "hard-sphere" type models. (a) The rectangular Hulburt-Hirschfelder model (infinite-heavy-central mass);[77] (b) the skewed (finite-central mass) potential;[78] (c) the hyperbolic model, due to Hulburt and Hirschfelder;[77] (d) a modified skewed potential;[82] (e) the circular hard-sphere potential.

and the wavefunction in region III is presented in terms of standing waves,

$$\psi_{III}(x,y) = \frac{1}{\sqrt{2\pi}} \sum_{n'=0}^{\infty} [A_{n'n_o} \phi^{I}_{n'}(x)\sin(k^{I}_{n'}y) + B_{n'n_o} \phi^{II}_{n'}(y)\sin(k^{II}_{n'}x)] \tag{170}$$

Here

$$k^{\ell}_{n'} = \left(\frac{2\mu}{\hbar^2}(E - \epsilon^{\ell}_{n'})\right)^{1/2} \quad ; \quad \ell = I,II \tag{171}$$

where μ is the mass of the system, E is the total energy ($\epsilon^{\ell}_{n'}$), and $\phi^{\ell}_{n'}(x)$; ℓ = I,II, are, respectively, the n-th eigenstate and the n-th eigenfunction, both in the ℓ-th channel. HH chose the vibrational potential to be an infinitely deep, square-well-type potential, and consequently:

$$\phi_n^{\ell}(z) = \sin\left(\frac{n\pi}{L_\ell} z\right) \quad ; \quad z = x,y \quad : \quad \ell = I,II \tag{172}$$

R_{nn_o} is the n-th *nonreactive* **T**-matrix element starting at an initial state (n_o), and T_{mn_o} is the m-th *reactive* **T**-matrix element. The conservation of particles is guaranteed by the relation:

$$\sum_n |R_{nn_o}|^2 + \sum_m |T_{mn_o}|^2 = 1 \tag{173}$$

The coefficients T_{mn_o} and R_{nn_o}, as well as A_{nn_o} and B_{nn_o}, are determined by continuity requirements (''matching'') at the boundaries of each region. Thus, at $x = L_I$ we require:

$$\psi_I(x,y) = \psi_{III}(x,y) \quad ; \quad \left.\frac{\partial\psi_I}{\partial x}\right|_{x=L_I} = \left.\frac{\partial\psi_{III}}{\partial x}\right|_{x=L_I} \tag{174}$$

and similar conditions must be met at $y = L_{II}$.

Now if the summations in Equations 168 to 170 are truncated at $n = N$, the continuity conditions can be shown to yield 4N nonhomogeneous algebraic equations for each of the 4N unknowns $A_{n'n_o}$, $B_{n'n_o}$, R_{nn_o}, and T_{mn_o}.

Two modifications of this treatment were suggested about 25 years later (see Figure 7B). Tang, Kleinman, and Karplus (TKK)[78] and (independently) Wilson[79] (showed how one can treat the same model, but for any mass combination and Baer and Kouri[80] removed the assumption concerning the infinite depth of the square-well potentials. TKK[78] followed the scheme offered by HH, but used a different system of coordinates in region III. Instead of applying Cartesian coordinates, they used polar coordinates defined with respect to the origin:

$$x = R\cos\phi \quad : \quad y = R\sin\phi \tag{175}$$

If β is the transformation angle between the two arrangement channel coordinates, (R_I,r_I) and (R_{II},r_{II}) given by Equation 12′, then $\psi_{III}(R,\phi)$ is defined as:

$$\psi_{III}(R,\phi) = \sum_n (A_nF_n + B_nG_n) \tag{176}$$

where

$$\begin{aligned} F_n(R,\phi) &= J_{nq}(\alpha R)\sin(nq\phi) \\ G_n(R,\ ) &= J_{(n-1/2)q}(\alpha R)\sin((n - 1/2)q\phi) \\ &+ \frac{J_{(n-1/2)q}(\alpha R_\ell)}{J_{(n+1/2)q}(\alpha R_\ell)} J_{(n+1/2)q}(\alpha R)\sin((n + 1/2)q\phi) \end{aligned} \tag{177}$$

Here $q = 2\pi/\beta$: $\alpha = 2\mu/\hbar^2 (E - V_I)$ (V_I is a step function-type potential defined in region III only; see Figure 7b), $R_\ell = \ell\sqrt{1 + S^2}$ (see Figure 7e), and J_k is the Bessel function of order k. The functions F_n and G_n are the two independent solutions of the SE in region III. In addition, they satisfy the following boundary conditions:

$$
\begin{aligned}
F_n(R,0) &= G_n(R,0) &= 0 \\
F_n(0,\phi) &= G_n(0,\phi) &= 0 \\
F_n(R,\beta) &= G_n(R,\beta) &= 0 \\
F_n\left(R_\ell, \frac{\beta}{2}\right) &= G_n\left(R_\ell, \frac{\beta}{2}\right) &= 0
\end{aligned}
\tag{178}
$$

TKK studied this model for two mass combinations, i.e., $m_A = m_C, m_\beta = \infty$ ($\beta = \pi/2$), and $m_A = m_B = m_C$ ($\pi = 2\pi/3$) and for three different values of V_I, namely, $V_I = 0$ (a flat interacting region), $V_I = 2.5[\pi/L]^2$ (an interaction region with a barrier), and $V_I = -1.5[\pi/L]^2$ (an interaction with a well). The first mass combination will be termed Light-Heavy-Light (LHL) and the second Light-Light-Light (LLL).

The main findings were:

1. The LLL and LHL mass combination cases yield oscillatory transition probability functions having different energy dependence. In the LHL case, the oscillatory behavior is caused by Feschbach-type resonances,[81] whereas in the LLL case, due to the broadening of the interaction region, additional interference effects enhance the oscillatory behavior.
2. In case of a barrier, we notice the existence of the quantum mechanical tunneling process which becomes larger as the angle β decreases.
3. With regard to the threshold behavior, the case of a barrier and the case with a potential well exhibit very similar features,[81] where the initial slope is small but increases gradually as the energy approaches the opening of the second vibrational state.

A treatment of the finite mass case, which differs from the TKK treatment and is much closer in spirit to the HH treatment, was given by Wilson.[79] He suggested using the ordinary (vibrational and translational) coordinates in regions I and II, and the interatomic distances in region III. Consequently, he was able to represent the wavefunction in region III in terms of trigonometric functions, thereby avoiding the relatively complicated Bessel functions.

The finite (deep) channel case (as opposed to the infinite channel case) was handled by Baer and Kouri[80] who applied their coupled τ-method (see Volume I, Chapter 4). Since this method circumvents the need for matching of wavefunctions, one can treat a finite deep potential case as easily as the infinite deep case. The results show that the effect of the channel depths is secondary as long as the energy is far from a resonance. The treatment should be carried out with more care when the energy is in the resonance region.

In 1943, HH recognized[77] that in order to increase the flexibility and therefore the reliability of the hard sphere model, a circular-type interaction region is more appropriate than a rectangular one. They suggested using a channel between two confocal hyperbolas where the potential energy is assumed to be zero inside and infinite outside (see Figure 7d). By introducing the two elliptical coordinates θ and ψ, which are related to the Cartesian coordinates x and y as:

$$
\begin{aligned}
x &= R \cosh\theta \sin\phi \\
y &= R \sinh\theta \cos\phi
\end{aligned}
\tag{179}
$$

SE takes the form:

$$
-\frac{\hbar^2}{2\mu}\left(\frac{\partial^2\psi}{\partial\theta^2} + \frac{\partial^2\psi}{\partial\ell^2}\right) + R^2(\cosh 2\theta + \cos 2\ell)\frac{E}{2}\psi = 0 \tag{180}
$$

In this model, ℓ serves as a vibrational-type coordinate and θ as the translational one. Having these coordinates, HH found a way to construct a reaction coordinate s:

$$s = R \int^{\theta} (\sinh^2\theta + \cos^2\ell)^{1/2} d\theta \tag{181}$$

If $\chi(s)$ is now defined as the translational part of the wavefunction, the corresponding SE becomes:

$$-\frac{\hbar^2}{2\mu}\frac{d^2\chi}{ds^2} + (V(s) - E)\chi = 0 \tag{182}$$

where V(s) is a given function of s.

Another model of "circular" shape, but where Cartesian coordinates were used instead of circular ones, was studied by the author.[82-83] Regions I and II were as before, but region III was now defined by the lines $R_I = S\ell$, $R_{II} = S\ell$, and $r_I = S(R_I - S\ell)$ (see Figure 7e). The Cartesian coordinates used in region III were ζ and η, defined as:

$$\zeta = \frac{R_I - Sr_I}{(1 + S^2)^{1/2}} \quad ; \quad \eta = \frac{SR_I - r_I}{(1 + S^2)^{1/2}} \tag{183}$$

and the corresponding wavefunction is

$$\psi_{III} = \left(\frac{2}{L}\right)^{1/2} \sum_{n'} \left(A_{n'} e^{ik_{n'}\zeta} + B_{n'} e^{-ik_{n'}\zeta}\right) \sin \frac{n'\pi}{L} (\eta - t) \tag{184}$$

where t is

$$t = \frac{1}{(1 + S^2)^{1/2}} L \tag{185}$$

The model where region III is bounded by two circular arcs, as shown in Figure 7c,[84] is very similar. In this case one employs polar coordinates (see Equation 53) and the Hamiltonian takes the same form as given in Equation 40. The corresponding wavefunction is

$$\psi_{III} = \frac{1}{\sqrt{\rho}} \sum_{n} \left(A_n e^{i\alpha_n\theta} + B_n e^{-i\alpha_n\theta}\right) \sin \frac{n\pi}{L} (\rho - \rho_1) \tag{186}$$

where

$$L = \rho_2 - \rho_1 \tag{187}$$

and α_n are determined by solving an algebraic eigenvalue problem. Here ρ_1 may or may not differ from zero.

In all previously described models, the vibrational potential was chosen to be a square-well potential. Almost no significant difficulties are expected when the vibrational potential is chosen to be harmonic. Gross and Korsch[85] did this when they considered a model in which only two regions are distinguished (instead of three); these are the reagent channel and the product channel, in each of which the potential V(r,R) is assumed to be separable, namely,

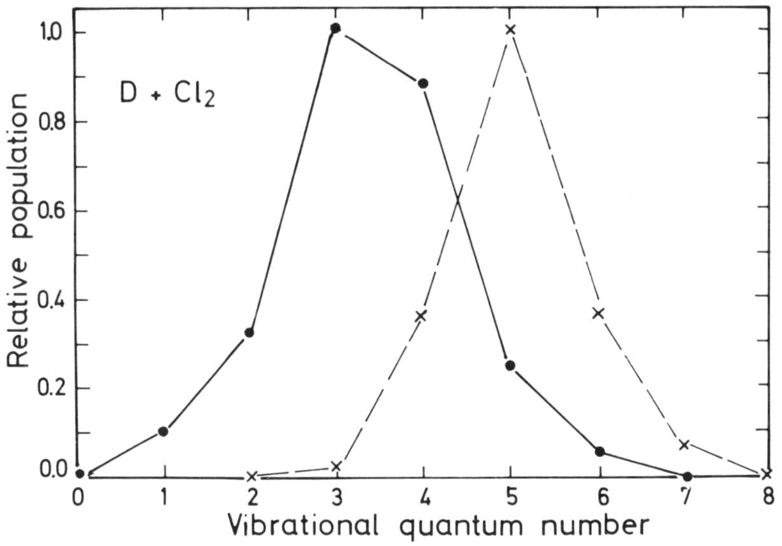

FIGURE 8. Vibrational distribution for the reaction $D + Cl_2$ ($v_i = 0$) $\rightarrow DCl(v_f) + Cl$ (———, experimental results;[91] -----, exact collinear quantum mechanical calculation.[90])

$$V(r,R) = V^R(R) + V^r(r);\ R = R_I, R_{II};\ r = r_I, r_{II} \tag{188}$$

The boundary between the two regions is defined by the line of intersection of the potentials:

$$V_I(r_I,R_I) = V_{II}(r_{II},R_{II}) \tag{189}$$

In general, one could assume any form for $V^R(R)$ and $V^r(r)$. Gross and Korsch, in the numerical part of their work, considered only the case where $V^R(R) = 0$, and $V^r(r)$ was assumed to be an harmonic oscillator.

D. The Inversion Process in Exothermic Reactions

1. Background

In 1928, Beutler and Polanyi[86] already speculated on the possibility that a simple chemical reaction yields highly vibrational excited products. The interest in such reactions was immensely enhanced in the 1960s and 1970s when it was recognized that they could lead to the formation of vibrational inverted populations appropriate for use with (chemical) lasers. It was soon recognized that the exothermic reactions:

$$\begin{aligned} &F + H_2 \rightarrow HF + H \\ &H + Cl_2 \rightarrow HCl + Cl \\ &Cl + HBr \rightarrow HCl + Br \\ &H + F_2 \rightarrow HF + F \end{aligned} \tag{190}$$

could not only yield the required inverted populations but in fact, under appropriate conditions, some of them could yield the most powerful and efficient lasers. The intensive experimental research that was carried out was followed by the theory. Using the classical trajectories technique, the Polanyi group[87-88] was the first to show that LEPS-type potential energy surfaces can yield results that are close to the experimental results. Regarding quantum mechanical treatments, it was only in 1973 that Wu et al.[89] obtained the inversion for $F + H_2$, followed by the work of Baer[90] for H(D) + Cl_2 (see results for $D + Cl_2$[91] in Figure 8) and $Cl + HBr$,[37] and Connor et al.[92] for $H + F_2$.

Much effort was directed at devising simplified models to explain the inversion process. In general, one can distinguish between dynamic models which relate the inversion to the curvature of the reaction path, and static models which are based on Franck-Condon (FC) overlap integrals between two shifted harmonic (or other) potentials. In the following sections, we discuss these two approaches to some extent.

2. Dynamic Models

The first dynamic model was suggested by Hofacker and Levine in 1971.[93] They considered an equation similar to the Marcus equation. For simplicity, we shall describe their finding using the Marcus equation as given in Equation 27. In general, Equation 27 is a coupled system of equations which becomes decoupled when τ^1, τ^2, and **M** are identically zero and $\boldsymbol{\eta}^2$ is equal to the unity matrix:

$$\boldsymbol{\eta}^{(2)} = \mathbf{I} \qquad (191)$$

They assumed that the source of the vibrational nonadiabatic transition is closely associated with η^2. If it is assumed that the curvature (k) is not too large, η^2 can be written:

$$\eta^2 \sim 1 + 2k\rho \qquad (192)$$

and consequently, the pertubation in Equation 27 is proportional to:

$$\Delta u_n \sim k(E - V_1(s) - \epsilon_n(s)) < \phi_n|\rho|\phi_{n\pm 1} > \qquad (193)$$

or

$$\Delta u_n = k\, E_{k_n}\bar{\rho}_n \qquad (194)$$

where E_{k_n} is the kinetic energy with respect to level n and $\bar{\rho}_n$ is closely related to the width of the oscillator. Since the coupling strength is measured with respect to the energy gap between the two considered states, Hofacker and Levine accordingly introduced a coupling term (g_n) defined as:

$$g_n = \Delta u_n/\hbar\omega_n = (E_{kn}/\hbar\omega_n)\, k\bar{\rho}_n \qquad (195)$$

On the basis of this treatment it was argued that the larger g_n, the more pronounced is the degree of inversion. The finding was confirmed to a certain extent by additional studies.[91,94-98]

A similar approach was taken by Basilevsky et al.[91,99-100] who also reached the same conclusion. However, since they considered the Marcus equation, they recognized the limitations of the theory, which was found to be applicable only for small curvatures (k), and had no possibility of being extended in a straightforward way to large curvatures. If we consider the definition of g (Equation 195) again, it can be seen that since k cannot become large, the only way to obtain a large value for g is to make the kinetic energy (E_{k_n}) large enough. This was confirmed to some extent by numerical studies.[90,100]

The difficulty involved with large curvatures is best seen from the Marcus equation discussed earlier. It is not just a matter of terms which are too large to be handled by low-order perturbation theories; essential singularities appear in the equation itself. The singularities are noticed when $\eta(s)$ becomes zero or when ρ, the vibrational coordinate, becomes equal to $k(s)^{-1}$ (see Equation 23). As long as k(s) is small, $k(s)^{-1}$ is large and consequently, $\eta(s)$ is different from zero in the region of physical significance. However, once k(s) becomes large enough, $\eta(s)$ becomes zero and the equation is singular in the region of physical importance. Consequently, both the approximate and the exact treatments of this equation become rather complicated.

One of the difficulties encountered in the study of dynamic models is concerned with the fact that the concept of the curvature is not uniquely defined. The curvature function is linked to the form of the reaction coordinate. Therefore, as long as a reaction coordinate cannot be defined uniquely, the curvature function is almost entirely arbitrary. Part of this difficulty can be overcome by defining an average curvature:

$$\bar{k}(s) = \int d\rho\, \psi^*(s,\rho) k(s,\rho) \psi(s,\rho) \tag{196}$$

where $\psi(s,\rho)$ is the total wavefunction. To obtain an analytic expression for $k(s,\rho)$, Baer and Beswick[20] approximated the reaction coordinate by a hyperbola in the vicinity of its maximal value: If **r** is the radius vector and ℓ is the corresponding polar angle, the equation of a hyperbola is given in the form:

$$\frac{1}{r} = u = -K(1 - e\cos\phi) \tag{197}$$

where

$$K = a/b^2 \tag{198}$$

Here a is the semitransverse axis, b is the semiconjugate axis, and e is the eccentricity. The curvature is defined as:[101]

$$k = \frac{u + (d^2u/d\phi^2)}{\left[1 + \left(\frac{1}{u}\frac{du}{d\phi}\right)^2\right]^{3/2}} \tag{199}$$

Substitution of Equation 197 in Equation 199 leads to:

$$k = K\frac{(1 - e\cos\phi)^3}{(1 - 2e\cos\phi + e^2)^{3/2}} \tag{200}$$

If β is the angle between the asymptotes given by Equation 12, it can be shown that:

$$e^{-1} = \cos(\beta/2) \tag{201}$$

and

$$K = \frac{1}{a}\,\mathrm{cotan}^2(\beta/2) \tag{202}$$

The meaning of a (and β) in a realistic case is shown in Figure 9. Equation 200 can be simplified for the case when $\phi \sim 0$, namely, in the vicinity where the curvature attains its maximal value. Following a few algebraic manipulations k becomes:

$$k(s) = K\left(1 - \frac{3}{2}e^2K^2s^2\right) \tag{203}$$

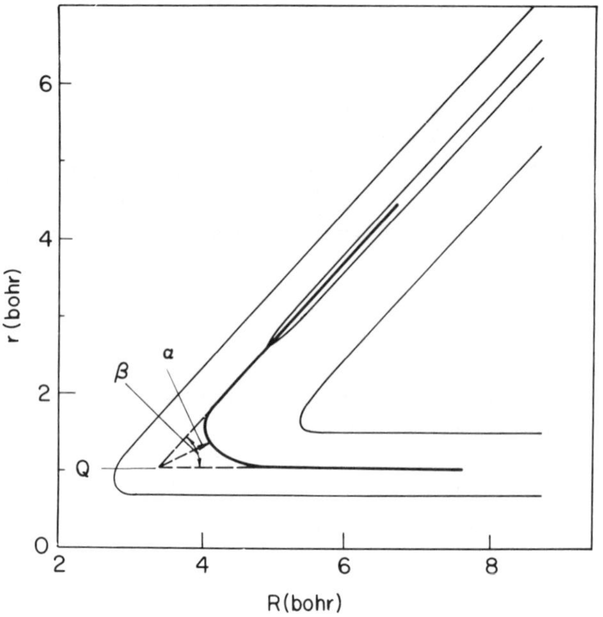

FIGURE 9. Parametric representation of the reaction coordinate in the vicinity of maximal curvature; α and β are parameters related to the hyperbola (see Equation 200).

where s is the reaction coordinate measured with respect to the line $\phi = 0$ and is related to ϕ as:

$$s = r(\phi = 0)\ell \tag{204}$$

where r is given by Equation 197. Equation 203 can be extrapolated by exponentiating the expression in the parenthesis:

$$k(s) = K \exp\left(-\frac{3}{2} e^2K^2s^2\right) \tag{205}$$

This expression has to be somewhat modified because it does not fulfill one of the basic requirements for k(s), i.e.,

$$J = \int_{-\infty}^{\infty} ds\, k(s) = \pi - \beta \tag{206}$$

Substituting Equation 205 in Equation 206, one obtains:

$$J = \cos(\beta/2)\sqrt{\frac{2\pi}{3}} \tag{207}$$

which, for small β values, is close to π/2 instead of close to π. Thus, Equation 205 underestimates the curvature values for larger s values.

In order to make k(s) also ρ-dependent, one has to replace a in Equation 202 by (a + ρ) so that k will become ρ-dependent, i.e.,

$$K = (a + \rho)^{-1} \cot^2(\beta/2) \tag{208}$$

Baer and Beswick[20] also suggested linking the inversion to the second derivative of k with respect to s and consequently defining an "inversion parameter" $\tilde{g}$ in the form:

$$\tilde{g} = \frac{\epsilon_k}{\hbar\omega} k\bar{\rho} \tag{209}$$

where

$$\epsilon_k = \frac{\hbar^2}{2\mu} \frac{1}{K} \frac{d^2k}{ds^2} = \frac{\hbar^2}{2\mu} \frac{\cos^2(\beta/2)}{a^2 \sin^4(\beta/2)} \tag{210}$$

It is seen that although $\tilde{g}$ looks similar to g as given in Equation 195, ϵ_k is no longer equal to the kinetic energy, but is related to the second derivative of the curvature function.

3. Static Models

The static models are related to the Distorted (or undistorted) Wave Born Approximation (DWBA), which for more than 30 years has been extensively applied in nuclear physics (see Volume II, Chapter 2). The first attempt to apply this method in the study of chemical reactions was probably by Baer and Wu in 1954,[57] followed 20 years later by a study by Micha.[102] However, it is only since the early 1970s that this method has become well established and thus, widely used. Tang and Karplus[103] rigorously followed the formal expression for the required **T** matrix element and made all the necessary theoretical and numerical efforts to obtain a meaningful result for the three-dimensional (3D) reactive cross-section for the H + H_2 reaction. The method was later modified by Choi and Tang[104-105] and extensively applied in their studies. This subject, as well as other related topics, are discussed in detail by Tang in Volume 2, Chapter 2, Section IV.

In the present chapter, we shall concern ourselves mainly with a type of models which, although closely related to the DWBA, can still be considered as an independent group, namely, the Franck-Condon (FC) type models. Whereas the DWBA treatment, as well as many other approximate treatments of reactive systems preceded the exact treatment and were actually devised as alternatives for the exact (but complicated) treatment, the FC models are a direct outcome of the exact treatment and were devised mainly for a qualitative explanation of the exact results. While analyzing the exact results for the reaction H(D) + $Cl_2 \rightarrow$ H(D)Cl + Cl, Baer[90] noted that the final vibrational distribution of the H(D)Cl molecule was strongly dependent on the initial vibrational state of the Cl_2 molecule and that a Gaussian-type vibrational distribution obtained for an initial ground state became strongly oscillatory once started in an excited state. This observation was later confirmed by Connor et al.[106] who applied a somewhat different surface. Since no similar effects were observed in the classical results,[107-109] one can consider this phenomenon to be a pure quantum mechanical effect. It was then suggested[90] that a FC model could account for these observations, and it was shown qualitatively that such a model correctly describes the findings of the exact calculations.

A more quantitative approach was taken by Halavee and Shapiro[110] who succeeded in deriving, from the DWBA integral, an expression for the **T** matrix elements which explicitly contained the corresponding FC overlap integral. Starting with the expression:

$$T_{n_\beta n_\alpha} = \iint dR_\alpha dr_\alpha \psi_{n_\beta}^{(-)}(R_\beta, r_\beta) \psi_{n_\alpha}(R_\alpha, r_\alpha) \tag{211}$$

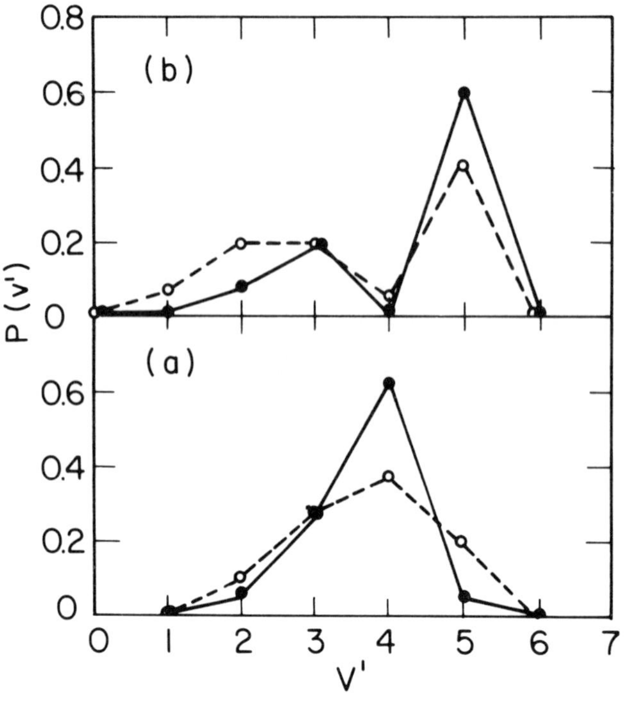

FIGURE 10. Collinear vibrational distribution for the reaction H + Cl(v_i) → HCl(v_f) + Cl. (a) v_i = 0; (b) v_i = 1. (———, exact quantum mechanical results;[90]-----, Franck-Condon model results[110]).

where α and β stand for the initial and final arrangement channels, they obtained, following various approximations, an expression which, for an adiabatic behavior in each arrangement channel, takes the form:

$$T_{n_\beta n_\alpha} = A_{n_\beta n_\alpha} \int dr_\alpha \phi_{n_\beta}(R_\alpha - \gamma_\alpha r_\alpha)\phi_{n_\alpha}(r_\alpha) \tag{212}$$

where $r_\beta = R_\alpha - \gamma_\alpha r_\alpha$ is the vibrational coordinate in the β-arrangement, γ_α stands for the mass ratio $m_C/(m_B + m_C)$, and $A_{n_\beta n_\alpha}$ is a coefficient which is connected with the translational part of the wavefunction. As can be seen, the overlap is between two shifted eigenfunctions, one related to the reagents and the other to the products. The shift (d) is by an amount:

$$d = \frac{R_\alpha}{\gamma_\alpha} + r_{\alpha_0} - \frac{r_{\beta_0}}{\gamma_\alpha} \tag{213}$$

where r_{α_0} and r_{β_0} are the equilibrium distances of the corresponding diatomics. In addition, the force constant (k_β) of the product molecule is changed and becomes $k_\beta\gamma_\alpha^2$. The exact and the model results for the reactions H + Cl_2 (v_i = 0,1) → HCl + Cl are compared in Figure 10.

Other groups, e.g., Schatz and Ross[111] and Fischer and Venzl[112] derived similar expression starting from the same Born integral. However, making other approximations, they obtained different values for $A_{n_\beta n_\alpha}$ and d. Schatz and Ross[113] also extended their treatment to three dimensions (3D) and obtained the corresponding expressions for the vibrational distributions.

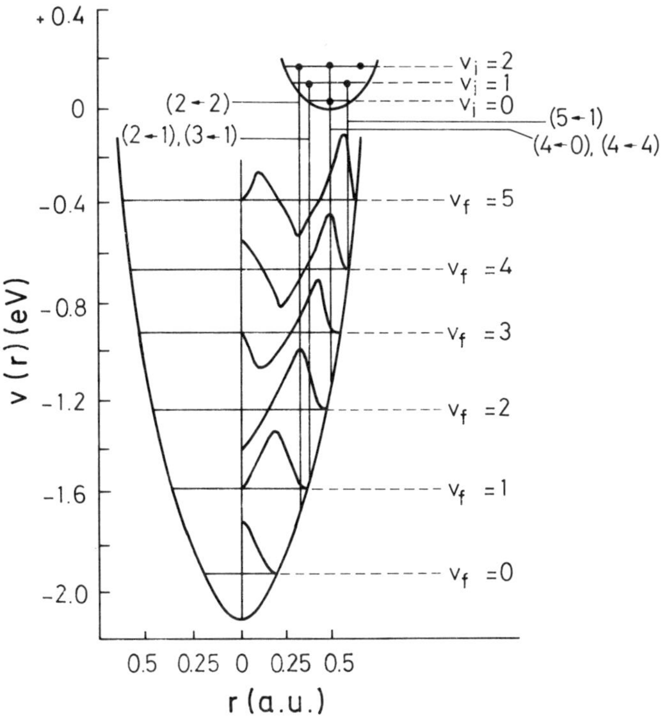

FIGURE 11. The two shifted harmonic potentials related to Cl_2 (the upper one) and HCl. The straight lines indicate the main possible Franck-Condon type transitions.[90]

Good agreement was obtained for the F + $H_2(D_2)$ systems, but the agreement for H(D) + Cl_2 systems was less satisfactory. Recently, Wong and Brumer[114] developed a FC-type model, applying the Faddeev decomposition of the scattering wavefunctions. Judging from the agreement with exact calculations, their approach is most promising. In this context, several semiclassical theories should be mentioned which, assuming the validity of the FC model, could also account for the observations of the exact calculations.[115-116]

Next we introduce a simplified approach[117] which is similar to a model presented by Hirschfelder and Wigner in 1939[13] (see Figure 4C). The Hirschfelder-Wigner model is applied, with one modification to account for vibrational nonadiabaticity: the two tubes (the reagent and the product channels) which are to be matched at R = 0 are shifted by an amount b (see Figure 11). In each of the tubes the system is assumed to be undisturbed and consequently the corresponding wavefunctions are as follows:

1. In the reactants tube the wavefunction is assumed to be of the form given by Equation 168 where x = R and y = r.
2. In the products tube the wavefunction is assumed to be of the form given by Equation 169 with y = R and x = r.

These two solutions and their first derivatives are matched at R = 0 (the equations to be solved are similar to Equation 174) and integrated over the vibrational coordinate. Consequently, the following equations for T_{mn_o} and R_{nn_o} are obtained:

$$T_{mn_o} - \sum_n \sigma^{(1)}_{mn} R_{nn_o} = \sigma^{(1)}_{mn_o} \quad ; \quad m = 0,1 \ldots M \qquad (214)$$

$$\sum_m \sigma^{(2)}_{nm} T_{mn_o} + R_{nn_o} = \delta_{nn_o} \quad ; \quad n = 0,1 \ldots M \tag{215}$$

where

$$\begin{aligned} \sigma^{(1)}_{mn} &= (k^{II}_m/k^{I}_n)^{1/2} S_{mn} \\ S_{mn} &= < \phi^{II}_m|\phi^{I}_n > \\ \sigma^{(2)}_{nm} &= \sigma^{(1)}_{mn} \end{aligned} \tag{216}$$

Thus, the S_{mn} are the overlap integrals between two shifted *harmonic oscillator* wavefunctions. Let us now consider the case where no reflection takes place, namely, $R_{nn_o} = 0$ for all n. Then it can be seen from Equation 214 that:

$$T_{mn_o} = \sigma^{(1)}_{mn_o} \tag{217}$$

which, for the corresponding reactive transition probabilities yields the expressions:

$$P_{mn_o} = \left(\frac{k^{II}_m}{k^{I}_{n_o}}\right) |S_{mn_o}|^2 \tag{218}$$

In case the eigenfunctions $\phi_n(r)$ are real, Equation 218 implies:

$$S_{mn_o} = \pm \left(\frac{k^{I}_{n_o}}{k^{II}_m}\right) P^{1/2}_{mn_o} \tag{219}$$

Equation 219 supplies the link between the *exact* results (the reactive transition probabilities) and the FC model. To check the validity of the model it is suggested that one examine whether the calculated probabilities fulfill certain relations that are known to exist between S_{mo} and S_{ml}; m = 0,1... It was shown that in the case of harmonic oscillator wavefunctions, the two kinds of overlaps are related as follows:[117]

$$S_{m1} = \tau \frac{(2\chi)^{1/2}}{1 + \chi} S_{m0} - 2 \frac{(m\chi)^{1/2}}{1 + \chi} S_{m-10} \tag{220}$$

where χ is the ratio between the two frequencies and τ is a mass-scaled shift (b), i.e.,

$$\chi = \omega/\Omega \quad ; \quad \tau = b/(\hbar/\mu\omega) \tag{221}$$

(μ is the reduced mass of the system).
Defining the two variables,

$$Y_m = (P_{m1}/P_{m0})^{1/2} \quad ; \quad Z_m = mk^{II}_m P_{m-10}/(k^{II}_{m-1} P_{m0}) \tag{222}$$

and recalling Equations 219 and 220, one can see that Y_m and Z_m fulfill a linear relationship:

$$\pm Y_m = \alpha - \beta(\pm Z_m) \tag{223}$$

where

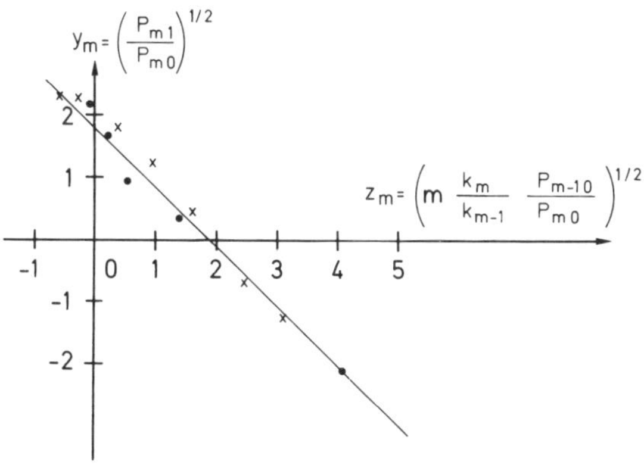

A

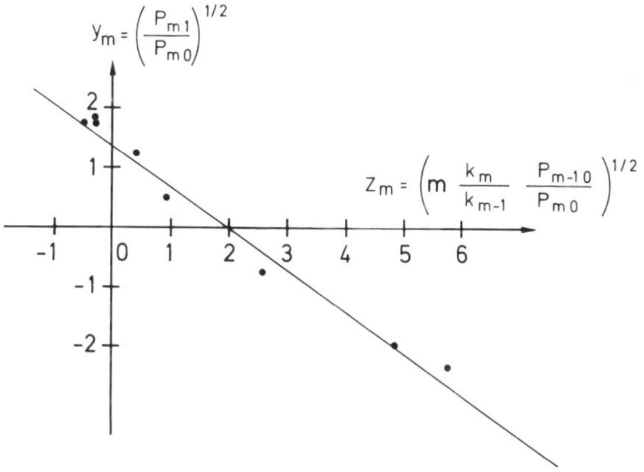

B

FIGURE 12. The values of $Y_m = \pm (P_{m1}/P_{m0})^{1/2}$ as a function of $Z_m = \pm \left|m(k_m P_{m-10})/(k_{m-1} P_{m0})\right|^{1/2}$ for m = 1,2, . . . The reactive transition probabilities P_{0m} and P_{1m} were obtained in the exact calculation.[117] (A) Results for H + Cl_2: o, the values for E = 0.268 eV; x, the values for E = 0.50 eV; (B) results for D + Cl_2 (the values are for E = 0.27 eV).

$$\alpha = \tau \frac{(2\chi)^{1/2}}{1 + \chi} \left(\frac{k_0^I}{k_1^I}\right)^{1/2}$$
$$\beta = 2 \frac{\chi^{1/2}}{1 + \chi} \left(\frac{k_0^I}{k_1^I}\right)^{1/2} \tag{224}$$

The signs in Equation 223 are not specified according to this theory, and therefore the relation between Y_m and Z_m is well defined up to the sign. This freedom was used to obtain the best possible straight line through the points. As input data to construct the variables Y_m and Z_m, the exact collinear result for H(D) + Cl_2[90] were applied. The final results are shown in Figure 12. The points for both systems are seen to follow straight lines quite nicely. Moreover, from the calculated α and β values we could also estimate τ and χ. Thus, for H + Cl_2 we obtained $\tau \sim 3$; $1 \leq \chi \leq 4$ and for D + Cl_2; $\tau \sim 3$ and $5 \leq \chi \leq 7$. Analyses of the results are given in Reference 117.

E. Resonances and Time Delay

The theory of resonances was fully developed in the field of nuclear physics and used in theoretical chemistry without any extension or modification. Although Child[118] was the first to consider the possibility of resonance effects in chemical reactions, it was only at the beginning of the 1970s that, following the first exact quantum mechanical calculations of the collinear reactive transition probabilities for $H + H_2$, resonances were studied in an extensive way. Levine et al.[89,119] gave the first satisfactory explanation for the resonances in the $H + H_2$ reactive transition probability. Their line of thinking was then followed and somewhat extended by Schatz and Kuppermann.[120]

The concept of the resonance in relation to scattering was introduced around the mid 1930s. The energy dependence of the cross-section in the neighborhood of a resonance is given by the remarkably accurate Breit-Wigner formula[121] which, when written for the corresponding **T** matrix element, takes the form:

$$T(E) = \frac{\Gamma/2}{E_o - E + i(\Gamma/2)} e^{i\delta_o} \tag{225}$$

Here E_o, Γ, and δ_o are assumed to be energy-independent constants, E_o being the resonance energy, Γ the width of the resonance curve at half its peak value, and δ_o a phase shift associated with the nonresonant scattering. Although the theories developed by Kopur and Peierls[123] and Adair[122] account for this behavior, it was only in 1958 that Feschbach[124] presented his approach, a theory free of parameters that led most straightforwardly from the compound nucleus assumption to the celebrated Breit-Wigner formula. The Feschbach theory is now the acceptable explanation for the resonance structure and the corresponding resonances are accordingly termed Feschbach resonances. In the following, we shall omit details of the theory, which Feschbach himself[124] has given in a most comprehensive way, and discuss its consequences and application with regard to collinear (chemical) reactive systems.

Since the findings of Levine and Wu,[119] the search for resonances was carried out mainly in two ways. Reconsidering the Breit-Wigner formula for the T-matrix element, one may write it either in the form:

$$T(E) = \sin\delta_R(E) e^{i(\delta_R(E) + \delta_0)} \tag{226}$$

where

$$\delta_R(E) = \tan^{-1} \frac{\Gamma/2}{E - E_0} \tag{227}$$

or in the form:

$$T(E) = \mathrm{Re}(T(E)) + i\,\mathrm{Im}(T(E)) \tag{228}$$

where

$$\mathrm{Re}(T(E)) = \frac{\Gamma/2}{(E - E_0)^2 + (\Gamma/2)^2} \left[(E_0 - E)\cos\delta_0 + \frac{\Gamma}{2} \sin\delta_0 \right] \tag{229a}$$

and

$$\mathrm{Im}(T(E)) = \frac{\Gamma/2}{(E - E_0)^2 + (\Gamma/2)^2} \left[(E_0 - E)\sin\delta_0 - \frac{\Gamma}{2} \cos\delta_0 \right] \tag{229b}$$

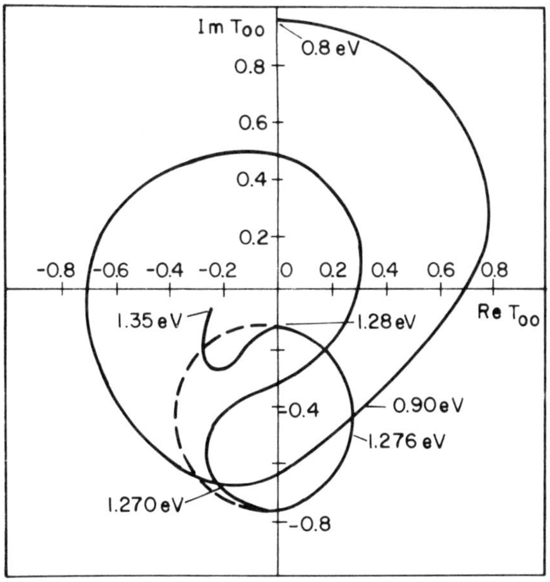

FIGURE 13. Argand plot of Im T^R_{00} vs. Re T^R_{00}, with the energy as a parameter for the H + H_2 system.[120] The dashed curve indicates the existence of a resonance circle in the energy region of E = 1.276 eV.

One way of determining the resonance is by applying Equation 227 and plotting δ_R (the resonance phase shift) as a function of the energy. In the vicinity of the resonance energy δ_R is expected to change from zero to π ($\delta_R(E = E_0) = \pi/2$). Another way is to apply Equation 229 and to plot Im(T) against Re(T) with the energy as a parameter (Argand plot). Thus, one can show that Equation 229 describes a circle with a radius of length 0.5 and center with the coordinates:

$$\begin{aligned} \mathrm{Re}(T)_c &= \sin\delta_0/2 \\ \mathrm{Im}(T)_c &= \cos\delta_0/2 \end{aligned} \tag{230}$$

Regarding the energy dependence, it can be shown that increasing E generates a counterclockwise circle. An Argand plot for the H + H_2 system is shown in Figure 13. In the vicinity of E = 1.276 eV, an almost perfect circle is formed, indicating the possible existence of a resonance.

Although in the early studies it was recognized that the reactive resonances are ordinary Feschbach resonances associated with closed adiabatic (vibrational) states (non-Feschbach resonances are the shape resonances which are closely associated with tunneling through potential barriers into potential wells) it was only several years later that attempts were made to show that this is really the case. Chapman and Hayes[125] treated the endothermic system I + H_2. While considering the reactive transition probability $P(\nu_i = 1 \rightarrow \nu_f = 0)$, they found a marked oscillatory behavior in the energy range 1.890 to 1.975 eV. They identified four peaks, all of which could be directly attributed to four compound states of the H_2I system. It turned out that when plotted as a function of the reaction coordinate, the vibrational state $\nu_f = 1$ in the product (HI) exit channel could support four bound states (see Figure 14). Since the asymptotic $\nu_f = 1$ state was still closed for the above energy range, then, based on the Feschbach theory and assuming a negligibly small interaction between the $\nu_f = 1$ and the $\nu_f = 0$ states, one would expect strong oscillatory behavior for energies in the vicinity

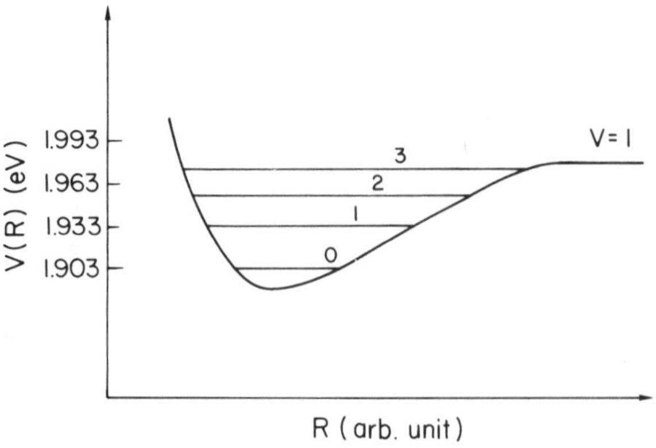

FIGURE 14. The four resonance states of the v = 1 vibrational state along the H + HI channel.[125] The energies were determined by a first-order perturbation calculation.

of these eigenstates. Applying a perturbative procedure, the corresponding eigenstates were calculated and compared with the scattering energies for which the probability function peaks. The deviations were less than 0.5%.

The study by Latham et al.[28] for the F + H_2 system can be considered to be somewhat similar in spirit to the above. Although the importance of wells along the reaction coordinate for different states had been clearly demonstrated, they could not attribute the various resonances in this system to any specific compound eigenstate. Rather, they showed that, due to the strong coupling between the different states in the interaction region, the final resonance was a result of many interacting states.

More recent studies were concentrated on resonances in 3D. Shoemaker and Wyatt[126-127] formulated the Feschbach projection operator theory for treating 3D reactive systems and applied it to a model with the aim of developing approximate treatments. Pollak and Wyatt[128] developed a semiclassical approach to study resonances in 3D and applied it to the H + H_2 and the F + H_2 system. More about the semiclassical method can be found Volume III, Chapter 3.

A concept closely related to that of resonance is time delay. Although time delays were treated in only a few cases, they are interesting enough to be discussed here. The concept of time delay originated in the work of Eisenbud and Wigner.[129-132] If we consider an incident beam which is a superposition of two monoenergetic beams of energy $h(\nu + \nu')$ and $h(\nu - \nu')$, where $\nu' << \nu$, the incident wavefunction is given by

$$\psi_{inc} = r^{-1}\,[\exp(-i((k + k')r - (\nu + \nu')t)) + \exp(-i((k - k')r - (\nu - \nu')t))] \tag{231}$$

and the outgoing wave is

$$\psi_{out} = r^{-1}\,[\exp(i((k + k')r - (\nu + \nu')t + (\delta_R + \delta_R'))) + \exp(i((k - k')r - (\nu - \nu')t + (\delta_R - \delta_R')))] \tag{232}$$

Here, $k + k'$ and $k - k'$ are the corresponding wave numbers ($k' << k$), and $(\delta_R + \delta'_R)$ and $(\delta_R - \delta'_R)$ are the phase shifts which correspond to the energies $h(\nu + \nu')$, and $h(\nu - \nu')$. The two components of the two waves considered are inphase when

$$2k'r + 2\nu't = 0 \tag{233}$$

for the incident wave, and

$$2k'r - 2\nu't + 2\delta_R' = 0 \tag{234}$$

for the outgoing wave. Since $(\nu'/k') = d\nu/dk$ is the velocity of the center of the wave, it can be seen that the incident beam moves with the correct velocity (see Equation 233), but the outgoing beam is retarded by a stretch $d\delta/dk$ and arrives at point $(r - d\delta/dk)$ at the time when it should have arrived at r, were it not for the action of the scattering center. In order to relate $(d\delta_R/dk)$ to time delay (τ_R), we divide it by v (the velocity of the incident beam) so that:

$$\tau_R = \frac{1}{v}\frac{d\delta_R}{dk} = \hbar \frac{d\delta_R}{dE} \tag{235}$$

For the case of a Feschbach resonance the phase shift is given in the form:

$$\delta_R = \tan^{-1}\left(\frac{\Gamma/2}{E_0 - E}\right) \tag{236}$$

and consequently the time delay is

$$\tau_R = -\hbar \frac{\Gamma/2}{(E_0 - E)^2 + (\Gamma/2)^2} \tag{237}$$

or

$$\tau_R = -P(E)\frac{2\hbar}{\Gamma} \tag{238}$$

where P(E) is the corresponding transition probability, i.e.,

$$P(E) = |T(E)|^2 = \frac{(\Gamma/2)^2}{(E - E_0)^2 + (\Gamma/2)^2} \tag{239}$$

An extension of the concept of time delay was given in 1960 by Smith.[133] Applying the definition of the time residence in a given region and defining the difference between the two residence times with and without interaction as the time delay (or lifetime), Smith was able to construct a general lifetime matrix **Q**. In the general case **Q** and **S** are related by:

$$\mathbf{Q} = i\hbar\, \mathbf{S} \frac{d\mathbf{S}^+}{dE} \tag{240}$$

which, in view of the conjugate relation between time and energy, takes the form:

$$\mathbf{Q} = \mathbf{S}\, t\, \mathbf{S}^+ \tag{241}$$

Smith not only showed that his general approach reduces to the Wigner-Eisenbud formulation[129] for the elastic case, but also proved that the diagonal element of **Q**, i.e., Q_{ii}, is the average time delay experienced by a particle starting at an initial state i:

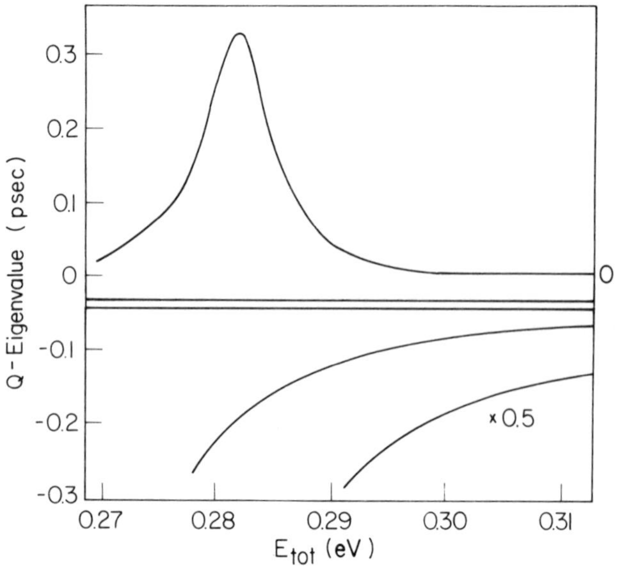

FIGURE 15. Eigenvalues of the time delay Q-matrix as a function of total energy E for the F + H_2 system.[134]

$$Q_{ii} = (\Delta t_i)_{ave} = \sum_j S_{ij}^* S_{ij} \Delta t_{ij} \tag{242}$$

where Δt_{ij} is the time delay in the Wigner-Eisenbud sense.

The Smith approach was recently tested with regard to the F + H_2 system and its isotopic analogs.[134] In Figure 15 the five eigenvalues of the **Q** matrix are shown as a function of the energy for the F + $H_2 \rightarrow HF(\nu)$ + H reaction. It is noticed that only one eigenvalue has a well-behaved functional shape that clearly indicates the existence of a resonance. The location of this particular resonance is at $E_{res} = 0.281$ eV which is very close to the position of the $\nu = 2$ (vibrational) state resonance. The corresponding time delay 0.35 ps which is about two orders of magnitude longer than the characteristic vibrational period.

III. THE THREE-DIMENSIONAL SYSTEM

A. Background

The transition from the 1D to the 3D system was found to be rather difficult. It took about 5 years to accomplish this step and even then the only system that was treated was the fully symmetric H + H_2 system, for which only a narrow energy range was exposed to calculations.[1] (The F + H_2 system was also treated in 3D but decoupling methods were employed, i.e., the Coupled States Approximation[135] or the Infinite Order Sudden Approximation.[136-137]) The first attempt to move into 3D was carried out by Wolken and Karplus.[138] In their study, they included only one vibrational state and a limited number of rotational states; consequently, unconverged results were obtained. The next treatment was by Schatz and Kupperman (SK)[139] who used a formalism developed by Kuppermann, Schatz, and Baer (KSB).[140] They were the first to arrive at a rate constant calculated from first principles. Almost at the same time Elkovitz and Wyatt (EW)[35] independently developed an alternative procedure which yielded results that somewhat differ from those of SK. Later, Walker, Stechel, and Light (WSL)[141] repeated the same calculation employing a method developed by Walker, Light, and Altenberger-Siczek[142] and obtained similar results to those of SK. A general 3D system is shown in Figure 16.

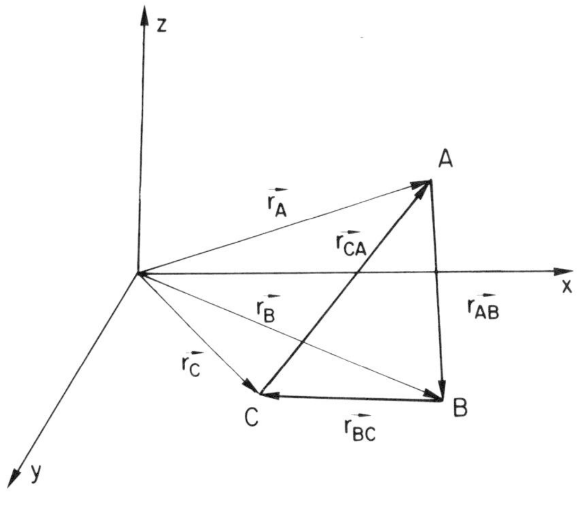

FIGURE 16. The three-dimensional system. The vectors $\mathbf{r}_x$, x = A,B,C are measured with respect to a fixed system of coordinates in space. The $\mathbf{r}_{xy}$; x,y = A,B,C are interatomic distances.

Since the first exact cross sections for the H + H_2 system were published, 7 years have passed and no exact treatment for any other system has been reported. It is now well accepted that neither of the above-mentioned methods can be converted into an everyday tool to study chemical reactions. Besides the complexity involved in handling the programs and the numerical instabilities in the calculations, these methods contain certain inherent difficulties which seem to be unsolvable when employed for nonsymmetric systems. These difficulties have to do with the matching procedure of the 3D wavefunctions in the interaction region and the bifurcation process[45] which becomes apparent when one has to propagate into more than one arrangement channel (see Figure 17). There are several methods available now which look rather promising as they avoid the matching. (One of them employs the Delves coordinates and was described in Section II.B.3 and the other is described in Chapter 4, but none were tested in a three-dimensional treatment). Since this is the situation, reference will not be made to any method in particular, but rather the reader will be exposed to the necessary background to carry out research in this area.

For an easy beginning, we first refer to the classical description of the 3D system and then continue with the quantum mechanical one. Within the quantum mechanical description we distinguish between the space- and the body-fixed systems of coordinates, discuss the connection and relationship between the various arrangement channels, and finally, refer to the scattering amplitude and the differential and integral cross sections.

B. The Classical Description of the System in Three Dimensions

The Hamiltonian for three moving particles in a 3D system is given by Equation 1′:

$$\mathcal{H} = \frac{1}{2m_A}\mathbf{P}_A^2 + \frac{1}{2m_B}\mathbf{P}_B^2 + \frac{1}{2m_C}\mathbf{P}_C^2 + V(\mathbf{r}_A, \mathbf{r}_B, \mathbf{r}_C) \tag{1'}$$

where $\mathbf{r}_A$, $\mathbf{r}_B$, and $\mathbf{r}_C$ are the distances of atoms A, B, and C from some reference point and $\mathbf{P}_A$, $\mathbf{P}_B$, and $\mathbf{P}_C$ are the corresponding moments. Since the potential depends on the distances between the three atoms only, we transform to a more convenient system of coordinates which is based on interatomic distances as well as on the coordinates and momenta of the center of mass (see Figure 16):

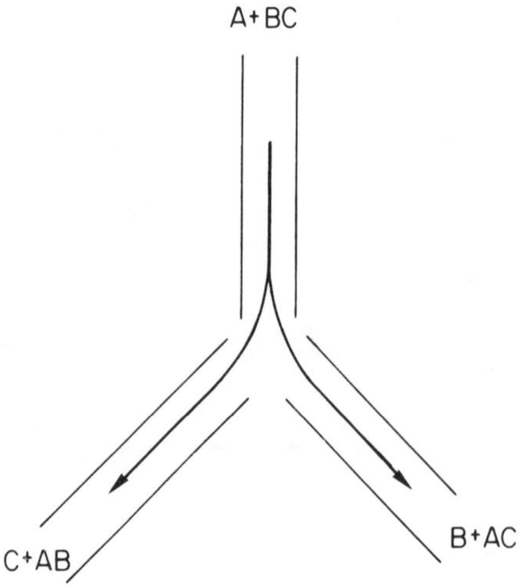

FIGURE 17. The bifurcation in the three channel problem. Here (A + BC) represents the reactants channel and (C + AB) and (B + AC) are the products channels.

$$\mathbf{r}_{AB} = \mathbf{r}_B - \mathbf{r}_A$$
$$\mathbf{r}_{BC} = \mathbf{r}_C - \mathbf{r}_B \tag{243}$$
$$\mathbf{R} = \frac{m_A\mathbf{r}_A + m_B\mathbf{r}_B + m_C\mathbf{r}_C}{M}$$

where M is the sum of the three masses:

$$M = m_A + m_B + m_C \tag{244}$$

By employing the usual contact transformation it can be shown that the Hamiltonian takes the form:

$$\mathcal{H} = \frac{1}{2\mu_{AB}}\mathbf{P}_{AB}^2 + \frac{1}{2\mu_{BC}}\mathbf{P}_{BC}^2 + \frac{\mathbf{P}^2}{2M} - \frac{1}{m_B}\mathbf{P}_{AB}\cdot\mathbf{P}_{BC} + V(|\mathbf{r}_{AB}|, |\mathbf{r}_{BC}|, |\mathbf{r}_{AC}|) \tag{245}$$

where $\mathbf{P}_{AB}$, $\mathbf{P}_{BC}$, and $\mathbf{P}$ are the conjugate momenta to $\mathbf{r}_{AB}$, $\mathbf{r}_{BC}$, and $\mathbf{R}$, respectively, and μ_{AB} and μ_{BC} are the reduced masses of the form:

$$\mu_{xy} = \frac{m_x m_y}{m_x + m_y} \quad ; \quad x,y = A,B,C \tag{246}$$

Since the potential does not depend on $\mathbf{R}$, the momentum $\mathbf{P}$ is constant and therefore, the corresponding term in Equation 245 can be ignored. It should be emphasized that Equation 245 is not unique and that other representations for $\mathcal{H}$ are possible. They are of the same functional form, but with the indexes A,B, and C permuted cyclically. The disadvantage of this representation is the crossed term which causes difficulties once the Hamiltonian becomes a (quantum mechanical) differential operator. Thus, introducing the coordinates $\mathbf{R}'$ and $\mathbf{r}'$ defined as:

$$\mathbf{r}' = \mathbf{r}_{BC} \tag{247}$$

$$\mathbf{R}' = \mathbf{r}_{AB} + \frac{m_C}{m_B + m_C} \mathbf{r}_{BC}$$

and again using the ordinary technique to form the corresponding conjugate momenta, the Hamiltonian $\mathcal{H}$ takes the form:

$$\mathcal{H} = \frac{1}{2\mu_R} \mathbf{P}^2_{R'} + \frac{1}{2\mu_{r'}} + V(\mathbf{R}',\mathbf{r}') \tag{248}$$

where

$$\mu_R = \frac{m_A(m_B + m_C)}{M} \quad ; \quad \mu_r = \mu_{BC} \tag{249}$$

Equation 248 already contains certain physical features of the particular arrangement of the three atoms. The coordinate $\mathbf{r}'$ stands for the distance between atoms B and C. Therefore, if B and C form a diatomic molecule, then $\mathbf{r}'$ is the axis vector of this molecule. The vector R' stands for the distance between atom A and the center of mass of BC (see Equation 247) and consequently, is a relevant coordinate for describing the interaction between A and diatomic molecule BC. Thus, the Hamiltonian given in Equation 248 describes the dynamics of the three-atom system for a given arrangement. In what follows this will be the reagents arrangement designated by λ. The two other arrangements are the ν-arrangement, where AB stands for the diatomic molecule and C is the isolated atom, and the κ-arrangement where AC is the diatomic molecule and B is the isolated atom. The arrangements ν and κ are considered to be the products arrangements (see Figure 18).

As a final transformation we scale the coordinates $\mathbf{r}'_\alpha$ and $\mathbf{R}'_\alpha$; $\alpha = \lambda, \nu, \kappa$ in order to make the presentation of the Hamiltonian as symmetric as possible with respect to all three arrangements (see also Equation 9).Thus

$$\begin{aligned} \mathbf{r}_\alpha &= a_\alpha \mathbf{r}'_\alpha \\ & \quad ; \quad a_\alpha = \left(\frac{\mu_{r\alpha}}{\mu_{R\alpha}}\right)^{1/4} \quad ; \quad \alpha = \lambda, \nu, \kappa \\ \mathbf{R}_\alpha &= a_\alpha^{-1\prime}\mathbf{R}'_\alpha \end{aligned} \tag{250}$$

Consequently, $\mathcal{H}$ becomes:

$$\mathcal{H} = \frac{1}{2\mu}\left(\mathbf{P}^2_{R\alpha} + \mathbf{P}^2_{r\alpha}\right) + V(\mathbf{R}_\alpha,\mathbf{r}_\alpha) \tag{251}$$

where μ is the mass of the system (see also Equation 11),

$$\mu = \left(\mu_{r_\alpha}\mu_{R_\alpha}\right)^{1/2} = \left(\frac{m_A m_B m_C}{M}\right)^{1/2} \tag{252}$$

Like in the collinear case, the transformation matrix from one set of coordinates to the other is an orthogonal matrix. For example, the transformation matrix from the λ to the ν arrangement is given in the form:

$$\begin{pmatrix} \mathbf{R}_\nu \\ \mathbf{r}_\nu \end{pmatrix} = \begin{pmatrix} \cos\beta & -\sin\beta \\ \sin\beta & \cos\beta \end{pmatrix} \begin{pmatrix} \mathbf{R}_\lambda \\ \mathbf{r}_\lambda \end{pmatrix} \tag{253}$$

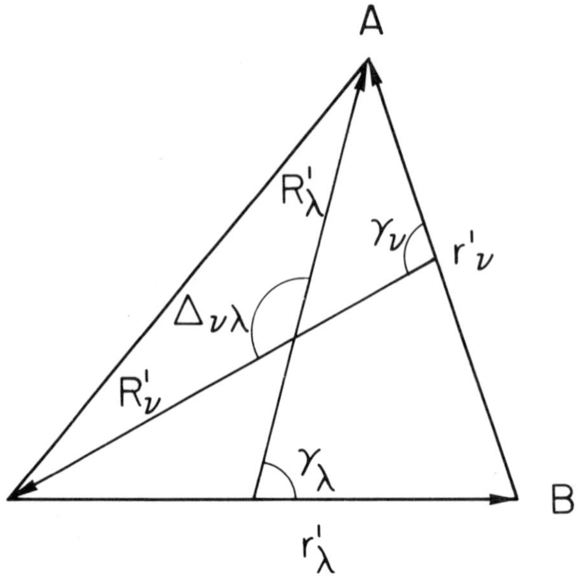

FIGURE 18. The three atom system; λ denotes the A + BC arrangement and ν stands for the C + AB arrangement.

where cosβ and sinβ are given by Equation 12′. As a final point in this discussion we refer to the potential. The potential is described by the three interatomic *distances* and consequently, $V(\mathbf{R}_\alpha, \mathbf{r}_\alpha)$ can be shown to be:

$$V(\mathbf{R}_\alpha, \mathbf{r}_\alpha) = V(R_\alpha, r_\alpha, \gamma_\alpha) \quad ; \quad \alpha = \lambda, \nu, \kappa \tag{254}$$

where R_α and r_α are equal to $|\mathbf{R}_\alpha|$ and $|\mathbf{r}_\alpha|$, respectively, and γ_α is the angle between **R** and **r** (see Figure 18), i.e.,

$$\gamma_\alpha = \cos^{-1}(R_\alpha \cdot r_\alpha) \quad ; \quad \alpha = \lambda, \nu, \kappa \tag{255}$$

C. The Quantum Mechanical Description of the System in Three Dimensions

In treating the three-body system in its full dimensionality, one distinguishes between two systems of coordinates.[143] The first is the space-fixed (SF) system[144] of coordinates which relates the four angular variables to one fixed system of coordinates in space. This system, as will be shown immediately, is a direct extension of the classical representation. The second system is the body-fixed (BF) system[145-148] of coordinates which relates the single left angular variable to a system which is fixed in the body but varies in space. The other three angles, the Euler Angles,[149] are responsible for the transformation from the SF to the BF system.

1. The Space-Fixed Representation

Once the (classical) Hamiltonian is given, the corresponding SE is

$$\mathcal{H}\psi = E\psi \tag{256}$$

where $\mathcal{H}$ is now a differential operator:

$$\mathcal{H} = -\frac{\hbar^2}{2\mu}(\nabla_R^2 + \nabla_r^2) + V(R,r,\gamma) \tag{257}$$

Here the index α is deleted to shorten the notation. The operators ∇_R^2 and ∇_r^2 are the Laplacian operators of the form:[150]

$$\nabla_R^2 = \frac{1}{R}\frac{\partial^2}{\partial R^2} R + \frac{\ell^2}{\hbar^2 R^2} \tag{258a}$$

$$\nabla_r^2 = \frac{1}{r}\frac{\partial^2}{\partial r^2} r + \frac{\mathbf{j}^2}{\hbar^2 r^2} \tag{258b}$$

where

$$\ell^2 = -\hbar^2\left(\frac{1}{\sin\theta}\frac{\partial}{\partial\theta}\sin\theta\frac{\partial}{\partial\theta} + \frac{1}{\sin^2\theta}\frac{\partial^2}{\partial\phi^2}\right) \tag{259}$$

and

$$\mathbf{j}^2 = -\hbar^2\left(\frac{1}{\sin\theta_r}\frac{\partial}{\partial\theta_r}\sin\theta_r\frac{\partial}{\partial\theta_r} + \frac{1}{\sin^2\theta_r}\frac{\partial^2}{\partial\phi_r^2}\right) \tag{260}$$

The angles θ and ϕ are the spherical angles associated with R, and θ_r and ϕ_r are spherical angles associated with r. The angle γ which appears in the potential function (see Equation 25) is related to these four angles as follows:

$$\cos\gamma = \cos\theta\cos\theta_r + \sin\theta\sin\theta_r\cos(\phi - \phi_r) \tag{261}$$

The representation of H as given by Equations 257 to 261 is called the SF representation because all four angles refer to a fixed system of coordinates in space.

The operators ℓ and $\mathbf{j}$ are the ordinary orbital and rotational (internal) angular momentum operators. The operator ℓ is given in the form:[150]

$$\begin{aligned}
\ell_x &= -i\hbar\left(-\cos\phi\cot\theta\frac{\partial}{\partial\phi} - \sin\phi\frac{\partial}{\partial\phi}\right)\\
\ell_y &= -i\hbar\left(-\sin\phi\cot\theta\frac{\partial}{\partial\phi} + \cos\phi\frac{\partial}{\partial\phi}\right)\\
\ell_z &= -i\hbar\frac{\partial}{\partial\phi}
\end{aligned} \tag{262}$$

and a similar representation exists for $\mathbf{j}$. The eigenvalues of ℓ^2 and $\mathbf{j}^2$ are $\ell(\ell + 1)$ and $j(j + 1)$, respectively, where ℓ and j are integers or zero. The corresponding eigenfunctions are the spherical harmonics $Y_{\ell m_\ell}(\theta,\phi)$ and $Y_{jm_j}(\theta_r,\phi_r)$,[150] where m_ℓ and m_j are the (integer) eigenvalues of ℓ_z and j_z, respectively.

This approach, introduced by Arthurs and Dalgarno[144] while studying inelastic collisions, proceeds as follows.

Employing the spherical harmonics mentioned previously, we are in a position to construct eigenfunctions, not only of ℓ^2 and $\mathbf{j}^2$, but also of $\mathbf{J}^2$ and $\mathbf{J}_z$. Here $\mathbf{J}$ is the total angular momentum:

$$\mathbf{J} = \ell + \mathbf{j} \tag{263}$$

This possibility is particularly important because in any collision process, whether inelastic or reactive, both $\mathbf{J}^2$ and $\mathbf{J}_z$ are conserved and consequently J and M serve as "good" quantum numbers.
Thus the corresponding eigenfunction is

$$\Gamma_{\ell j}^{JM}(\theta, \phi, \theta_r, \phi_r) = \sum_{m_\ell, m_j} C(jm_j\ell m_\ell|JM)\, Y_{\ell m_\ell}(\theta,\phi)\, Y_{jm_j}(\theta_r,\phi_r) \tag{264}$$

where $C(jm_j\ell m_\ell|JM)$ are the the Clebsch-Gordan coefficients.[149] The fact that we are able to form an eigenfunction of $\mathbf{J}^2$ and $\mathbf{J}_z$ from products of well-known functions is used, in the following way.

The total wavefunction $\psi(\mathbf{R},\mathbf{r})$, can be expressed in terms of a set of SF functions $[\psi_{JM}(\mathbf{R},\mathbf{r})]$ which are assumed to be eigenfunctions of H, J^2, and J_z. Thus,

$$\psi(\mathbf{R},\mathbf{r}) = \sum_{J=0} \sum_{M=-J}^{J} Z_{JM}\psi_{JM}(\mathbf{R},\mathbf{r}) \tag{265}$$

Here Z_{JM} are undetermined constants and $\psi_{JM}(\mathbf{R},\mathbf{r})$ are functions given in the form:

$$\psi_{JM}(\mathbf{R},\mathbf{r}) = \psi_{JM}(R,\theta,\phi,r,\theta_r,\phi_r) = \sum_{\ell j} G_{\ell j}^{JM}(R,r)\Gamma_{\ell j}^{JM}(\theta,\phi,\theta_r,\phi_r) \tag{266}$$

where $G_{\ell j}^{JM}(R,r)$ are yet to be determined. Substituting Equation 265 in Equation 256 and recalling Equations 257 to 260, 264, and 266, one arrives at the system of coupled equations for $G_{\ell j}^{JM}(R,r)$:

$$\left(-\frac{\hbar^2}{2\mu}\left(\frac{1}{R}\frac{\partial^2}{\partial R^2}R + \frac{1}{r}\frac{\partial^2}{\partial r^2}r\right) - \frac{j(j+1)}{2\mu r^2}\hbar^2 - \frac{\ell(\ell+1)}{2\mu R^2}\hbar^2 - E\right) G_{\ell j}^{JM}(R,r) = -\sum_{\ell' j'} <\ell j|V|j'\ell'> G_{\ell' j'}^{JM}(R,r) \tag{267}$$

where

$$<\ell j|V|j'\ell'> = \iint d\omega d\omega_r\, \Gamma_{\ell j}^{JM}(\omega,\omega_r)V(R,r,\gamma)\,\Gamma_{\ell' j'}^{JM}(\omega,\omega_r) \tag{268}$$

Here $\omega(\omega_r)$ stands for the pair of angles (θ,ϕ) $[(\theta_r,\phi_r)]$ and the differential $d\omega$ stands for:

$$d\omega = d(\cos\theta)d\phi \tag{269}$$

In the inelastic case Equation 267 is solved subject to boundary conditions and the solution leads to the corresponding S- or the T-matrix elements. These are used to form the desired differential and integral cross sections.

This approach is not employed in studies on exchange collisions because it is believed that the transformation from one arrangement to the other is more complicated than necessary. On the other hand, the BF system is believed to be more convenient for this transformation and therefore, we continue the discussion within this framework.

2. The Body-Fixed Representation

The transformation to the BF representation is made by employing the Euler angles and the coefficients of the irreducible representations of the rotation group. One chooses the three Euler angles $(\theta,\phi,0)$ which yield a new z-axis in the direction of R and a new y-axis in the old (x′y′) plane. This transformation also affects the two angles (θ_r,ϕ_r) which become γ and δ where γ is given by Equation 261 and δ is given by:

$$\delta = \cot^{-1}(-\sin\theta\cot\theta_r\cos(\phi_r - \phi) + \cos\theta\cot(\phi_r - \phi)) \tag{271}$$

The wavefunction ψ^{JM} is now explained in the form:

$$\psi^{JM} = \sum_{\Omega=-J}^{J}\sum_{j>|\Omega|} D^{J}_{\Omega M}(\theta,\phi,0)y_{j\Omega}(\gamma,\delta)G^{JM}_{j\Omega}(r,R) \tag{272}$$

The Hamiltonian which is compatible with this representation is obtained from the one given in Equations 257 to 258 by replacing ℓ^2 (the angular part as written in Equation 272 is no longer an eigenfunction of ℓ^2).

$$\ell^2 = (\mathbf{J} - \mathbf{j})^2 \tag{273}$$

Consequently, ℓ^2 becomes:

$$\ell^2 = \mathbf{J}^2 + \mathbf{j}^2 - J_-j_+ - J_+j_- - 2J_zj_z \tag{274}$$

where $J_+(j_+)$ and $J_-(j_-)$ are the raising and lowering operators:

$$J_\pm = J_x \pm iJ_y \quad ; \quad j_\pm = j_x \pm ij_y \tag{275}$$

Substitution of Equation 274 in Equations 257 to 258 yields:[151]

$$\mathcal{H} = -\frac{\hbar^2}{2\mu}\left(\frac{1}{R}\frac{\partial^2}{\partial R^2}R + \frac{1}{r}\frac{\partial^2}{\partial r^2}r\right) + \frac{1}{2\mu}\left(\frac{1}{r^2} + \frac{1}{R^2}\right)\mathbf{j}^2 + \frac{1}{2\mu R^2}(\mathbf{J}^2 - J_-j_+ - J_+j_- - 2J_zj_z) + V(r,R,\gamma) \tag{276}$$

where the differential representation of the various operators are:

$$\begin{aligned}
\mathbf{J}^2 &= -\hbar^2\left[\frac{\partial^2}{\partial\theta^2} + \cot\theta\frac{\partial}{\partial\theta} + \frac{1}{\sin^2\theta}\left(\frac{\partial^2}{\partial\delta^2} + \frac{\partial^2}{\partial\phi^2} - 2\cos\theta\frac{\partial^2}{\partial\delta\partial\phi}\right)\right] \\
\mathbf{j}^2 &= -\hbar^2\left(\frac{\partial^2}{\partial\gamma^2} + \cot\gamma\frac{\partial}{\partial\gamma} + \frac{1}{\sin^2\gamma}\frac{\partial^2}{\partial\gamma^2}\right) \\
J_\pm &= \hbar e^{\mp i\delta}\left(-i\cot\theta\frac{\partial}{\partial\delta} + \frac{i}{\sin\theta}\frac{\partial}{\partial\phi} \pm \frac{\partial}{\partial\theta}\right) \\
j_\pm &= \hbar\left(i\cot\gamma\frac{\partial}{\partial\delta} \pm \frac{\partial}{\partial\gamma}\right) \\
J_z &= j_z = -i\hbar\frac{\partial}{\partial\delta}
\end{aligned} \tag{277}$$

Substituting Equation 272 in Equation 276, employing the well-known properties of the $D^J_{M\Omega}(\theta,\phi,0)$ and $y_{j\Omega}(\gamma,\delta)$ functions,[149] multiplying through by $D^J_{M\Omega'}(\theta,\phi,0)y_{j'\Omega'}(\gamma,\delta)$, integrating over θ,ϕ,γ, and δ, and replacing Ω' and j' by Ω and j, one obtains the following system of coupled equations:

$$(T^{Jj}_{\Omega\Omega} - E)\, G^{JM}_{j\Omega}(R,r) + T_{\Omega\Omega+1}\, G^{JM}_{j\Omega+1} + T_{\Omega\Omega-1}\, G^{JM}_{j\Omega-1} + \sum_{j'>|\Omega|} V^{\Omega}_{jj'}\, G^{JM}_{j'\Omega}(R,r) = 0 \tag{278}$$

$$j > |\Omega|;\quad \Omega = -J. \ldots, 0, \ldots .J \;;\quad J = 0,1,2,\ldots$$

where

$$T^{JM}_{\Omega\Omega} = -\frac{\hbar^2}{2\mu}\left(\frac{1}{R}\frac{\partial^2}{\partial R^2}R + \frac{1}{r}\frac{\partial^2}{\partial r^2}r\right) + \frac{\hbar^2}{2\mu}\left(\frac{j(j+1)}{r^2} + \frac{1}{R^2}(J(J+1) + j(j+1) - 2\Omega^2)\right) \tag{279}$$

$$T^{JM}_{\Omega\Omega\pm1} = -\frac{\hbar^2}{2\mu R^2}[(J(J+1) - \Omega(\Omega \pm 1))\,(j(j+1) - \Omega(\Omega \pm 1))]^{1/2} \tag{280}$$

and

$$V^{\Omega}_{jj'} = < j\Omega|V(r,R,\gamma)|j'\Omega > = \int d(\cos\gamma)y_{j\Omega}(\gamma,0)V(r,R,\gamma)y_{j'\Omega}(\gamma,0) \tag{281}$$

Equation 278 can be further simplified by replacing $G^{JM}_{j\Omega}(R,r)$ by

$$G^{JM}_{j\Omega}(R,r) = \frac{1}{Rr}F^{JM}_{j\Omega}(R,r) \tag{282}$$

Substituting Equation 282 in Equation 278 one gets a similar equation to Equation 278 except that $T^{Jj}_{\Omega\Omega}$ takes a more convenient form:

$$T^{Jj}_{\Omega\Omega} = -\frac{\hbar^2}{2\mu}\left(\frac{\partial^2}{\partial R^2} + \frac{\partial^2}{\partial r^2}\right) + \frac{\hbar^2}{2\mu}\left\{\frac{j(j+1)}{r^2} + \frac{1}{R_2}\left[J(J+1) + j(j+1) - 2\Omega^2\right]\right\} \tag{279'}$$

Our next step is to determine the asymptotic physical boundary conditions in each arrangement channel. We first expand $F^{JM}_{j\Omega}(R,r)$ in terms of the vibrational asymptotic basis set which contains the eigenfunctions of the corresponding diatomic molecule. Thus

$$F^{JM}_{j\Omega}(R,r) = \sum_{n} \zeta^{JM}_{nj\Omega}(R)\phi_{nj}(r) \tag{283}$$

where $\phi_{nj}(r)$ are solutions of the eigenvalue problem:

$$\left|-\frac{\hbar^2}{2\mu}\left(\frac{\partial^2}{\partial r^2} + v(r)\right) + \frac{j(j+1)\hbar^2}{2\mu r^2} - \epsilon_{nj}\right|\phi_{nj}(r) = 0 \tag{284}$$

Here ϵ_{nj} are the vibrational eigenvalues and v(r) is the (asymptotic) diatomic potential:

$$v(r) = \lim_{R\to\infty} V(R,r,\gamma) \tag{285}$$

In Equation 283 the functions ζ^{JM}_{njr} (R) are the translational scattering functions, which (like in the collinear case) have to be assigned asymptotic physical expressions. However, first we have to consider Equations 278 with the definitions given in equations 279′ to 285. Substituting Equation 283 in Equation 278, we obtain, for this limiting case, the following coupled system of equations:

$$\left\{\frac{\partial^2}{\partial R^2} - \frac{1}{R^2}(J(J+1) + j(j+1) - 2\Omega^2) + k^2_{nj}\right\} \zeta^{JM}_{nj\Omega} - \frac{1}{R^2}\left(t^{Jj}_{\Omega\Omega+1}\,\zeta^{JM}_{nj\Omega+1}(R) + t^{Jj}_{\Omega\Omega-1}\,\zeta^{JM}_{nj\Omega-1}(R)\right) = 0 \tag{286}$$

where

$$k^2_{nj} = \frac{2\mu}{\hbar^2}(E - \epsilon_{nj}) \tag{287}$$

and

$$t^{Jj}_{\Omega\Omega\pm1} = R^2\frac{2\mu}{\hbar^2}T^{Jj}_{\Omega\Omega\pm1} \tag{288}$$

The solutions of this system of equations can be found by diagonalizing the (symmetric) interaction matrix **t** defined as[146]

$$t_{\Omega\Omega'} = \begin{cases} 2\Omega^2 & ; \quad \Omega' = \Omega \\ t_{\Omega\Omega\pm1} & ; \quad \Omega' = \Omega \pm 1 \\ 0 & ; \quad \text{otherwise} \end{cases} \tag{289}$$

More details on this subject are given by Walker and Light.[148]

A simpler and much more straightforward approach is to consider the far asymptotic solution (when the R^{-2} terms are also neglected) which can be shown to be:

$$\lim_{R\to\infty} \zeta^{JM}_{nj\Omega}(R) = \frac{1}{k_{nj}}\left[\sin\left(k_{nj}R - (J+j)\frac{\pi}{2}\right)\delta_{nn_o}\delta_{jj_o}\delta_{\Omega\Omega_o} + K(nj\Omega|n_oj_o\Omega_o|JM)\cos\left(K_{nj}R - (j+J)\frac{\pi}{2}\right)\right] \tag{290}$$

where (n_o,j_o,Ω_o) stands for some initial state. Equation 290 is for an open state. A similar expression exists for the closed states (see also Equation 78). The coefficient $\mathbf{K}(nj\Omega|n_oj_o\Omega_o|JM)$ stands for the corresponding **K** matrix elements. Later we elaborate on these matrix elements and their relations to the **S** and **T** matrixes.

3. The $\lambda \rightarrow \nu$ (Rearrangement) Transformation

One of the features which makes exchange collisions different for other kinds of collisions is the existence of more than one arrangement channel. The transformation from one arrangement to the other was a major subject during 1970s. It is assumed that the transformation is best done in the BF system of coordinates. This is also the main reason why all quantum mechanical reactive collision studies were carried out in this framework.

To perform the transformation we have to study the relation between the various sets of coordinates which are assigned to the different arrangements. Equation 253 yields the relation between the coordinates of the λ and the ν channels. From this we obtain[140]

$$R_\nu^2 = \cos^2\beta\, R_\lambda^2 + \sin^2\beta\, r_\lambda^2 - \sin2\beta\, r_\lambda R_\lambda \cos\gamma_\lambda \tag{291a}$$

$$r_\nu^2 = \sin^2\beta\, R_\lambda^2 + \cos^2\beta\, R_\lambda^2 + \sin2\beta\, r_\lambda R_\lambda \cos\gamma_\lambda \tag{291b}$$

$$r_\nu^2 + R_\nu^2 = r_\lambda^2 + R_\lambda^2 \tag{291c}$$

$$\cos\gamma_\nu = \frac{1}{R_\nu r_\nu}\left[{}^1\!/_2 \sin2\beta\,(R_\lambda^2 - r_\lambda^2) + \cos2\beta\,\cos\gamma_\lambda\, R_\lambda r_\lambda\right] \tag{291d}$$

An expression is needed for the important angle $\Delta_{\lambda\nu}$ which is the angle between $\mathbf{R}_\lambda$ and $\mathbf{R}_\nu$ (see Figure 18):

$$\cos\Delta_{\lambda\nu} = \mathbf{R}_\lambda \cdot \mathbf{R}_\nu = \cos\beta\,\frac{R_\lambda}{r_\lambda} - \sin\beta\,\cos\gamma_\lambda\,\frac{r_\lambda}{R_\lambda} \tag{291e}$$

Like in the collinear case where the transformation from one arrangement to the other is carried out along a line in the interaction region (see Equation 70), here also the matching is carried out in the interaction region, but over a surface. A surface of this kind can be generated by imposing a linear relation between r_λ and r_ν, i.e.,

$$r_\nu = Br_\lambda \tag{292}$$

The main physical feature of this surface is that any continuous path which starts at some point in the λ-arrangement and ends in the ν-arrangements has to cross this surface. Since the matching of the wavefunctions of the two arrangements is done on this surface, we also list the relations between the two sets of coordinates as calculated on this surface.

The equation of the surface in the λ-coordinates is found by replacing r_ν by Br_λ in Equation 291b.[152-153] Thus,

$$R_\lambda^2 \sin^2\beta + r_\lambda^2(\cos^2\beta - B^2) + r_\lambda R_\lambda \cos\gamma_\lambda\, \sin2\beta = 0 \tag{293}$$

The nice feature of this surface is that if $\phi_{\lambda 0}$ is defined as

$$\cot\phi_{\lambda 0} = r_\lambda / R_\lambda \tag{294}$$

then Equation 293 becomes

$$\cot\phi_{\lambda 0} = \frac{\sin\beta}{B^2 - \cos^2\beta}\left(\cos\beta\cos\gamma_\lambda + \sqrt{B^2 - \sin^2\gamma_\lambda \cos^2\beta}\right) \tag{295}$$

With these definitions in hand we can immediately also express γ_ν and $\Delta_{\lambda\nu}$:

$$\cos\gamma_\nu = -\frac{\cos\gamma_\lambda + (1 - B^2)\cot\phi_{\lambda 0}\cot\beta}{B(1 + (1 - B^2)\cot^2\phi_{\lambda 0})^{1/2}} \tag{296a}$$

$$\cos\Delta_{\lambda\nu} = \frac{\cos\beta - \cos\gamma_\lambda \sin\beta\ \cot\phi_{\lambda 0}}{B(1 + (1 - B^2)\cot^2\phi_{\lambda 0})^{1/2}} \tag{296b}$$

As a final expression we would like to give the general relation between γ_λ, γ_ν, and B which is[153]

$$\begin{aligned} F(\gamma_\lambda,\gamma_\nu,B) &= \cos\gamma_\lambda(B^2 - \cos^2\beta\ \sin^2\gamma_\lambda)^{1/2} \\ &+ B\cos\gamma_\nu(1 - B^2\cos^2\beta\ \sin^2\gamma_\nu)^{1/2} \\ &+ \cos\beta(\sin^2\gamma_\lambda - B^2\sin^2\gamma_\nu) = 0 \end{aligned} \tag{297}$$

The other subject that is closely connected with the transformation has to do with the representation of the wavefunctions in the two arrangement channels. We are mainly concerned with the transformation involved to obtain the ν-representation of the wavefunction, assuming the λ-representation is given.

Since J and M are good quantum numbers in any of the arrangements, it can be assumed without losing the generality of the problem that

$$\psi^{JM}(\mathbf{R}_\lambda,\mathbf{r}_\lambda) = \psi^{JM}(\mathbf{R}_\nu,\mathbf{r}_\nu) \tag{298}$$

where ($\mathbf{R}_\lambda$, $\mathbf{r}_\lambda$) and ($\mathbf{R}_\nu$, $\mathbf{r}_\nu$) are related to each other as given in Equation 253. Employing Equation 272, we obtain:

$$\begin{aligned} &\sum_{\Omega_\lambda=-J}^{J} \sum_{j_\lambda>|\Omega_\lambda|} D^J_{\Omega_\lambda M}(\theta_\lambda,\phi_\lambda,\delta_\lambda) y_{j_\lambda\Omega_\lambda}(\gamma_\lambda,0) G^{JM}_{j_\lambda\Omega_\lambda}(R_\lambda,r_\lambda) = \\ &\sum_{\Omega_\nu=-J}^{J} \sum_{j_\nu>|\Omega_\lambda|} D^J_{\Omega_\nu M}(\theta_\nu,\phi_\nu,\delta_\nu) y_{j_\nu\Omega_\nu}(\gamma_\nu,0) G^{JM}_{j_\nu\Omega_\nu}(R_\nu,r_\nu) \end{aligned} \tag{299}$$

The expressions on each side of Equation 299 are not seen to be identical to that in Equation 272; but they are the same due to the relation:

$$D^J_{\Omega M}(\theta,\phi,0) y_{j\Omega}(\gamma,\delta) = D^J_{\Omega M}(\theta,\phi,\delta) y_{j\Omega}(\gamma,0) \tag{300}$$

To simplify the treatment we introduce an internal wavefunction:

$$\overline{G}^{JM}_{\Omega}(r,R,\gamma) = \sum_{j>|\Omega|} y_{j\Omega}(\gamma,0) G^{JM}_{j\Omega}(r,R) \tag{301}$$

so that Equation 299 becomes:

$$\sum_{\Omega_\lambda=-J}^{J} D^J_{\Omega_\lambda M}(\theta_\lambda,\phi_\lambda,\delta_\lambda) \overline{G}^{JM}_{\Omega_\lambda}(r_\lambda,R_\lambda,\gamma_\lambda) = \sum_{\Omega_\nu=-J}^{J} D^J_{\Omega_\nu M}(\theta_\nu,\phi_\nu,\delta_\nu) \overline{G}^{JM}_{\Omega_\nu}(r_\nu,R_\nu,\gamma_\nu) \tag{302}$$

In order to determine the relation between (θ_λ,ϕ_λ,δ_λ) and (θ_ν,ϕ_ν,δ_ν) it is important to remember that the angles (θ_λ,ϕ_λ,δ_λ) are defined with respect to $\mathbf{R}_\lambda$ and the angles (θ_ν,ϕ_ν,γ_ν)

are defined with respect to $\mathbf{R}_\nu$. Since the angle between $\mathbf{R}_\lambda$ and $\mathbf{R}_\nu$ is $\Delta_{\lambda\nu}$ (see Figure 18) it can be seen that

$$(\theta_\nu,\phi_\nu,\delta_\nu) \equiv (\theta_\lambda + \Delta_{\lambda\nu}, \ell_\lambda, \delta_\lambda) \tag{303}$$

and consequently there exists:

$$\mathbf{D}^J(\theta_\nu,\phi_\nu,\delta_\nu) = \mathbf{D}^J(\theta_\lambda,\phi_\lambda,\delta_\lambda)\mathbf{d}^J(\Delta_{\lambda\nu}) \tag{304}$$

where $\mathbf{D}^J$ are matrixes with the elements $D^J_{\Omega M}$ and $\mathbf{d}^J$ is the "small" d matrix defined as

$$\mathbf{d}^J(\Delta_{\lambda\nu}) = \mathbf{D}^J(0,\Delta_{\lambda\nu},0) \tag{305}$$

Substituting Equation 304 in Equation 302, one finds the relation:

$$G^J_{\Omega_\nu M}(r_\nu,R_\nu,\gamma_\nu) = \sum_{\Omega_\lambda} d^J_{\Omega_\nu\Omega_\lambda}(\Delta_{\lambda\nu})G^J_{\Omega_\lambda M}(r_\lambda,R_\lambda,\gamma_\lambda) \tag{306}$$

or in matrix notation:

$$\mathbf{G}^J_\nu = \mathbf{d}^J_{\nu\lambda}\mathbf{G}^J_\lambda \tag{307}$$

While considering this expression it is important to remember that $\Delta_{\lambda\nu}$ is only dependent on λ-coordinates as given by Equation 291e or, if needed, on the matching surface, as given by Equation 296b.

Before ending this subject we refer specifically to the matching equations. Equation 307, when applied on the matching surface, is the first matching equation. The second is the equation which relates the normal derivatives of the $\mathbf{G}^J_\alpha$ functions on the matching surface. Thus, if $\mathbf{n}_{\lambda\nu}$ is a unit vector along the normal to the surface, then the second matching equation is

$$\frac{\partial}{\partial \mathbf{n}_{\lambda\nu}}\mathbf{G}^J_\nu = \frac{\partial}{\partial \mathbf{n}_{\lambda\nu}}(\mathbf{d}^J_{\nu\lambda}\mathbf{G}^J_\lambda) \tag{308}$$

For the special case where B = 1, Equation 308 can be shown to reduce to:[140]

$$\frac{\partial}{\partial \mathbf{n}_{\lambda\nu}}\mathbf{G}^J_\nu = \mathbf{d}^J_{\nu\lambda}\frac{\partial}{\partial \mathbf{n}_{\lambda\nu}}\mathbf{G}^J_\lambda \tag{309}$$

The relations given in Equations 307, 308, or 309 have to be fulfilled for any two arrangement channels. Thus, in the three-atom case, three sets of equations of this type have to be satisfied.

4. Reactive, Differential, and Integral Cross-Sections

The numerical treatments of reactive collisions have so far been carried out in the BF system only. Therefore, the discussion in this section will be limited to the case.

In Section III.C.2, the derivation of the $\mathbf{K}^{JM}$ matrix elements was presented. Like in the collinear case (see Equation 79), the relation between $\mathbf{K}^{JM}$ and $\mathbf{S}^{JM}$ is given by the Heitler Damping relation, i.e.,[46]

$$\mathbf{S}^{JM} = \frac{1 + i\mathbf{K}^{JM}}{1 - i\mathbf{K}^{JM}} \tag{310}$$

In this notation, $\mathbf{S}^{JM}$ and $\mathbf{K}^{JM}$ still have two indexes J and M, but since it can be shown that M has to be equal to the initial m_λ (λ is the initial channel), from now on each of the matrixes will be designated by J only. To apply the Heitler relation one is forced to calculate a full $\mathbf{K}^J$ matrix which means solving for initial conditions in all open states of the three arrangement channels. The $\mathbf{K}^J$ and $\mathbf{S}^J$ matrixes have the following structure:

$\lambda \rightarrow \lambda$	$\nu \rightarrow \lambda$	$\kappa \rightarrow \lambda$
$\lambda \rightarrow \nu$	$\nu \rightarrow \nu$	$\kappa \rightarrow \nu$
$\lambda \rightarrow \kappa$	$\nu \rightarrow \kappa$	$\kappa \rightarrow \kappa$

where the notation $\alpha \longrightarrow \beta$ means a transition from a given initial state in the α-arrangement to a final open state in the β-arrangement. Once the $\mathbf{S}^J$ matrix elements are calculated, the rest of the treatment is similar to that given for the nonreactive case. Although most of the asymptotic analysis was carried out by applying Pack's procedure,[146] the final steps in this treatment will be according to the procedure of Jacob and Wick,[147] who employed the helicity representation in which the BF z-axis is chosen to be along the translational wave vector and not along the translational radius vector. Now if the SF z-axis is chosen so that it is in the opposite direction from the initial wavevector, then we have:

$$(\theta_r,\phi_r)\big|_{\text{initial}} \equiv (\gamma\ \delta)\big|_{\text{initial}}$$

and consequently the asymptotic physical wavefunction takes the form:

$$\psi^{\alpha}_{n_\lambda j_\lambda m_\lambda}(\mathbf{r}_\alpha,\mathbf{R}_\alpha) = \frac{1}{R_\lambda r_\lambda} e^{ik_{n_\lambda j_\lambda}R_\lambda}\phi_{n_\lambda j_\lambda}(r_\lambda) y_{j_\lambda m_\lambda}(0_{r_\lambda},\phi_{r_\lambda})\delta'_{\alpha\lambda}$$
$$+ \sum_{n_\alpha j_\alpha m_\alpha} \frac{1}{r_\alpha} f(n_\alpha,j_\alpha,m_\alpha \leftarrow n_\lambda,j_\lambda,m_\lambda|\theta_\alpha,\phi_\alpha) y_{j_\alpha m_\alpha}(\gamma_\alpha,\delta_\alpha)\phi_{n_\alpha j_\alpha}(r_\alpha)\frac{e^{ik_{n_\alpha j_\alpha}R_\alpha}}{R_\alpha} \tag{311}$$

where f are the scattering amplitudes, and $\delta'_{\alpha\lambda}$ stands for

$$\delta'_{\alpha\lambda} = \delta_{\alpha\lambda}\delta_{n_\alpha n_\lambda}\delta_{j_\alpha}\delta_{j_\lambda}\delta_{m_\alpha m_\lambda} \tag{312}$$

Following the usual procedures (as in the nonreactive case) it can be shown that the relation between the **S** matrix elements and the f values is

$$f(n_\alpha,j_\alpha,m_\alpha \leftarrow n_\lambda,j_\lambda,m_\lambda|\theta_\alpha,\phi_\alpha) = \frac{i^{j_\lambda - j_\alpha + 1}}{2k_{n_\lambda j_\lambda}} e^{im_\alpha\phi_\alpha}$$
$$\sum_J (2J + 1) d^J_{m_\alpha m_\lambda}(\theta_\alpha) T(n_\alpha j_\alpha m_\alpha|n_\lambda j_\lambda m_\lambda|J) \tag{313}$$

where

$$T(n_\alpha j_\alpha m_\alpha|n_\lambda j_\lambda m_\lambda|J) = \delta'_{\alpha\lambda} - S(n_\alpha j_\alpha m_\alpha|n_\lambda j_\lambda m_\lambda|J) \tag{314}$$

It can be seen that the quantum numbers that characterize the S-matrix element as given in Equation 314 are identical to those that characterize the K-matrix elements as given in Equation 290 if and only if $M = -\Omega_0 = m_\lambda$ and $\Omega = m_\alpha$. In Equation 313 the symbol $K_{n_\lambda j_\lambda}$ stands for the unscaled wave number.

The differential cross-section is given in the form:[154]

$$\frac{d\sigma}{d\theta} = \pi|f|^2 \tag{315}$$

whereas the integral cross-section, which is found by integrating over θ, becomes:

$$\sigma(n_\alpha j_\alpha m_\alpha | n_\lambda j_\lambda m_\lambda) = \frac{\pi}{k^2_{n_\lambda j_\lambda}} \sum_{J=O}^{\infty} (2J + 1)|T(n_\alpha j_\alpha m_\alpha | n_\lambda j_\lambda m_\lambda | J)|^2 \tag{316}$$

Both $d\sigma/d\theta$ and σ may be averaged over initial m_λ and summarized over m_α to give the degeneracy-averaged corresponding differential and integral cross-sections; σ may be written as:

$$\sigma(n_\alpha j_\alpha \leftarrow n_\lambda j_\lambda) = \frac{\pi}{k^2_{n_\lambda j_\lambda}} \sum_J (2J + 1)P(n_\alpha, j_\alpha | n_\lambda j_\lambda | J) \tag{317}$$

where

$$P(n_\alpha j_\alpha | n_\lambda j_\lambda) = \frac{1}{2j_\lambda + 1} \sum_{m_\alpha} \sum_{m_\lambda} | T(n_\alpha j_\alpha m_\alpha | n_\lambda j_\lambda m_\lambda | J)|^2 \tag{318}$$

Here it is understood that $|m_\alpha| \leqslant \min(J, j_\alpha)$ and $|m_\lambda| \leqslant \min(J, j_\lambda)$.

D. Numerical Results

As discussed elsewhere, the methods available for obtaining exact 3D results are very inefficient and therefore were applied to the low-energy region of the $H + H_2$ system only.[139,141,159] In this section a few selected results are presented with the aim of showing the similarities and the differences between classical (CL) and quantum mechanical (QM) results.

In Table 1 several (distinguishable partial) total QM integral cross-sections for the reactions:

$$H' + H_2(\nu_i = 0, j_i = 0) \rightarrow H'H + H \tag{319}$$

are shown as a function of total energy. The results are compared with the corresponding CL results; thus, the SK[139] and EW[159] calculations that were carried out on the Porter Karplus surface[156] are compared with the corresponding KPS CL result,[155] and the WSL QM results[141] obtained for the Siegbahn-Liu-Truhlar-Horowitz potential[157-158] are compared with the corresponding CL ones of Mayne and Toennies.[160] The CL and the QM values overlap reasonably well in the higher energy region, but differ appreciably in the lower region. This is due to tunneling effects which are not incorporated in the CL treatment.

Reactive rate constants are shown as a function of T^{-1} in Figure 19. Here the differences between the QM and the CL results are quite significant over the entire temperature range, emphasizing the importance of tunneling for this physical magnitude. In the low temperature region the deviations are of a few orders of magnitude, but they become smaller as the temperature is increased. An interesting feature is the fact that the CL rate constant curve

Table 1
QUANTUM MECHANICAL CROSS SECTIONS (a_0^2) FOR THE H + H_2 (v = 0, j = 0) REACTION CALCULATED EMPLOYING THE THREE AVAILABLE (QUANTUM) METHODS; THE CORRESPONDING CLASSICAL RESULTS ARE GIVEN FOR COMPARISON

	PK[156] potential			SLTH[157,158] potential	
		Quantum			
E_t (eV)	Classical KPS[155]	SK[139]	EW[159]	Classical MT[160]	Quantum WSL[141]
0.45	0.0	1.8(−4)[a]	—	0.0	1.2(−5)
0.50	0.0	5.0(−3)	—	0.0	2.8(−4)
0.55	0.0	6.0(−2)	—	0.0	4.6(−3)
0.58	—	—	—	8.0(−3)	—
0.60	0.4	0.35	0.4	—	0.053
0.63	—	—	—	0.21	—
0.65	0.9	0.9	1.44	—	0.30
0.70	1.4	1.5	2.08	0.71	—

[a] Numbers in parentheses stand for powers of ten.

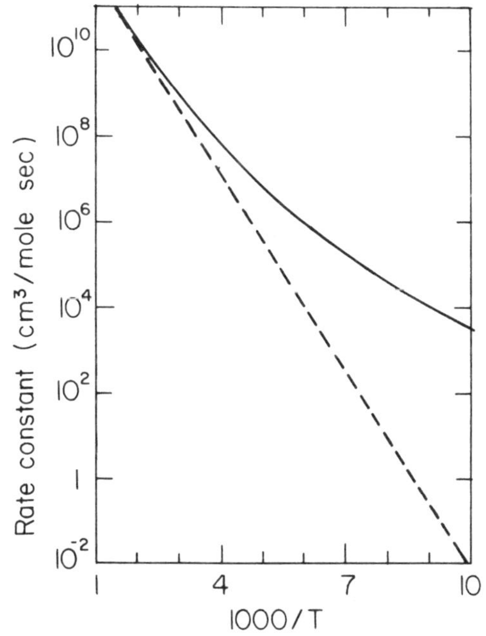

FIGURE 19. Arrhenius plot for the (distinguishable) H + H_2 reaction (———QM results due to SK[139]---CL results due to KPS[155]).

is linear whereas the QM one is nonlinear, a feature extensively discussed by Bell and others already in the early 1930s (see Section II.C.1).

IV. SUMMARY

This review was divided into two main parts: one that deals with the collinear system and the other that treats the system in its full dimensionality. The research is more or less evenly distributed between the two subjects. However, during the last few years there has been a significant shift in interest with most of the effort now being directed towards the 3D system, and this includes both the *exact* and the *approximate* treatments. The collinear arrangement serves only as test case for new computational methods and is no longer a subject for research per se.

ACKNOWLEDGMENT

I am most grateful to Ms. R. Plottel for editing this chapter, Mrs. S. Saphier for drawing the figures, and to Mrs. L. Wolff for typing the manuscript.

REFERENCES

1. **Baer, M.,** A review of quantum mechanical approximate treatments of three-body reactive systems, in *Advances of Chemical Physics,* Vol. 49. Prigogine, I. and Rice, S. A., Eds., John Wiley & Sons, N.Y., 1982, 191.
2. **Eyring, H. and Rolanyi, M.,** Uber einfache Gasreaktionen, *Z. Physikal. Chem.,* B12, 279, 1932.
3. **Pelzer, H. and Wigner, E.,** Uber die Geschwindigkeitskonstante von Austausreaktionen, *Z. Physikal. Chem. B,* 15, 445, 1932.
4. **Rosen, N.,** Lifetimes of unstable molecules, *J. Chem. Phys.,* 1, 319, 1933.
5. **Whittaker, E.,** *A Treatise on the Analytical Dynamics of Particles and Rigid Bodies,* 4th ed., Cambridge University Press, 1937.
6. **Smith, F. T.,** Participation of vibration in exchange reactions, *J. Chem. Phys.,* 32, 1352, 1959.
7. **Delves, L. M.,** Tertiary and general order collisions, *Nucl. Phys.,* 9, 391, 1959.
8. **Langer, R. M.,** The quantum mechanics of chemical reactions, *Phys. Rev.,* 34, 92, 1929.
9. **Eyring, H.,** The energy of activation for bimolecular reactions involving hydrogen and halogens, according to the quantum mechanics, *J. Am. Chem. Soc.,* 53, 2537, 1931.
10. **Evans, M. G. and Polanyi, M.,** Inertia and driving force of chemical reactions, *Trans. Faraday Soc.,* 34, 11, 1938.
11. **Eyring, H.,** The calculation of activation energies, *Trans. Faraday Soc.,* 34, 3, 1938.
12. **Wigner, E. P.,** The transition state method, *Trans. Faraday Soc.,* 34, 29, 1938.
13. **Hirschfelder, J. O. and Wigner, E. P.,** Some quantum mechanical considerations in the theory of reactions involving an activation energy, *J. Chem. Phys.,* 7, 616, 1939.
14. **Hofacker, G. L.,** Quantentheorie Chemischer Reaktionen, *Natarforschung,* 189, 607, 1963.
15. **Marcus, R. A.,** On the analytical mechanics of chemical reactions. Quantum mechanics of linear collisions, *J. Chem. Phys.,* 45, 4493, 1966.
16. **Marcus, R. A.,** On the analytical mechanics of chemical reactions. Classical mechanics of linear collisions, *J. Chem. Phys.,* 45, 4500, 1966.
17. **Gorben, H. C. and Stehle, P.,** *Classical Mechanics,* 2nd ed., John Wiley & Sons, N.Y., 1960, 319.
18. **Podolsky, B.,** Quantum mechanically correct form of hamiltonian function for conservative system, *Phys. Rev.,* 32, 812, 1928.
19. **Rankin, C. C. and Light, J. C.,** Quantum solution of collinear reactive systems: $H + Cl_2 \rightarrow HCl + Cl$, *J. Chem. Phys.,* 51, 1701, 1969.
20. **Baer, M. and Beswick, J. A.,** A theoretical study of exothermic reactions, *Chem. Phys.,* 21, 443, 1977.
21. **Basilevsky, M. V.,** Qualitative interpretation of the dynamics of collinear exchange reactions, *Chem. Phys.,* 12, 315, 1976.
22. **Wyatt, R. E.,** Quantum mechanics of the $H + H_2$ reaction: investigation of vibrational diabatic models, *J. Chem. Phys.,* 51, 3489, 1969.
23. **Wu, S-F. and Levine, R. D.,** Quantum mechanical computational studies of chemical reactions. I. Close coupling method for the collinear $H + H_2$ reaction, *Mol. Phys.,* 22, 881, 1971.
24. **Child, M. S.,** *Molecular Collisions Theory,* Academic Press, London, 1974.
25. **Madden, P. A. and Murrel, J. N.,** Quantum mechanical calculations on collinear reactive collisions over potential wells, *Mol. Phys.,* 31, 1643, 1976.

26. **Adams, J. T., Smith, R. L., and Hays, R. F.,** Integral equation approach to collinear reactive scattering: $A + BC \rightarrow AB + C$, *J. Chem. Phys.*, 61, 2193, 1974.
27. **Light, J. C. and Walker, R. B.,** Hermitian quantum equations for scattering in reaction coordinates, *J. Chem. Phys.*, 65, 1598, 1976.
28. **Latham, S. L., McNutt, J. F., Wyatt, R. E., and Redmon, M. J.,** Quantum dynamics of the $F + H_2$ reaction: resonance models and energy and flux distributions in the transition state, *J. Chem. Phys.*, 69, 3746, 1978.
29. **Smith, F. T.,** Diabatic and adiabatic representations for atom collision problems, *Phys. Rev.*, 179, 111, 1969.
30. **Baer, M.,** Adiabatic and diabatic representations for atom-molecule collisions: treatment of the collinear arrangement, *Chem. Phys. Lett.*, 35, 112, 1975.
31. **Baer, M.,** Adiabatic and diabatic representations for atom-diatom collisions: treatment of the three-dimensional case, *Chem. Phys.*, 15, 49, 1976.
32. **Baer, M., Drolshagen, G., and Toennies, J. P.,** The adiabatic-diabatic approach to vibrational inelastic scattering. I. Theory and study of simple collinear model, *J. Chem. Phys.*, 73, 1690, 1980.
33. **Wu, S-F, Johnson, B. R., and Levine, R. D.,** Quantum mechanical computational studies of chemical reaction. II. Isotope exchange reactions for collinear $H + H_2$ system, *Mol. Phys.*, 25, 609, 1973; III. Collinear $A + BC$ reactions with some model potential energy surfaces, *Mol. Phys.*, 25, 839, 1973.
34. **Saxon, R. P. and Light, J. C.,** Quantum calculations of planar reactive $H + H_2$. I. Theory, *J. Chem. Phys.*, 56, 3874, 1972; II. Application, *J. Chem. Phys.*, 56, 3885, 1972.
35. **Elkovitz, A. B. and Wyatt, R. E.,** Three dimensional natural coordinate asymmetric top theory of reactions: application to $H + H_2$, *J. Chem. Phys.*, 63, 702, 1975.
36. **Johnson, B. R.,** Reaction-path technique in close coupling computations: $H + H_2$, *Chem. Phys. Lett.*, 13, 172, 1972.
37. **Baer, M.,** A collinear quantum mechanical treatment of the heavy-light-heavy mass combination: $Cl + HBr \rightarrow HCl + Br$, *J. Chem. Phys.*, 62, 305, 1975.
38. **Hauke, G., Manz, J., and Romelt, J.,** Collinear triatomic reactions described by Delves Coordinates, *J. Chem. Phys.*, 73, 5040, 1980.
39. **Kuppermann, A., Kay, J. A., and Dwyer, H. P.,** Hypershperical coordinates in quantum mechanical collinear reactive scattering, *Chem. Phys. Lett.*, 74, 257, 1980.
40. **Manz, J. and Romelt, J.,** Dissociative collinear reactions evaluated by S-matrix propagation along Delve's radial coordinates, *Chem. Phys. Lett.*, 77, 172, 1981.
41. **Kaye, J. A. and Kuppermann, A.,** Quantum mechanical coupled channel collision-induced dissociation calculations with hyperspherical coordinates, *Chem. Phys. Lett.*, 78, 546, 1981.
42. **Kaye, J. A. and Kuppermann, A.,** Collinear quantum mechanical probabilities and rate constants for the $Br + HCl$ ($\nu = 2,3,4$) reaction using hyperspherical coordinates, *Chem. Phys. Lett.*, 92, 574, 1982.
43. **Pollak, E.,** Classical analysis of collinear light atom transfer reactions, *J. Chem. Phys.*, 78, 1228, 1983.
44. **Manz, J., Meyer, R., Pollak, E., and Romelt, J.,** A new possibility of chemical bonding: vibration stabilization of IHI, *Chem. Phys. Lett.*, 93, 184, 1982.
45. **Middleton, P. B. and Wyatt, R. E.,** Quantum mechanical study of a reaction path bifurcation model, *Chem. Phys. Lett.*, 21, 57, 1973.
46. **Mott, N. F. and Massey, H. S. W.,** *The Theory of Atomic Collisions*, 3rd ed., Clarendon Press, Oxford, 1965, 391.
47. **Light, J. C. and Walker, R. B.,** An R-matrix approach to the solution of coupled equations for atom molecule reactive scattering, *J. Chem. Phys.*, 65, 4272, 1976.
48. **Hildebrand, F. B.,** *Introduction to Numerical Analysis*, McGraw-Hill, N.Y., 1956, chap. 5.
49. **Truhlar, D. G.,** Finite difference value method for solving one dimensional eigenvalue equations, *J. Comp. Phys.*, 10, 123, 1972.
50. **Neilsen, W. B. and Gordon, R. G.,** On a semiclassical study of molecular collisions. I. General method, *J. Chem. Phys.*, 58, 4131, 1973.
51. **Miller, G. and Light, J. C.,** Quantum calculations of collinear reactive triatomic systems. II. Theory, *J. Chem. Phys.*, 54, 1635, 1971.
52. **Billing, G. D. and Baer, M.,** A propagator method for integration of classical trajectory equations, *Chem. Phys. Lett.*, 48, 372, 1977.
53. **Gordon, R. G.,** New method for constructing wave functions for bound states and scattering, *J. Chem. Phys.*, 51, 14, 1969.
54. **Manz, J.,** The relation of chemical potentials and reactivity studied by state path sum. I. Formal theory, *Mol. Phys.*, 28, 399, 1974.
55. **Diestler, D. J. and McKoy, V.** Quantum mechanical treatment of inelastic collisions. II. Exchange reactions, *J. Chem. Phys.*, 48, 2951, 1968.

56. **Askar, A., Cakmak, A. S., and Rabitz, H. A.,** Finite element methods for reactive scattering, *Chem. Phys.*, 33, 267, 1978.
57. **Baur, E. and Wu, T. Y.,** A quantum mechanical calculation of the rates of some chemical reactions, *J. Chem. Phys.*, 21, 726, 1953.
58. **Eckart, C.,** The penetration of a potential barrier by electrons, *Phys. Rev.*, 35, 1303, 1930.
59. **Bell, R. P.,** The application of quantum mechanics to chemical kinetics, *Proc. R. Soc. London Ser. A*, 139, 466, 1933.
60. **Wigner, E.,** Uber das Uberschreiten von potentialschwellen bei chemische Reaktionen, *Z. Phys. Chem. B*, 19, 203, 1933.
61. **Jeffreys, H.,** On certain approximate solutions of linear differential equations of the second order, *Proc. Math. Soc.*, 23, 428, 1927.
62. **Bell, R. P.,** Quantum mechanical effects in reactions involving hydrogen, *Proc. R. Soc. London Ser. A*, 148, 24, 1935.
63. **Bell, R. P.,** The tunnel effect correction for parabolic potential barriers, *Trans. Faraday Soc.*, 55, 1, 1959.
64. **Shavitt, I.,** A calculation of the rates of the ortho-Para conversions and isotope exchanges in hydrogen, *J. Chem. Phys.*, 31, 1359, 1959.
65. **Boys, S. F. and Shavitt, I.,** *Univ. of Wisconsin Naval Res. Lab. Tech. Rep.* WIS F-13, 1959.
66. **Weston, W. E. Jr.,** H_3 activated complex and the rate of reaction of hydrogen atoms with hydrogen molecules, *J. Chem. Phys.*, 31, 892, 1959.
67. **Shavitt, I.,** Correlation of experimental rate constants of the hydrogen exchange reactions with a theoretical H_3 potential surface, using transition state theory, *J. Chem. Phys.*,49, 4048, 1968.
68. **Shavitt, I., Stevens, R. M., Minn, F. L., and Karplus, M.,** Potential energy surface for H_3, *J. Chem. Phys.*, 48, 2700, 1968.
69. **Truhlar, D. G. and Kuppermann, A.,** Exact tunneling calculations, *J. Am. Chem. Soc.*, 93, 1840, 1971.
70. **Pollak, E.,** A classical determination of vibrationally adiabatic barriers and wells of a collinear potential energy surface, *J. Chem. Phys.*,74, 5586, 1981.
71. **Pollak, E.,** Isotope effects in the hydrogen exchange reaction, *Chem. Phys. Lett.*, 80, 45, 1981.
72. **Jellinek, J. and Pollak, E.,** An adiabatic analysis of the reactive infinite order sudden approximation, *J. Chem. Phys.*, 78, 3014, 1983.
73. **Pollak, E. and Wyatt, R. E.,** Semiclassical determination of adiabatic barriers on a three dimensional potential energy surface, *J. Chem. Phys.*, in press.
74. **Johnson, H. S. and Rapp, D.,** Large tunneling corrections in chemical reactions rates. II, *J. Am. Chem. Soc.*, 83, 1, 1961.
75. **Child, M. S.,** Measurable consequences of a dip in the activation barrier for an adiabatic chemical reaction, *Mol. Phys.*, 12, 401, 1967.
76. **Walker, R. B. and Hayes, E. F.,** Theoretical analysis of the quantum contributions to the reactions: $H_2(\nu = 1) + H \rightarrow H + H_2(\nu' = 0,1)$ and $H_2(\nu = 1) + D \rightarrow HD(\nu' = 0,1) + H$, *J. Phys. Chem.*, 87, 1255, 1983.
77. **Hulbert, H. H. and Hirschfelder, J. O.,** The transmission coefficient in the theory of absolute reaction rates, *J. Chem. Phys.*, 11, 276, 1943.
78. **Tang, K. T., Kleinman, B., and Karplus, M.,** Solvable quantum mechanical model for three body rearrangement scattering, *J. Chem. Phys.*, 50, 1119, 1969.
79. **Wilson, D. J.,** Quantum vibrational transition probabilities in atom diatomic molecule collsions. III. Reactive scattering, *J. Chem. Phys.*, 51, 5008, 1969.
80. **Baer, M. and Douri, D. J.,** Theory of reactive scattering. II. Application of the τ operator formalism to a linear model for three body rearrangement, *J. Chem. Phys.*, 56, 4840, 1972.
81. **Hirschfelder, J. O. and Tang, K. T.,** Quantum mechanical streamlines. III. Idealized reactive atom-diatom molecule collision, *J. Chem. Phys.*, 64, 760, 1976.
82. **Baer, M.,** Comparison between quantum mechanical and classical treatment of hard sphere model for collinear three-body rearrangement collisions, *J. Chem. Phys.*, 54, 3670, 1971.
83. **Mahan, B. M.,** Collinear collision chemistry. II. Energy disposition in reactive collisions, *J. Chem. Educ.*, 51, 377, 1974.
84. **Baer, M.,** unpublished results.
85. **Cross, M. and Korsch, H. J.,** Product state distribution in an idealized collinear reaction, *Chem. Phys. Lett.*, 28, 573, 1974.
86. **Beutler, H. and Polanyi, M.,** Uber hochverdunte Flamen. I. Flamen im einfachen Rohr. Vorlantige Analyse des Reaktions mechanismus. Reaktions geschwindigkeit, Leuchtvergang, *Z. Phys. Chem.*, *B*, 1, 3, 1928.
87. **Kuntz, P. J., Nemeth, E. M., Polanyi, J. C., Rosner, S. D., and Young, C. E.,** Energy distribution among products of exothermic reactions. II. Repulsive, mixed and attractive energy release, *J. Chem. Phys.*, 44, 1168, 1966.

88. **Anlauf, K. G., Kuntz, P. J., Maylotte, D. H., Pacey, P. D., and Polanyi, J. C.,** Energy distribution among reaction products. II. H + XY and X + HY, *Disc. Faraday Chem. Soc.*, 44, 83, 1967.
89. **Wu, S-F, Johnson, B. R., and Levine, R. D.,** Quantum mechanical computational studies of chemical reaction. III. Collinear A + BC reactions with some potential energy surfaces, *Mol. Phys.*, 25, 839, 1973.
90. **Baer, M.,** The isotope reactive systems H + Cl_2 and D + Cl_2. A quantum mechanical treatment of the collinear arrangement, *J. Chem. Phys.*, 60, 1057, 1974.
91. **Anlauf, K. G., Horne, D. S., MacDonald, R. G., Polanyi, J. C., and Woodal, K. B.,** Energy distribution among reaction products V : H + X_2 (X = Cl,Br), D + Cl_2, *J. Chem. Phys.*, 57, 1561, 1972.
92. **Connor, J. N. L., Jakubetz, W., and Manz, J.,** Exact quantum mechanical transition probabilities for the collinear reaction H + $F_2(\nu - 0) \rightarrow HF(\nu \leq 11)$ + F, *Chem. Phys. Lett.*, 39, 75, 1976.
93. **Hofacker, G. L. and Levine, R. D.,** A nonadiabatic model for population inversion in molecular collisions, *Chem. Phys. Lett.*, 9, 617, 1971.
94. **Hofacker, G. L. and Rosch, N.,** Semiclassical evaluation of the translational vibrational coupling from the potential energy surface of an atom diatom reaction, *Ber. Bunsenges. Phys. Chem.*, 77, 661, 1973.
95. **Hofacker, G. L. and Michel, K. W.,** Prediction of vibrational inversion among products of exothermic radical reactions by a classical model, *Ber. Bunsenges. Phys. Chem.*, 78, 174, 1974.
96. **Hofacker, G. L. and Levine, R. D.,** The evolution of entropy along the reaction path in an atom-diatom collision, *Chem. Phys. Lett.*, 33, 404, 1975.
97. **Haasel, K. D.,** A semiclassical theory of translational vibration coupling in nonadiabatic chemical reactions, *Ber. Bunsenges. Phys. Chem.*, 79, 285, 1975.
98. **Duff, J. W. and Truhlar, D. G.,** Effect of curvature of the reaction path as dynamic effects in endothermic chemical reactions and product energies in exothermic reactions, *J. Chem. Phys.*, 62, 2477, 1975.
99. **Basilevsky, M.,** Vibrational transitions induced by a chemical reaction: a quantum model for the strong coupling region, *Mol. Phys.*, 28, 617, 1974.
100. **Basilevsky, M. and Rayaboy, V. M.,** Quantum investigation of linear triatomic exchange reactions. A computational method, *Chem. Phys.*, 41, 461, 1979; Dynamics of linear exchange reactions. Quasiclassical model for high energy vibrational inversion, *Chem. Phys.*, 41, 477, 1979; Exact solution for the quasiclassical model of vibrational population inversion in exothermal exchange reactions, *Chem. Phys.*, 41, 484, 1979.
101. **Chrisholm, J. S. R. and Morris, R. M.,** *Mathematical Methods in Physics*, Vol. 2, Elsevier/North Holland, Amsterdam, 371, 1964.
102. **Micha, D. A.,** A quantum mechanical model for simple molecular reactions, *Ark. Fys.*, 30, 411, 1965; The exchange reaction of H and H_2, *Ark. Fys.*, 30, 425, 1965; Angular distribution of products of hydrogen atom-hydrogen molecule reactions, *Ark. Fys.*, 30, 437, 1965.
103. **Tang, K. T. and Karplus, M.,** Quantum theory of (H,H_2) scattering: approximate treatment of reactive scattering, *Phys. Rev.* A, 4, 1844, 1971.
104. **Choi, B. H. and Tang, K. T.,** Three-dimensional quantum mechanical studies of D + $H_2 \rightarrow$ HD + H reactive scattering, *J. Chem. Phys.*, 62, 3642, 1975.
105. **Shan, Y., Choi, B. H., Poe, R. T., and Tang, K. T.,** Three-dimensional quantum mechanical study of the F + H_2 reactive scattering, *Chem. Phys. Lett.*, 57, 379, 1978.
106. **Connor, J. N. L., Lagana, A., Whitehead, J. C., Jakubetz, W., and Manz, J.,** Uni and bimodel product energy distributions for the reactions H + $Cl_2(\nu = 1)$ and D + $Cl_2(\nu = 1)$, *Chem. Phys. Lett.*, 62, 479, 1979.
107. **Essen, H., Billing, G. D., and Baer, M.,** Comparison of quantum mechanical and quasiclassical calculations of collinear reaction rate constants for the H + Cl_2 and D + Cl_2 system, *Chem. Phys.*, 17, 443, 1976.
108. **Truhlar, D. G., Merrick, J. A., and Duff, J. W.,** Comparison of Trajectory calculations, transition state theory, quantum mechanical reaction probabilities and rate constants for the collinear reaction H + $Cl_2 \rightarrow$ HCl + Cl, *J. Am. Chem. Soc.*, 98, 677, 1971.
109. **Gray, J. C., Truhlar, D. G., and Baer, M.,** Test of trajectory calculations against quantum mechanical state-to-state and thermal collinear reaction rates for H + $Cl_2 \rightarrow$ HCl + Cl, *J. Phys. Chem.*, 85, 1045, 1979.
110. **Halavee, U. and Shapiro, M.,** A collinear analytic model for atom-diatom chemical reactions, *J. Chem. Phys.*, 64, 2826, 1976.
111. **Schatz, G. C. and Ross, J.,** Franck Condon factors in studies of dynamics of chemical reactions. I. General theory, application to collinear atom-diatom reactions, *J. Chem. Phys.*, 66, 1021, 1977.
112. **Fisher, S. and Venzel, G.,** On the dynamics of exothermic diatomic exchange reactions, *J. Chem. Phys.*, 67, 1335, 1977.
113. **Schatz, G. C. and Ross, J.,** Franck-Condon factors in studies of dynamics of chemical reactions. II. Vibration-rotation distributions in atom-diatom reactions, *J. Chem. Phys.*, 66, 1037, 1977.
114. **Wong, J. K. C. and Brumer, P.,** The Faddeev equations and Franck-Condon models for chemical reactions, *Chem. Phys. Lett.*, 68, 517, 1979.

115. **Child, M. S. and Whaley, K. B.,** Observation of a condon reflection products state distribution in the collinear H + Cl_2 reaction, *Faraday Disc. Chem. Soc.*, 67, 57, 1979.
116. **Billing, G. D., Eu, B. C., Garisto-Zaritsky, N., and Neyland, C.,** A stochastic-collision complex model theory of chemical reactions, *J. Chem. Phys.*, 73, 1627, 1980.
117. **Baer, M.,** Franck-Condon model for collinear reactive systems. Factorization of the reactive vibrational amplitudes and probabilities, *J. Phys. Chem.*, 85, 3974, 1981.
118. **Child, M. S.,** Measurable consequences of a dip in the activation barrier for an adiabatic chemical reaction, *Mol. Phys.*, 12, 401, 1962.
119. **Levine, R. D. and Wu, S-F.,** Resonances in reactive collisions: computational study of the H + H_2 collision, *Chem. Phys. Lett.*, 11, 557, 1971.
120. **Schatz, G. C. and Kuppermann, A.,** Role of direct and resonant (compound state) processes and of their interferences in the quantum dynamics of the collinear H + H_2 exchange reaction, *J. Chem. Phys.*, 59, 964, 1973.
121. **Breit, G. and Wigner, E.,** Capture of slow neutrons, *Phys. Rev.*, 49, 519, 1936.
122. **Adair, R. K.,** High energy maxima in the π − P cross section, *Phys. Rev.*, 113, 338, 1959.
123. **Kapur, P. I. and Peierls,** The dispersion formula for nuclear reaction, *Proc. R. Soc. London A.*, 166, 277, 1938.
124. **Feschbach, H.,** Unified theory of nuclear reactions, *Ann. Phys.*, 5, 357, 1958.
125. **Chapman, F. M. and Hayes, E. F.,** Quantum mechanical study of resonances effects in the collinear reaction H_2 + I → HI + H, *J. Chem. Phys.*, 66, 2554, 1977.
126. **Shoemaker, C. L. and Wyatt, R. E.,** Resonances in three-dimensional chemical reactions. I. Feshbach analysis and computational techniques, *J. Chem. Phys.*, 77, 4982, 1982.
127. **Shoemaker, C. L. and Wyatt, R. E.,** Resonances in three-dimensional chemical reactions. II. Applications to a model atom-diatom reaction, *J. Chem. Phys.*, 77, 4994, 1982.
128. **Pollak, E. and Wyatt, R. E.,** Semiclassical prediction of resonance energies in three-dimensional reactive collisions, *J. Chem. Phys.*, 77, 2689, 1982.
129. **Wigner, E. P. and Eisenbud, L.,** Higher angular momenta and long range interaction in resonance reactions, *Phys. Rev.*, 72, 29, 1947.
130. **Eisenbud, L.,** Ph.D. thesis, Princeton University, N.J., 1948.
131. **Wigner, E. P.,** Lower limit for the energy derivative of the scattering phase shift, *Phys. Rev.*, 98, 145, 1955.
132. **Bohm, D.,** *Quantum Theory*, Prentiss-Hall, Englewood Cliffs, 1951, 260.
133. **Smith, F. T.,** Lifetime matrix in collision theory, *Phys. Rev.*, 118, 349, 1960.
134. **Kuppermann, A. and Kay, J. A.,** Collision lifetime matrix analysis of the first resonance in the collinear F + H_2 reaction and its isotopically analogues, *J. Phys. Chem.*, 85, 1964, 1981.
135. **Redmon, M. J. and Wyatt, R. E.,** Quantum resonances structure in the three-dimensional F + H_2 reaction, *Chem. Phys. Lett.*, 63, 209, 1979; Wyatt, R. E. and Redmon, M. J., Quantum mechanical differential reaction cross sections for the F + $H_2(\nu = 0) \rightarrow HF(\nu = 2,3)$ + H reactions, *Chem. Phys. Lett.*, 96, 284, 1983.
136. **Jellinek, J., Baer, M., and Kouri, D. J.,** Quantum mechanical state-to-state differential cross sections for the reaction F + H_2 → H + HF, *Phys. Rev. Lett.*, 47, 1588, 1981.
137. **Baer, M., Jellinek, J., and Kouri, D. J.,** Quantum mechanical treatment of the F + H_2 → HF + H reaction, *J. Chem. Phys.*, 78, 2962, 1982.
138. **Wolken, G., Jr. and Karplus, M.,** Theoretical studies of H + H_2 reactive scattering, *J. Chem. Phys.*, 60, 351, 1974.
139. **Schatz, G. C. and Kuppermann, A.,** Quantum mechanical reactive scattering for three dimensional atom plus diatom systems. II. Accurate cross sections for H + H_2, *J. Chem. Phys.*, 65, 4668, 1976.
140. **Kupperman, A., Schatz, G. C., and Baer, M.,** Quantum mechanical reactive scattering: theory for planar atom plus diatom systems, *J. Chem. Phys.*, 65, 4596, 1976.
141. **Walker, R. B., Stechel, E. B., and Light, J. C.,** Accurate H_3 dynamics on an accurate H_3 surface, *J. Chem. Phys.*, 69, 2922, 1978.
142. **Walker, R. B., Light, J. C., and Altenberger-Siczek, A.,** Chemical reaction theory of asymmetric atom molecule collisions, *J. Chem. Phys.*, 64, 1166, 1976.
143. **Kouri, D. J.,** Rotational excitation. II. Approximation methods, in *Atom-Molecule Collision Theory. A Guide for the Experimentalist*, Bernstein, R. B., Ed., Plenum Press, N.Y., 1979, chap. 9.
144. **Arthurs, A. M. and Dalgarno, A.,** The theory of scattering by rigid rotator, *Proc. R. Soc. London A*, 256, 540, 1960.
145. **Curtiss, C. F., Hirschfelder, J. O., and Adler, F. T.,** The separation of the rotational coordinates from the N-particle Schrodinger equation, *J. Chem. Phys.*, 18, 1638, 1950.
146. **Pack, R. T.,** Space fixed vs body fixed axes in atom-diatomic molecule scattering. Sudden approximation, *J. Chem. Phys.*, 60, 633, 1974.

147. **Jacob, M. and Wick, G. C.,** On the general theory of collisions of particles with spin, *Ann. Phys. N.Y.*, 7, 404, 1959.
148. **Walker, R. B. and Light, J. C.,** Body-fixed quations for atom-molecule scattering: exact and centrifulgal decoupling methods, *Chem. Phys.*, 7, 84, 1975.
149. **Rose, M. E.,** *Elementary Theory of Angular Momentum*, John Wiley & Sons, N.Y., 1967.
150. **Schiff, L. L.,** *Quantum Mechanics*, McGraw-Hill, N.Y., 1968, chap. 4.
151. **Vezzetti, D. J. and Rubinow, S. I.,** Asymptotic solution of the Schrodinger equation for the three-body problem, *Ann. Phys.*, 35, 373, 1965.
152. **Khare, V., Kouri, D. J., and Baer, M.,** Infinite order sudden approximation for reactive scatering. I. Basic l-labeled formulation, *J. Chem. Phys.*, 71, 1188, 1979.
153. **Jellinek, J. and Baer, M.,** Infinite order sudden approximation for reactive scattering within classical mechanics. I. Theory, *J. Chem. Phys.*, 76, 4883, 1982.
154. **Schiff, L. L.,** *Quantum Mechanics*, McGraw-Hill, N.Y., 1968, chap. 5.
155. **Karplus, M., Porter, R. N., and Sharma, R. D.,** Exchange reactions with activation energy. I. Simple barrier potential for (H,H_2), *J. Chem. Phys.*, 43, 3259, 1965.
156. **Porter, R. N. and Karplus, M.,** Potential energy surface for H_3, *J. Chem. Phys.*, 40, 1105, 1964.
157. **Siegbahn, P. and Liu, B.,** An accurate three dimensional potential energy surface for H_3, *J. Chem. Phys.*, 68, 2457, 1978.
158. **Truhlar, D. G. and Horowitz, C. J.,** Functional representation of Liu and Siegbahn's accurate *ab initio* potential energy calculations for $H + H_2$, *J. Chem. Phys.*, 68, 2466, 1978; *J. Chem. Phys.*, 71, 1514, 1979.
159. **Elkowitz, A. B. and Wyatt, R. E.,** Quantum mechanical reaction cross sections for the three dimensional hydrogen exchange reaction, *J. Chem. Phys.*, 62, 2504, 1975.
160. **Mayne, H. R. and Toennies, J. P.,** Quasiclassical trajectory studies of the $H + H_2$ reaction on an accurate potential energy surface. III. Comparison of rate constants and cross sections with experiment, *J. Chem. Phys.*, 75, 1794, 1981.

Chapter 4

THE GENERAL THEORY OF REACTIVE SCATTERING: THE INTEGRAL EQUATION APPROACH

Donald J. Kouri

TABLE OF CONTENTS

I. INTRODUCTION

The use of integral equations for treating collision problems has a long history,[1-6] but it has only been relatively recently that numerical methods were developed which enabled their efficient noniterative solution.[7-13] The approach has been used successfully to treat a variety of inelastic (nonreactive) collision systems. However, the situation with reactive scattering is considerably more complicated for several reasons. First, the standard Lippmann-Schwinger (LS) integral equation is, in general, unsuitable for dealing with reactive systems due to difficulties associated with the implicit manner in which the various asymptotic boundary conditions enter that equation.[14-16] Second is the fact that the scattering coordinates appropriate in one arrangement are not convenient in other arrangements. Although one may circumvent this to some degree through the use of so-called ''natural collision coordinates'',[17] the form of the resulting integral equations is not as simple or convenient as the nonreactive LS equation and it is nontrivial to extend this to real three-dimensional (3D) reactive collisions.

A few years ago, an alternative formulation of the three-body problem was presented by Faddeev[14] which solved the fundamental formal problems with the LS equation for reactions by replacing the single integral equation by a system of coupled integral equations (CIE) (one for each possible arrangement of the system). This approach built on ideas due to Eyges[18] in which the three-body wavefunction was decomposed into components — each associated with a possible subclustering of the system. Subsequently, other sets of CIEs have been presented.[19-70] All these formalisms are examples of a general approach to quantum mechanics known as arrangement channel quantum mechanics (ACQM)[46,54,61,62,64,70] which has application not only to reactive scattering theory but also, in general, to bound-state problems[40,47,49,50,69] and to the quantum statistical mechanics of reactive fluids.[54,67] There are now several reviews which discuss most of the formalisms that are of current interest.[71-75] Of these formalisms, the ones which we believe are most easily utilized for chemical problems are the channel coupling array version of the ACQM equations and the Baer-Kouri-Levin-Tobocman (BKLT) coupled T equations. The former are essentially the differential equation form of the latter. The ACQM form of the equations (with the channel permuting array[23-25,30,32-35,36,38,41] choice of the channel coupling array) is currently being used successfully for reactive scattering by Shapiro and Top.[64]

However, as this chapter deals with integral equation methods, we shall not dwell on this method but rather concentrate primarily on the BKLT equation. In addition to the BKLT equations, the only other CIEs which have been utilized for chemical reaction systems are those of Faddeev.[14] This has been in work carried out by Micha and co-workers[76-82] and by Brumer and Shapiro.[83-84] Part of this work has been focused on high-energy collisions and has utilized perturbation theory. However, part of it has dealt with application of the Faddeev equations to the $H + H_2$ reaction in 3D.[82] Finally, work on numerical applications of the BKLT equations to nuclear reactions has recently been reported.[75] Those studies, while interesting, will be outside the scope of the chapter. We instead will concentrate only on those CIE methods which have been applied to chemical reactions (both in the collinear and 3D domains). However, even then, we will concentrate solely on nonperturbation solution methods because we believe this is the crucial aspect of current research on integral equation methods for reactive scattering.

This chapter is organized as follows. In Section II, we discuss some formal and numerical background using as our framework the ordinary LS equation for nonreactive scattering. Our purpose is twofold. First, these equations have a formal structure which will be strongly paralleled in the CIE formalism. Second, the solution methods adapted for the reactive scattering CIE will be patterned after similar methods for the nonreactive case. We shall also see how the complications of reactive systems prevent straightforward use of the simplest solution method normally used in noniterative numerical solutions of the ordinary LS equation. Next, in Section III, we discuss some of the problems arising when one tries to apply the ordinary LS equation to reactive scattering. This serves to motivate the introduction of CIEs for reactive scattering. In Section IV we then present the formal aspects of the CIE approach to reactive scattering. We derive not only the BKLT equations, but also a variety of other related equations in order to demonstrate the similarities among all these equations. In Section V we present a detailed discussion of the method used to solve the BKLT equations and present some sample results demonstrating its utility. We also discuss some of the implications of the structure of the BKLT equations for their efficient solution. In Section VI we discuss briefly how one can adapt the Faddeev equations to the consideration of chemical systems in which the simple assumption of pairwise interactions is not valid. Finally, in Section VII, we present our views on the important problems to be addressed in future applications of the CIEs to reactive scattering.

II. INTEGRAL EQUATIONS FOR NONREACTIVE SCATTERING

In order to introduce many of the formal and computational techniques which are used in an integral equation approach to reactive scattering, it is convenient first to consider the analogous approach to nonreactive scattering. We shall begin with a heuristic derivation of the LS integral equation from the Schrödinger equation (SE).[1] We use this to introduce some of the standard manipulations of Green's operators in order to express the equation in various forms. Next, we introduce the integral equations satisfied by the scattering (S), transition (T), and reactance (R) operators, and explore interrelations among them. We show how one obtains the coordinate representation of these equations in order to display explicitly their integral equation character. We then introduce a distorted wave form of the Green's operator and discuss how the preceding results are modified in order to obtain an exact solution in terms of distorted waves. Finally, we discuss solution methods solving the LS equation.

A. Lippmann-Schwinger Equation

In abstract form, we may write the Schrödinger equation as:

$$(E - H)|\psi> = 0 \tag{1}$$

where E is the total energy of the system, H is the total Hamiltonian operator (taken as a sum of an unperturbed H_o and the perturbation V producing the scattering between different states of H_o),

$$H = H_o + V \tag{2}$$

and $|\psi>$ is the full scattering state vector. The state $|\psi>$ describes a situation with well-defined boundary conditions which will be made explicit in the integrated form of Equation 1. We may arrange Equation 1 according to:

$$(E - H_o)|\psi> = V|\psi> \tag{3}$$

and the solution is obtained by applying the inverse $(E - H_o)^{-1}$ to both sides:

$$|\psi> = (E - H_o)^{-1} V|\psi> \tag{4}$$

Provided $(E - H_o)^{-1}$ exists, this clearly satisfies Equation 1. However, if the strength of the perturbation tends to zero, Equation 4 produces a null state which cannot correspond to the true physical situation. Rather, as V tends to zero, $|\psi>$ should tend to some unperturbed eigenket of H_o corresponding to the system in the absence of any forces leading to scattering. We can add to $|\psi>$ in Equation 4 an eigenstate of H_o,

$$(E - H_o)|\phi> = 0 \tag{5}$$

since

$$|\psi> = |\phi> + (E - H_o)^{-1} V |\psi> \tag{6}$$

also satisfies the Schrödinger equation provided only that the energy of the states $|\phi>$ and $|\psi>$ are equal. Equation 6 is the celebrated LS equation.[1] The only problem with this analysis is that $(E - H_o)^{-1}$ will not exist if the spectrum of H_o includes E, and $|\phi>$ added to Equation 4 only yields a solution to the Schrödinger equation if the eigenstate of $H_o(|\phi>)$ has energy (E). The resolution of the problem is achieved by use of a limit procedure in which the energy (E) in $(E - H_o)^{-1}$ is displaced off the real axis by an infinitesmal amount ($\pm i\epsilon$). Then for H_o having only a real spectrum $(E - H_o \pm i\epsilon)^{-1}$ manifestly exists, and the solution is

$$\lim_{\epsilon \to 0_+} (|\psi^{\pm}> = |\phi> + (E - H_o \pm i\epsilon)^{-1} V|\psi^{\pm}>) \tag{7}$$

The limit $\epsilon \to 0_+$ is to be taken only after $(E - H \pm i\epsilon)^{-1}$ is allowed to act on the ket $V|\psi^{\pm}>$. In fact, the $i\epsilon$ can be interpreted as switching off the interaction in the distant past and future (i.e., if one looks at the Fourier transform of $(E - H_o \pm i\epsilon)^{-1}$ acting on $V|\psi^{\pm}>$) and corresponds to the process of preparing wavepackets for the projectile and target which are subsequently allowed to collide under the action of the full Hamiltonian H. The $+i\epsilon$ corresponds to so called "causal" boundary conditions in which for $t \to -\infty$, no scattered waves are present. It will be this "solution" of the Schrödinger equation with which we shall be mostly concerned. We also remark that Equation 6 represents a solution of the Schrödinger equation in a formal sense only (the unknown ket appears on both sides of the equation). Equation 6 however, contains not only the information of the Schrödinger equation, but also the boundary conditions and is global in the sense that it provides the complete state vector. This point will be clearer later when we discuss the coordinate representive of $|\psi\pm>$.

B. Perturbation Theory, Diagrammatic Sums, and Related Equations

One procedure to try and solve Equation 6 explicity for $|\psi^+>$ is to iterate the equation. Thus, we substitute Equation 6 for $|\psi^+>$ in the right hand side to obtain:

$$|\psi^+> = |\phi> + G_o^+ V|\phi> + G_o^+ V G_o^+ V|\psi^+> \tag{8}$$

where G_o^+ is a short notation for $(E - H_o + i\epsilon)^{-1}$,

$$G_o^+ = (E - H_o + i\epsilon)^{-1} \tag{9}$$

If we repeat this *ad infinitum,* we obtain a power series for $|\psi^+>$ in terms of V:

$$|\psi^+> = |\phi> + G_o^+ V|\phi> + G_o^+ V G_o^+ V|\phi> + G_o^+ V G_o^+ V G_o^+ V|\phi> + \ldots \tag{10}$$

This can be viewed as a perturbation expansion for $|\psi^+>$ in terms of powers of the perturbation (V), and its convergence properties are strongly dependent on the nature of V.[4]

An alternative viewpoint is to represent the terms in the sum by diagrams in which the system begins in state $|\phi>$, interacts via V, propagates freely under H_o (the G_o^+ term), interacts again via V, propagates under H_o, and so on. If we assume that the free particle motion of the center of mass of the system has already been separated, then the interaction V (for a two-body system) produces an effect on *all* particle motions left in the system and such a sequence of graphs is said to be ''connected''. That is, after a finite number of terms in Equation 10, all diagrams will contain interactions involving all particles (in the present nonreactive, two-body collision with center of mass separated, this occurs after the first term so that $G_o^+ V|\phi>$ and all higher terms possess connected diagrams). For a suitable force law, this implies that the kernel of the original integral Equation 6 becomes compact after a finite number of iterations. This, in turn, guarantees that finite rank (matrix) approximations to the kernel can be found which are elements of a convergent sequence. Of course, although convergence is ensured, nothing is implied as to the rate of convergence.

We now factor $|\phi>$ out to the right in Equation 10 to obtain:

$$|\psi^+ > = (1 + G_o^+V + G_o^+VG_o^+V + \ldots)\,|\phi > \tag{11}$$

We let x represent G_o^+ and write this as:

$$|\psi^+ > = \left(\sum_{n=0}^{\infty} X^n\right) |\phi > \tag{12}$$

The series $\Sigma_{n=o}^{\infty} x^n$ is recognized as the Taylor expansion of $(1 - x)^{-1}$ so we can also write:

$$|\psi^+ > = (1 - G_o^+V)^{-1}\,|\phi > \tag{13}$$

which is a formal solution of the LS equation. We can rearrange it by noting:

$$(1 - G_o^+V)^{-1} = (G_o^+\,(G_o^+)^{-1} - G_o^+V)^{-1} \tag{14}$$

$$= ((G_o^+)^{-1} - V)^{-1}\,(G_o^+)^{-1} \tag{15}$$

Now

$$(G_o^+)^{-1} = E - H_o + i\epsilon \tag{16}$$

and so we obtain:

$$|\psi^+ > = (E - H_o - V + i\epsilon)^{-1}\,(E - H_o + i\epsilon)|\phi > \tag{17}$$

$$= G^+\,(i\epsilon)|\phi > \tag{18}$$

by Equation 5. The quantity G^+, given by:

$$G^+ = (E - H + i\epsilon)^{-1} \tag{19}$$

is known as the full Green's operator for the system. We can rewrite Equation 18 as:

$$|\psi^+ > = G^+(E - H + i\epsilon - E + H)|\phi > \tag{20}$$

by just adding and subtracting E − H appropriately. Now, by Equations 19, 2, and 5, this yields:

$$|\psi^+ > = |\phi > + G^+V\,|\phi > \tag{21}$$

This represents a formal solution of the LS equation. The procedure of summing the Taylor series is an example of summing of diagrams in an infinite perturbation expansion. There are other ways to go between Equations 6 and 21 that are more compact. Consider the identity:

$$E - H + i\epsilon = E - H_o + i\epsilon - V \tag{22}$$

and apply G^+ from the left and G_o^+ from the right to obtain:

$$G^+ = G_o^+ + G^+VG_o^+ \tag{23}$$

Alternatively, one applies G^+ from the right and G_o^+ from the left to obtain:

$$G^+ = G_o^+ + G_o^+VG^+ \tag{24}$$

These represent LS equations for the full (interacting) Green's operator G^+ in terms of the unperturbed one G_o^+. If we now replace G_o^+ in Equation 6 with Equation 23, we find:

$$|\psi^+ > = |\phi > + G^+V|\psi^+ > - G^+VG_o^+V|\psi^+ > \tag{25}$$

We then eliminate $G_o^+V|\psi^+>$ in the last term using the LS equation to obtain again:

$$|\psi^+ > = |\phi> + G^+V|\phi> \tag{26}$$

It may not be obvious that the LS equation can be represented as an integral equation. To see this, it is convenient to make our notation somewhat more explicit. We consider scattering of a structureless particle off an infinitely massive particle producing a force field. The initial particle has momentum ($\mathbf{k}_i$) and energy (E) so that the unperturbed eigenket and full scattering ket will be labeled by those quantum numbers. (In fact, we shall suppress the energy label E except where it is crucial). Then the LS equation reads:

$$|\psi^+(\mathbf{k}_i)> = |\phi(\mathbf{k}_i) > + G_o^+V|\phi^+(\mathbf{k}_i)> \tag{27}$$

We now project this state onto the coordinate representative $<\mathbf{r}|$ to obtain:

$$<\mathbf{r}|\psi^+(\mathbf{k}_i)> = <\mathbf{r}|\phi(\mathbf{k}_i)> + <\mathbf{r}|G_o^+V|\psi^+(\mathbf{k}_i)> \tag{28}$$

The inner product ($<\mathbf{r}|\psi^+(\mathbf{k}_i)>$) is a scalar (number) and in fact, the set of all such numbers constitutes the scattering wavefunction, $\psi^+(\mathbf{k}|\mathbf{r})$. Similarly, $<\mathbf{r}|\phi(\mathbf{k}_i)>$ is the unperturbed wavefunction $\phi(\mathbf{k}_i|\mathbf{r})$, which, for the system under consideration, is simply a plane wave exp ($i\mathbf{k}_i \cdot \mathbf{r}$). The normalization is that associated with unit incident flux. The flux vector is given by:

$$\mathbf{j} = \frac{-i\hbar}{2m}\,[\psi^*\nabla\psi - \psi\nabla\psi^*] \tag{29}$$

for an arbitrary wavefunction ψ. Finally, we must determine what $<\mathbf{r}|G_o^+ V|\psi^+(\mathbf{k}_i)>$ is. This is most readily done by inserting resolutions of the identity in the coordinate representation according to:

$$<\mathbf{r}|G_o^+ V|\psi^+(\mathbf{k}_i)> = \int d\mathbf{r}' \int d\mathbf{r}''<\mathbf{r}|G_o^+|\mathbf{r}'><\mathbf{r}'|V|\mathbf{r}''><\mathbf{r}''|\psi^+(\mathbf{k}_i)> \tag{30}$$

In simple systems (like those normally occurring in chemistry) the interaction is *local* (diagonal) in the coordinate representation so that

$$<\mathbf{r}'|V|\mathbf{r}''> = \delta(\mathbf{r}' - \mathbf{r}'')\, V(\mathbf{r}') \tag{31}$$

Then we have:

$$<\mathbf{r}|G_o^+ V|\psi^+ (\mathbf{k}_i)> = \int d\mathbf{r}'< \mathbf{r}|G_o^+|\mathbf{r}'>V(\mathbf{r}')\psi^+(\mathbf{k}_i|\mathbf{r}') \tag{32}$$

Now, if $<\mathbf{r}|G^+{}_o|\mathbf{r}'>$ were also local in the coordinate representation, the above equation would immediately simplify and the LS equation would *not* be an integral equation. In fact, $<\mathbf{r}|G^+{}_o|\mathbf{r}'>$ is not diagonal (i.e., not proportional to $\delta(\mathbf{r} - \mathbf{r}')$) and thus, we find:

$$\psi^+(\mathbf{k}_i|\mathbf{r}) = \phi(\mathbf{k}_i|\mathbf{r}) + \int d\mathbf{r}'\ G_o^+(\mathbf{r}|\mathbf{r}')V(\mathbf{r}')\psi^+(\mathbf{k}_i|\mathbf{r}') \tag{33}$$

where $G_o^+(\mathbf{r}|\mathbf{r}')$ is the unperturbed Green's function. It is readily evaluated by inserting the resolution of the identity in the momentum representation so that:

$$G_o^+(\mathbf{r}|\mathbf{r}') = \int d\mathbf{k} \int d\mathbf{k}'\ <\mathbf{r}|\mathbf{k}><\mathbf{k}|\ \frac{1}{E - H_o + i\epsilon}\ |\mathbf{k}'><\mathbf{k}'|\mathbf{r}'> \tag{34}$$

We require then $<\mathbf{k}|(E - H_o + i\epsilon)^{-1}|\mathbf{k}'>$ and we note that

$$(E - H_o + i\epsilon)|\mathbf{k}'> = \left(E - \frac{\hbar^2 k'^2}{2m} + i\epsilon\right)|\mathbf{k}'> \tag{35}$$

Apply $(E - H_o + i\epsilon)^{-1}$ to both sides to obtain:

$$|\mathbf{k}'> = (E - H_o + i\epsilon)^{-1}\left(E - \frac{\hbar^2 k'^2}{2m} + i\epsilon\right)|\mathbf{k}'> \tag{36}$$

and note that the scalar $E - \hbar^2 k'^2/2m + i\epsilon$ commutes with $(E - H_o + i\epsilon^{-1}$. Then we divide both sides of Equation 36 by the scalar $E - \hbar^2 k^2/2m + i\epsilon$ to obtain:

$$<\mathbf{k}|(E - H_o + i\epsilon)^{-1}|\mathbf{k}'> = \frac{<\mathbf{k}|\mathbf{k}'>}{E - \dfrac{\hbar^2 k'^2}{2m} + i\epsilon} \tag{37}$$

However, we choose to normalize the unperturbed states such that:

$$<\mathbf{k}|\mathbf{k}'> = \delta(\mathbf{k} - \mathbf{k}') \tag{38}$$

and we then find that

$$G_o^+(\mathbf{r}|\mathbf{r}') = \int d\mathbf{k} \frac{<\mathbf{r}|\mathbf{k}><\mathbf{k}|\mathbf{r}'>}{E - \frac{\hbar^2 k^2}{2m} + i\epsilon} \tag{39}$$

Finally, we recognize $<\mathbf{r}|\mathbf{k}>$ and $<\mathbf{k}|\mathbf{r}'>$ as normalized plane waves (the second complex conjugated) so that:

$$G_o^+(\mathbf{r}|\mathbf{r}') = \int d\mathbf{k} \frac{\exp[i\mathbf{k} \cdot (\mathbf{r} - \mathbf{r}')]}{(2\pi)^3\left(E - \frac{\hbar^2 k^2}{2m} + i\epsilon\right)} \tag{40}$$

The integral over $\mathbf{k}$ is done in spherical polar coordinates by means of Cauchy's residue theorem with the final result being

$$\psi^+(\mathbf{k}_i|\mathbf{r}) = \phi(\mathbf{k}_i|\mathbf{r}) - \frac{m}{2\pi\hbar^2} \int d\mathbf{r}' \frac{\exp(ik|\mathbf{r} - \mathbf{r}'|)}{|\mathbf{r} - \mathbf{r}'|} V(\mathbf{r}')\psi^+(\mathbf{k}_i|\mathbf{r}') \tag{41}$$

It is obvious that this is an integral equation for the scattering wavefunction $\psi^+(\mathbf{k}_i|\mathbf{r})$. The $G_o^+(\mathbf{r}|\mathbf{r}')$ is nonlocal, basically because it contains the inverse of the Hamiltonian H_o. This contains momenta which involve derivative operations when expressed in the coordinate representation. The inverse of a differential operator clearly must involve some type of integral operator.

One important question that can be asked at this stage is how does one obtain the scattering information given the wavefunction $\psi^+(\mathbf{k}_i|\mathbf{r})$? In addressing this, we recall that the cross-section is defined as the proportionality constant relating the number of projectiles scattered into a given solid angle per unit time by A targets to the number (N) of incident projectiles per unit time. The probability per second of observing a scattered particle in solid angle $(d\Omega)$ is given in terms of the scattered flux by $\mathbf{j} \cdot d\hat{S}$, where $d\hat{S}$ is a vector surface element subtended by the solid angle $d\Omega$. This must equal the product of the number of scatterers (in our case, one) times the magnitude of the incident flux times the cross-section, expressed as $d\sigma/d\Omega(\mathbf{k}_f, \mathbf{k}_i)d\Omega$. Using the relation:

$$dS = r\, r^2\, d\Omega \tag{42}$$

and Equation 29 to evaluate the scattered flux, one obtains:

$$\frac{d\sigma}{d\Omega}(\mathbf{k}_f,\mathbf{k}_i) = |f(\mathbf{k}_f,\mathbf{k}_i)|^2 \tag{43}$$

where the differential scattering amplitude is given by:

$$f(\mathbf{k}_f,\mathbf{k}_i) = \frac{-m}{2\pi\hbar^2} \int d\mathbf{r}\, \exp(-i\mathbf{k}_f \cdot \mathbf{r})V(\mathbf{r})\psi^+(\mathbf{k}_i|\mathbf{r}) \tag{44}$$

It is seen that the scattering amplitude is proportional to $<\phi(\mathbf{k}_f)|V|\psi^+(\mathbf{k}_i)>$. It is also instructive to examine once again the formal LS equation:

$$|\psi^+(\mathbf{k}_i)> = |\phi(\mathbf{k}_i)> + G_o^+ V|\psi^+(\mathbf{k}_i)> \tag{45}$$

We note again that:

$$G_o^+ = \int d\mathbf{k} \frac{|\phi(\mathbf{k})><\phi(\mathbf{k})|}{E - \frac{\hbar^2 k^2}{2m} + i\epsilon} \tag{46}$$

and we project Equation 45 with a final state $<\phi(\mathbf{k}_f)|$ to obtain:

$$<\phi(\mathbf{k}_f)|\psi^+(\mathbf{k}_i)> = <\phi(\mathbf{k}_f)|\phi(\mathbf{k}_i)> + \int d\mathbf{k} \frac{<\phi(\mathbf{k}_f)|\phi(\mathbf{k})>}{E - \frac{\hbar^2 k^2}{2m} + i\epsilon} <\phi(\mathbf{k})|V|\psi^+(\mathbf{k}_i)> \tag{47}$$

The first term on the r.h.s. of Equation 47 vanishes for nonforward scattering and by Equation 38 we obtain:

$$<\phi(\mathbf{k}_f)|\psi^+(\mathbf{k}_i)> = <\phi(\mathbf{k}_f)|V|\psi^+(\mathbf{k}_i)>/i\epsilon \tag{48}$$

Here we assume that the energy of the state $|\phi(\mathbf{k}_f)>$ equals E. It then follows that:

$$<\phi(\mathbf{k}_f)|V|\psi^+(\mathbf{k}_i)> = \lim_{\epsilon \to 0_+} i\epsilon <\phi(\mathbf{k}_f)|\psi^+(\mathbf{k}_i)> \tag{49}$$

which provides an alternative prescription for the scattering amplitude. We now are prepared to introduce the transition and related operators.

C. Equations for the T, S, and R Operators

From the preceding section we see that the relevant scattering information is contained in the scalar $<\phi(\mathbf{k}_f)|V|\psi^+(\mathbf{k}_i)>$ and this is just the projection onto $<\phi(\mathbf{k}_f)|$ of the ket produced by the action of the perturbation (V) on the scattering ket $|\psi^+(\mathbf{k}_i)>$. It is convenient to introduce an alternative operator in place of the potential (V) by the equation:

$$T|\phi(\mathbf{k}_i)> = V|\psi^+(\mathbf{k}_i)> \tag{50}$$

The operator T is called the transition operator and it is that operator whose action on the unperturbed state $|\phi(\mathbf{k}_i)>$ produces the same ket as the perturbation acting on the full scattering state. We note that in $V|\psi^+(\mathbf{k}_i)>$, the potential is contained to infinite order in $|\psi^+(\mathbf{k}_i)>$ and obviously to first order in the factor V. By contrast, the potential is present to zero-th order in $|\phi(\mathbf{k}_i)>$ and to infinite order in T. T can be thought of as an effective potential and we note then that:

$$<\phi(\mathbf{k}_f)|T|\phi(\mathbf{k}_i)> = <\phi(\mathbf{k}_f)|V|\psi^+(\mathbf{k}_i)> \tag{51}$$

Indeed, the Born approximate expression for the transition rate (the Fermi golden rule) involves the matrix element $<\phi(\mathbf{k}_f)|V|\phi(\mathbf{k}_i)>$ and replacing V by T yields the *exact* transition rate. Equation 50 enables us to easily derive the LS equation satisfied by T. We substitute Equation 50 into Equation 45 to obtain:

$$|\psi^+(\mathbf{k}_i)> = |\phi(\mathbf{k}_i)> + G_o^+ T|\phi(\mathbf{k}_i)> \tag{52}$$

then apply V to both sides and use Equation 51 again to obtain:

$$T|\phi(\mathbf{k}_i)> = [V + VG_o^+ T]|\phi(\mathbf{k}_i)> \tag{53}$$

This must hold for all state vectors $|\phi(\mathbf{k}_i)>$ and we therefore identify the operator equation for T as:

$$T = V + VG_o^+T \tag{54}$$

We could also have manipulated directly with Equation 21 by substituting with Equation 24 to obtain:

$$|\psi^+\rangle = |\phi\rangle + (G_o^+ + G_o^+VG^+)V\,|\phi\rangle \tag{55}$$

$$= |\phi\rangle + G_o^+(V + VG^+V)\,|\phi\rangle \tag{56}$$

Then again follow the analysis of Equations 45 to 48 so that[2,85]

$$\langle\phi|\psi^+\rangle = \langle\phi|(V + VG^+V)\,|\phi\rangle\,/i\epsilon \tag{57}$$

or

$$\lim_{\epsilon\to 0_+} i\epsilon\,\langle\phi|\psi^+\rangle = \langle\phi|(V + VG^+V)|\phi\rangle \tag{58}$$

We therefore identify

$$T = V + VG^+V \tag{59}$$

as an alternative expression for the T operator. One can verify that Equations 59 and 54 are equivalent by the use of Equation 24 to write:

$$T = V + V(G_o^+ + G_o^+VG^+)\,V \tag{60}$$

$$= V + VG_o^+T \tag{61}$$

Alternatively, the use of Equation 23 leads to:

$$T = V + TG_o^+V \tag{62}$$

All these equations yield equivalent results in the case of nonreactive scattering but we shall see there are interesting differences occurring in reactive T operators defined via the analog of Equation 50 and those defined by the approach of Equations 55 to 59. Before going on, it is important to note that the comparison of Equation 59, 61, and 62 yields the relations:

$$G^+V = G_o^+T \tag{63}$$

and

$$VG^+ = TG_o^+ \tag{64}$$

We now introduce other types of operators of significance in scattering. We begin by noting that the Green's operator G_0^+ can be written as:

$$\frac{1}{E - H_o + i\epsilon} = \frac{P}{E - H_o} - i\pi\delta(E - H_o) \tag{65}$$

where $P/(E - H_o)$ is the principal value Green's operator G_0^P. It is real as opposed to G_0^+. Now we write:

$$T = V + VG_o^p T - i\pi V\delta(E - H_o)T \tag{66}$$

and we define a new operator, R, by the LS equation:

$$R = V + VG_o^p R \tag{67}$$

We form the difference of Equations 66 and 67 to write:

$$T - R = VG_o^p (T - R) - i\pi V\delta(E - H_o)T \tag{68}$$

Solving this for T − R, we find

$$T - R = -i\pi(1 - VG_o^p)^{-1}V\,\delta(E - H_o)T \tag{69}$$

and note that by Equation 67, this can also be written as:

$$T = R - i\pi\, R\, \delta(E - H_o)T \tag{70}$$

This is the Heitler damping equation[86] and it provides a means of obtaining T from R. The equation satisfied by R (the reactance operator) has the advantage of being real and it is only after solving Equation 67 that any complex arithmetic is required. The appearance of the operator $\delta(E - H_o)$ forces the equation to include only those matrix elements of T and R between states at the same energy (E). It should be clear that beginning with the equivalent expression for T, Equation 62 and an analogous one for R,

$$R = V + RG_o^p V \tag{71}$$

We can also establish that

$$T = R - i\pi\, T\, \delta(E - H_o)R \tag{72}$$

We now introduce the scattering operator (S) whose matrix elements are the probability amplitudes for a given transition. We define S by the relation:

$$S = 1 - 2\pi i\, \delta(E - H_o)T \tag{73}$$

The significance of this equation is most easily appreciated by expressing S in terms of R. This is achieved by solving Equation 72 for T and substituting the result into Equation 73 to yield:

$$S = 1 - 2\pi i\, \delta(E - H_o)R(1 + i\pi\delta(E - H_o)R)^{-1} \tag{74}$$

$$= (1 - i\pi\, \delta(E - H_o)R)(1 + i\pi\delta(E - H_o)R)^{-1} \tag{75}$$

The structure of this result implies that for Hermitian R, S will be unitary.

In order to illustrate more clearly the relationship between matrix elements of the T, R, and S operators, it is convenient to consider the LS equation in a little more detail. Rather than use a plane wave basis (as was done in arriving at Equation 41), it is convenient to expand in terms of angular partial waves. Thus, we make use of

$$-\frac{m}{2\pi\hbar^2}\frac{\exp(ik|\mathbf{r} - \mathbf{r}'|)}{|\mathbf{r} - \mathbf{r}'|} = -\frac{2mk}{\hbar^2}\sum_{\ell\mu} Y_{\ell\mu}(\mathrm{r})Y^*_{\ell\mu}(\hat{\mathrm{r}}')h_\ell^+(kr_>)j_\ell(kr_<) \tag{76}$$

$$\psi^{+}(\mathbf{k}_i|\mathbf{r}) = 4\pi\sum_{\ell\mu} i^{\ell}\, Y^{*}_{\ell\mu}(k_i)Y_{\ell\mu}(\hat{r})g_{\ell}(r) \tag{77}$$

and

$$\exp(i\mathbf{k}_i \cdot \mathbf{r}) = 4\pi\sum_{\ell\mu} i^{\ell}\, Y^{*}_{\ell\mu}(k_i)Y_{\ell\mu}(\hat{r})j_{\ell}(kr) \tag{78}$$

where h_{ℓ}^{+} is a spherical Hankel function of order ℓ, j_{ℓ} is a spherical Bessel function of order ℓ, and in writing Equation 77 we have made use of the fact that the scattering has symmetry with respect to the azimuthal angle. Substituting these expressions into the LS equation and simplifying yields the radial integral equation:

$$g_{\ell}(r) = j_{\ell}(kr) - \frac{2mk}{\hbar^2}\int_0^{\infty} dr' r'^2 h_{\ell}^{+}(kr_{>})j_{\ell}(kr_{<})V(r')g_{\ell}(r') \tag{79}$$

The spherical Bessel, Hankel, and Neumann functions asymptotically behave like:

$$\lim_{r\to\text{large}} j_{\ell}(kr) = \sin(kr - \ell\pi/2)/kr \tag{80}$$

$$\lim_{r\to\text{large}} n_{\ell}(kr) = \cos(kr - \ell\pi/2)/kr \tag{81}$$

and

$$\lim_{r\to\text{large}} h_{\ell}^{+}(kr) = \exp(i[kr - \ell\pi/2])/kr \tag{82}$$

The partial wave T matrix is found from the large r form of g_{ℓ} (r) and is given by:

$$T_{\ell} = \frac{4mik}{\hbar^2}\int_0^{\infty} dr\, r^2 j_{\ell}(kr)V(r)g_{\ell}(r) \tag{83}$$

In terms of T_{ℓ}, one then has:

$$\lim_{r\to\text{large}} g_{\ell}(r) = j_{\ell}(kr) + \frac{i}{2kr}\exp(i[kr - \ell\pi/2])T_{\ell} \tag{84}$$

If we instead rewrite g_{ℓ} (r) in terms of incoming and outgoing waves, we find:

$$\lim_{r\to\text{large}} g_{\ell}(r) = -\exp(-i[kr - \ell\pi/2]) + (1 - T_{\ell})\exp(i[kr - \ell\pi/2])/2ikr \tag{85}$$

and the S matrix is then given by:

$$S_{\ell} = 1 - T_{\ell} \tag{86}$$

Analogously, we could write g_{ℓ} (r) in terms of sine and cosine functions, and find:

$$g_{\ell}(r) = j_{\ell}(kr)(1 - T_{\ell}/2) - \frac{2mk}{\hbar^2}\int_0^{\infty} dr' r'^2 n_{\ell}(kr_{>})j_{\ell}(kr_{<})V(r')g_{\ell}(r') \tag{87}$$

Renormalizing g_ℓ (r) by:

$$g_\ell(r) = U_\ell(r)\,(1 - T_\ell/2) \tag{88}$$

where

$$U_\ell(r) = j_\ell(kr) - \frac{2mk}{\hbar^2}\int_0^\infty dr'r'^2 n_\ell(kr_>)j_\ell(kr_<)V(r')U_\ell(r') \tag{89}$$

and defining R_ℓ by:

$$R_\ell = -\frac{2mk}{\hbar^2}\int_0^\infty dr\ r^2\ j_\ell(kr)V(r)U_\ell(r) \tag{90}$$

one may easily verify, using Equation 83, that:

$$T_\ell = -2i\ R_\ell\ (1 - T_\ell/2) \tag{91}$$

Solving for T_ℓ yields:

$$T_\ell = -2i\ R_\ell/\ (1 - iR_\ell) \tag{92}$$

If we substitute this into Equation 86, we finally obtain:

$$S_\ell = (1 + i\ R_\ell)/(1 - i\ R_\ell) \tag{93}$$

and comparison with Equation 75 shows the obvious connection of the simple one-channel problem to the operator relation. Finally, we note from Equations 89 and 90 that R_ℓ is real. Therefore, from Equation 93 we have:

$$|S_\ell|^2 = 1 \tag{94}$$

and we can write:

$$S_\ell = \exp(2in_\ell) \tag{95}$$

where η_ℓ is the phase shift for the ℓ-th partial wave. It is then readily verified that:

$$T_\ell = -2i\ \sin\eta_\ell\ \exp(i\eta_\ell) \tag{96}$$

and

$$R_\ell = \tan\eta_\ell \tag{97}$$

We next consider how these relations change when one uses distorted wave states as a reference rather than the Bessel functions. We add and subtract a distortion potential(U) in the Schrödinger equation so that our LS equation now reads:

$$|\psi^+> = |g^+> + \overline{G}_o^+\ (V - U)|\psi^+> \tag{98}$$

The state $|g^+>$ is an eigenstate of $H_o + U$; i.e., it is a distorted wave. The $|g^+>$ is chosen because $|\psi^+>$ must have only outgoing waves and the $|g^+>$ provides the scattering due to the distortion potential. The quantity $\overline{G}_0^+$ is $(E - H_0 - U + i\epsilon^{-1})$ and $\overline{G}_0^+(V - U)|\psi^+>$ provides the additional scattering due to the rest of the potential. The full scattering amplitude is given, for a particular partial wave ℓ, by:

$$T_\ell = \overline{T}_\ell + T_\ell' \tag{99}$$

where $\overline{T}_\ell$ is the transition amplitude due to the distortion U, and T_ℓ' is the transition amplitude due to V − U. In fact,

$$T' = (V - U) + (V - U)(E - H + i\epsilon)^{-1}(V - U) \tag{100}$$

in general, and:

$$T'\,|g^+> = (V - U)|\psi^+> \tag{101}$$

The amplitude due to U is given, for the ℓ-th partial wave, by Equation 96 so that:

$$\overline{T}_\ell = -2i\,\sin\overline{\eta}_\ell\,\exp(i\overline{\eta}_\ell) \tag{102}$$

where $\overline{\eta}_\ell$ is the phase shift due to U. The reference states g_ℓ^+ satisfy the asymptotic boundary condition:

$$\lim_{r\to \text{large}} g_\ell^+(r) = \exp(i\overline{\eta}_\ell)\,\sin\left(kr - \frac{\ell\pi}{2} + \overline{\eta}_\ell\right)/kr \tag{103}$$

in place of Equation 80 for $j_\ell(kr)$. The relevant matrix element of T_ℓ' is[2]

$$T'_\ell = <g_\ell^-\,|T'|g_\ell^+> \tag{104}$$

and we note that:

$$g_\ell^- = (g_\ell^+)^* \tag{105}$$

It follows that:

$$T'_\ell = T_\ell\,\exp(2i\overline{\eta}_\ell) \tag{106}$$

where $\widetilde{T}_\ell$ is given by:

$$T_\ell = <g_\ell|T'|g_\ell> \tag{107}$$

and

$$\lim_{r\to\infty} g_\ell(r) = \sin\left(kr - \frac{\ell\pi}{2} + \overline{\eta}_\ell\right)/kr \tag{108}$$

We now recall Equation 86 which states that:

$$S_\ell = 1 - [-2i\,\sin\overline{\eta}_\ell\,\exp(i\overline{\eta}_\ell) + T_\ell\,\exp(2i\overline{\eta}_\ell)] \tag{109}$$

This is easily rearranged to yield:

$$S_\ell = [1 - T_\ell] \exp(2i\bar{\eta}_\ell) \tag{110}$$

$$= \exp(i\bar{\eta}_\ell)\, \tilde{S}_\ell \exp(i\bar{\eta}_\ell) \tag{111}$$

This result shows that the full S matrix is constructed from the product of the S matrix due only to V-U and phase factors which build in the scattering due to the distortion U. The $\bar{S}_\ell$, in turn, can be determined by calculating the matrix elements of the reactance operator associated with V − U,

$$R = (V - U) + (V - U)\left(\frac{P}{E - H_o - U}\right)\tilde{R} \tag{112}$$

and using the damping equation 93. The form of Equation 111 has an appropriate generalization to the reactive case.

We now wish to consider some of the numerical aspects of integral equations, again using the nonreactive LS equation to illustrate both what we can expect in the case of reactive scattering, and points of difference between reactive and nonreactive collisions. We shall restrict our discussion to the partial wave version of the LS equation (Equation 79). This is an inhomogeneous Fredholm integral equation of the second kind,[87-88] and one procedure for solving it is to convert it into algebraic equations by expanding the solution in some basis of linearly independent functions, $\{\chi_n\}$. It is convenient here to assume these are elements of an orthonormal set under weight r^2 but as we shall see later, this is not essential. We develop $g_\ell(r)$ in the set $\{\chi_n\}$ such that:

$$g_\ell(r) = \sum_n a_n^\ell\, \chi_n(r) \tag{113}$$

Substitute this into Equation 79, multiply by $r^2\, \chi_m{}^*(r)$, and integrate over r to obtain:

$$a_m^\ell = \int_0^\infty dr\, r^2\, \chi^*{}_m(r) j_\ell(kr) - \frac{2mk}{\hbar^2}\int_0^\infty dr\, r^2 \int_0^\infty dr'\, r'^2\, \chi^*{}_m(r) h_\ell^+(kr_>) j_\ell(kr_<) V(r') \sum_n a_n^\ell \chi_n(r') \tag{114}$$

This is of the form:

$$a_m^\ell = I_m^\ell + \sum_n C_{mn}^\ell\, a_n^\ell \tag{115}$$

where the definitions of I_m^ℓ and C_{mn}^ℓ are obvious. Since the equations are inhomogeneous, the solution exists provided the coefficient matrix, $\mathbf{1}^\ell - \mathbf{C}^\ell$ possesses an inverse. However, the I_m^ℓ involve an integral of $j_\ell(kr)$ times $\chi_m{}^*(r)$ weighted by r^2, and similarly, the α_n^ℓ are integrals involving the $g_\ell(r)$. Neither integrand tends to zero as $r \rightarrow \infty$ for the obvious choices of $\{\chi_n\}$. As a result, these integrals are ill-defined as written. This whole question can be avoided by working with a related integral equation which can be obtained from Equation 79 by applying V(r) to both sides and defining the amplitude density[7-10,89] $\zeta_\ell(r)$ by:

$$\zeta_\ell(r) = V(r) g_\ell(r) \tag{116}$$

$$= V(r) j_\ell(r) - \frac{2mk}{\hbar^2} V(r) \int_0^\infty dr' r'^2 h_\ell^+(kr_>) j_\ell(kr_<) \zeta_\ell(r') \tag{117}$$

The amplitude density obviously tends to zero as $r \rightarrow \infty$ (and for well-behaved potentials, is finite at the origin). We then obtain Equation 115 but with the definitions,

$$a^{\ell}_{m} = \int_0^{\infty} dr\, r^2 \chi^*_m(r)\zeta_\ell(r) \tag{118}$$

$$I^{\ell}_{m} = \int_0^{\infty} dr\, r^2 \chi^*_m(r)V(r)j_\ell(kr) \tag{119}$$

and

$$C^{\ell}_{mn} = -\frac{2mk}{\hbar^2}\int_0^{\infty} dr\, r^2 \int_0^{\infty} dr\, r'^2 \chi^*_m(r)V(r)h^+_\ell(kr_>)j_\ell(kr_<\chi_n(r') \tag{120}$$

Because $\zeta_\ell(r)$ is contained, one can now use L^2 basis functions for the $\{\chi_n\}$.

An alternative procedure is to introduce a quadrature approximation to the integral in Equation 79. This generally leads to much larger sets of equations and, at least in its most straightforward form, is not computationally feasible.[7,87] However, quadrature solutions of closely related integral equations can be easily carried out. To examine this, we convert Equation 79 into an inhomogeneous Volterra integral equation of the second kind.[9-10] We begin by eliminating the r< and r> variables by splitting the integral over r′ into two pieces such that:

$$g_\ell(r) = j_\ell(kr) - \frac{2mk}{\hbar^2} h^+_\ell(kr)\int_0^{r} dr' r'^2 j_\ell(kr')V(r')g_\ell(r')$$
$$-\frac{2mk}{\hbar^2} j_\ell(kr)\int_r^{\infty} dr' r'^2 h^+_\ell(kr')V(r')g_\ell(r') \tag{121}$$

We then add and substract the quantity:

$$I = -\frac{2mk}{\hbar^2} j_\ell(kr)\int_0^{r} dr' r'^2 h^+_\ell(kr')V(r')g_\ell(r') \tag{122}$$

and write the result as:

$$g_\ell(r) = j_\ell(kr)\,(1 - C_\ell)$$
$$+\frac{2mk}{\hbar^2}\int_0^{r} dr' r'^2 [j_\ell(kr)h^+_\ell(kr') - h^+_\ell(kr)j_\ell(kr')]V(r')g_\ell(r') \tag{123}$$

with

$$C_\ell = \frac{2mk}{\hbar^2}\int_0^{\infty} dr\, r^2 h^+_\ell(kr)V(r)g_\ell(r) \tag{124}$$

Then we define $u_\ell(r)$ by:

$$g_\ell(r) = U_\ell(r)\,(1 - C_\ell) \tag{125}$$

and find that:

$$U_\ell(r) = j_\ell(kr) + \frac{2mk}{\hbar^2}\int_0^r dr'r'^2\, K(r,r')V(r')U_\ell(r') \tag{126}$$

We note that Equation 124 can be written as:

$$C_\ell = \frac{2mk}{\hbar^2}\int_0^\infty dr\; r^2 h_\ell^+(kr)V(r)U_\ell(r)\;(1 - C_\ell) \tag{127}$$

so that:

$$1 - C_\ell = 1/\{1 + \frac{2mk}{\hbar^2}\int_0^\infty dr\; r^2 h_\ell^+(kr)V(r)U_\ell(r)\} \tag{128}$$

Thus, once Equation 126 for $u_\ell(r)$ has been solved, $1 - C_\ell$ will be known and one may then calculate $g_\ell(r)$. However, the real quantity of interest is T_ℓ (cf. Equation 83) which is given by:

$$T_\ell = \frac{4mik}{\hbar^2}\int_0^\infty dr\; r^2 j_\ell(kr)\; V(r)U_\ell(r)\;(1 - C_\ell) \tag{129}$$

Again, once Equation 126 has been solved out to sufficiently large r, the integral in Equation 129 will be known along with $1 - C_\ell$ and, therefore, T_ℓ can be computed. Completely analogous Volterra equations can be derived for computing R_ℓ rather than T_ℓ. Now an important feature of the kernel of Equation 126 arises due to the nature of $K(r,r')$ when evaluated at $r = r'$:

$$K(r,r) \equiv 0 \tag{130}$$

As a consequence, the kernel of the Volterra equation is much better behaved than that of the corresponding Fredholm equation[4] and is used in much of the formal analysis of scattering problems. For our purpose here, we wish to illustrate the simpler behavior of Equation 126 when solved by a Newton-Cotes quadrature. We replace the integral by such a quadrature to obtain:

$$U_\ell(r_t) = j_\ell(kr_t) + \frac{2mk}{\hbar^2}\sum_{s=1}^{t} W_s r_s^2\, K(r_t,r_s)V(r_s)U_\ell(r_s) \tag{131}$$

However, by Equation 130, $K(r_t,r_t)$ vanishes so that Equation 131 becomes a simple recursion formula:

$$U_\ell(r_t) = j_\ell(kr_t) + \frac{2mk}{\hbar^2}\sum_{s=1}^{t-1} W_s r_s^2\, K(r_t,r_s)V(r_s)U_\ell(r_s) \tag{132}$$

If one writes out this expression more explicitly, it is readily seen that in the course of recurring $u_\ell(r_t)$ out to r_N (where we assume $V(r_N) = 0$ for $r > r_N$), one in fact generates quadrature approximations to the integrals occurring in Equations 128 and 129. This procedure has been used in the study of a number of scattering problems.[3,7-13] At this point we have concluded our introduction to integral equation methods within the context of nonreactive scattering.

III. LIPPMANN-SCHWINGER EQUATION FOR REACTIVE SCATTERING

A. The Lippmann-Schwinger Equation with Reaction

We now shall consider the LS equation in the case where there can be rearrangements. Again the Hamiltonian H can be split into an unperturbed portion and an interaction, but now this splitting can be done in more than one way since the system can exist in several partitions. We thus write:

$$H = H_\alpha + V_\alpha \qquad \alpha = 1, 2, \ldots \tag{133}$$

where α labels a given partition of the system into an atom and a molecule, H_α describes the atom-diatom pair in the absence of any interaction between them, and V_α is the interaction between the atom and diatom in partition α. We note that:

$$H_\alpha = \lim_{R_\alpha \to \text{large}} H \tag{134}$$

where R_α is the distance from the atom to the diatom center of mass in partition α. Now we assume the system is initially in partition α with a full set of quantum numbers i describing the initial molecular internal state and relative momentum. Then the Schrödinger equation is written:

$$(E - H_\alpha)|\psi(\alpha i)\rangle = V_\alpha |\psi(\alpha i)\rangle \tag{135}$$

and the formal solution or LS equation is

$$|\psi^+(\alpha i)\rangle = |\phi(\alpha i)\rangle + G_\alpha^+ V_\alpha |\psi^+(\alpha i)\rangle \tag{136}$$

where now the Green's operator is

$$G_\alpha^+ = (E - H_\alpha + i\epsilon)^{-1} \tag{137}$$

We can again use the LS equation for G^+,

$$G^+ = G_\alpha^+ + G_\alpha^+ V_\alpha G^+ \tag{138}$$

$$= G_\alpha^+ + G^+ V_\alpha G_\alpha^+ \tag{139}$$

and rewrite Equation 121 as:

$$|\psi^+(\alpha i)\rangle = |\phi(\alpha i)\rangle + G^+ V_\alpha |\phi(\alpha i)\rangle \tag{140}$$

Now Equation 136 has the property that, as R_α tends to large values, the Green's operator G_α^+ produces the outgoing nonreactive scattered waves. However, for reactions, there must also be outgoing scattered waves in all other partitions which are energetically accessible. One may ask the question whether these are also contained in the LS formal solution (Equation 136) or equivalently in Equation 140. As we shall see, this is the case. We note that in general,

$$E - H + i\epsilon = E - H_\lambda + i\epsilon - V_\lambda \tag{141}$$

for *any* partition λ.

This immediately implies that:

$$G^{+} = G_{\lambda}^{+} + G_{\lambda}^{+} V_{\lambda} G^{+} \tag{142}$$

$$= G_{\lambda}^{+} + G^{+} V_{\lambda} G_{\lambda}^{+} \tag{143}$$

for *all* partitions λ. We substitute Equation 142 into Equation 140 and rearrange to obtain the expression:

$$|\psi^{+}(\alpha i)> = |\phi(\alpha i)> + G_{\lambda}^{+}(V_{\alpha} + V_{\lambda}G^{+}V_{\alpha})|\phi(\alpha i)> \tag{144}$$

We then see that G_{λ}^{+} will produce outgoing waves in arrangement λ as is required. However, we also can identify $V_{\alpha}+V_{\lambda}G^{+}V_{\alpha}$ as the transition amplitude for going from arrangement α to λ. Thus,

$$T_{\lambda\alpha} = V_{\alpha} + V_{\lambda}G^{+}V_{\alpha} \tag{145}$$

is one possible generalization of the transition operator for reactive scattering into partition λ from α. This procedure is the analog of that leading to Equations 55 and 56. We also can express:

$$G_{\lambda}^{+} = \sum_{j} \frac{|\phi(\lambda j)> <\phi(\lambda j)|}{E - E(\lambda j) + i\epsilon} \tag{146}$$

and compute:

$$\lim_{\epsilon \to 0_{+}} i\epsilon < \phi(\lambda t)|\psi^{+}(\alpha i) > = < \phi(\lambda t)|T_{\lambda\alpha}|\phi(\alpha i)> \tag{147}$$

just like in the nonreactive case considered earlier.

Alternatively, we can note that:

$$E - H_{\alpha} + i\epsilon = E - H_{\lambda} + i\epsilon - H_{\alpha} + H_{\lambda} \tag{148}$$

which implies that:

$$G_{\lambda}^{+} = G_{\alpha}^{+} + G_{\lambda}^{+}(H_{\lambda} - H_{\alpha})\, G_{\alpha}^{+} \tag{149}$$

If we substitute this into Equation 136, we obtain:

$$|\psi^{+}(\alpha i)> = |\phi(\alpha i)> + G_{\lambda}^{+}V_{\alpha}|\psi^{+}(\alpha i) > - G_{\lambda}^{+}(H_{\lambda} - H_{\alpha})G_{\alpha}^{+}V_{\alpha}|\psi^{+}(\alpha i)> \tag{150}$$

Then using Equation 146 again yields:

$$|\psi^{+}(\alpha i)> = |\phi(\alpha i)> + G_{\lambda}^{+}V_{\alpha}|\psi^{+}(\alpha i)> + G_{\lambda}^{+}(H_{\lambda} - E)|\phi(\alpha i)> \tag{151}$$

$$- G_{\lambda}^{+}(H_{\lambda} - H_{\alpha})|\psi^{+}(\alpha i)> \tag{152}$$

$$= -i\epsilon|\phi(\alpha i)> + G_{\lambda}^{+}V_{\lambda}|\psi^{+}(\alpha i)>$$

If one again follows the prescription of computing Lim $i\epsilon<\phi(\lambda t)|\psi^{+}(\alpha i)>$, $\epsilon \to o_{+}$ the result is that the scattering amplitude for going from state α, i to state λ, t is

$$f(\lambda t|\alpha i) = <\phi(\lambda t)|V_\lambda|\psi^+(\alpha i)> \tag{153}$$

It follows that we can also introduce a transition operator for the reaction $\alpha i \rightarrow \lambda t$ by the analog of Equation 50 so that:

$$T_{\lambda\alpha}|\phi(\alpha i)> = V_\lambda|\psi^+(\alpha i)> \tag{154}$$

Then using Equation 140 we obtain the expression:

$$T_{\lambda\alpha}|\phi(\alpha i)> = (V_\lambda + V_\lambda G^+ V_\alpha)|\phi(\alpha i)> \tag{155}$$

which holds for all possible initial states $|\phi(\alpha i)>$. Thus, the result is

$$T_{\lambda\alpha} = V_\lambda + V_\lambda G^+ V_\alpha \tag{156}$$

It is extremely important to note that, unlike the nonreactive case, $\tilde{T}_{\lambda\alpha}$ above and $\hat{T}_{\lambda\alpha}$ in Equation 145 are different operators! However, it is easy to show that the physical transition amplitude matrix elements of the two operators are equal. We do this by considering:

$$<\phi(\lambda t)|(T_{\lambda\alpha} - T_{\lambda\alpha})|\phi(\alpha i)> = <\phi(\lambda t)|(V_\lambda - V_\alpha)|\phi(\alpha i)> \tag{157}$$

But

$$<\phi(\lambda t)|(V_\lambda - V_\alpha)|\phi(\alpha i)> = <\phi(\lambda t)|(E + V_\lambda - E - V_\alpha)|\phi(\alpha i)> \tag{158}$$

$$= <\phi(\lambda t)|(H_\lambda + V_\lambda - H_\alpha - V_\alpha)|\phi(\alpha i)> \tag{159}$$

$$= 0 \tag{160}$$

This proves the equality of the physical matrix elements of the two transition operators. The operator $\hat{T}\lambda\alpha$ is termed the prior and $\tilde{T}_{\lambda\alpha}$ the post form of the transition operator.

B. Problems with the Reactive Lippmann-Schwinger Equation

It is now instructive to ask how one would try to use the LS equation for reactive collisions. We note that either Equation 146 or 152 is in a form which is analogous to the nonreactive case and indeed, one can attempt to solve the equations by converting the coordinate representation (Fredholm) integral equations into Volterra equations.[9-10] These can then be solved recursively (noniteratively) just as in the nonreactive case. Rather than do this, it is instead convenient to look at the equations for the transition operators. We shall use the prior form $\hat{T}_{\lambda\alpha}$ and note that use of the LS equation:

$$G^+ = G_\lambda^+ + G_\lambda^+ V_\lambda G^+ \tag{161}$$

in Equation 145 yields:

$$T_{\lambda\alpha} = V_\alpha + V_\lambda G_\lambda^+ (V_\alpha + G_\lambda^+ V_\alpha) \tag{162}$$

$$= V_\alpha + V_\lambda G_\lambda^+ T_{\lambda\alpha} \tag{163}$$

This is precisely a LS equation for $T_{\lambda\alpha}$ and it can be solved for the reactive transition T-matrix elements by the noniterative procedure. Analogously, if we consider the nonreactive process, Equation 163 yields (for $\lambda = \alpha$) the LS equation:

$$T_{\alpha\alpha} = V_{\alpha} + V_{\alpha}G_{\alpha}^{+} T_{\alpha\alpha} \quad (164)$$

Now solving these two equations will give us (in principle) the nonreactive and reactive T-matrixes. However, a little thought easily shows that unless great care is taken, erroneous results will be obtained. The problems stem from the fact that the eigenstate decomposition of the Green's operator (G_{α}^{+}) actually involves both discrete and continuous states of the diatom in arrangement α (or equivalently, G_{λ}^{+} involves continuum states of the arrangement λ molecule). Normally, these eigenfunction expansions are truncated to a finite sum and the continuum portion is neglected. This is justified by the fact that at low energies, the dissociative continuum states of the various diatoms are not energetically accessible (they are said to be ''closed'' channels). However, the information that there are partitions of the system possible other than α is contained in Equation 164 in the continuum portion of the operator G_{α}^{+}. As a result, truncation of the Green's operator to include only discrete diatom states does not build in the influence of the other arrangements properly. The most disastrous consequence is that solution of the truncated version of Equation 164 yields an S-matrix that conserves *all* the flux in the nonreactive channel. This is because associated with Equation 164 is the companion equation for a reactance operator:

$$R_{\alpha\alpha} = V_{\alpha} + V_{\alpha}G_{\alpha}^{P} R_{\alpha\alpha} \quad (165)$$

The $\hat{T}_{\alpha\alpha}$ and $\hat{R}_{\alpha\alpha}$ are related via a damping relation just like Equation 70. When the dissociative continuum is neglected, these become simple matrix equations which force the resulting approximate $\hat{T}_{\alpha\alpha}$ matrix elements to conserve all the flux. However, the reactive transition amplitudes are being computed separately via Equation 163 (also with a truncated Green's operator expansion). Again, there is no communication as to flux entering other partitions. The net result is that calculations using Equations 163 and 164 or their reactance operator equivalents yield total probabilities that are always bigger than one. No practical method of including the continuum portions of the Green's operators has yet been devised so that these equations are not useful. There are similar problems which arise when one tries to solve Equations 146 or 152 noniteratively.

Another aspect of the LS equation can be seen when one iterates any of the equations given above. For example, we can iterate Equation 163 to obtain:

$$T_{\lambda\alpha} = V_{\alpha} + V_{\lambda}G_{\lambda}^{+} V_{\alpha} + V_{\lambda}G_{\lambda}^{+} V_{\lambda}G_{\lambda}^{+}V_{\alpha} + \ldots \quad (166)$$

In order to illustrate our point, we consider a specific system of three particles which interact by pairwise additive forces. Then we label the particles 1, 2, 3 and partitions also 1, 2, 3. We take partition i as that in which particle i is free and particles j and k are bound as a diatom. Then:

$$V = V^{1} + V^{2} + V^{3} \quad (167)$$

where V^{i} equals the interaction between j and k, and thus:

$$V_{i} = \sum_{j \neq i} V^{j} \quad (168)$$

$$H_{i} = T + V^{i} \quad (169)$$

with $j \neq i$, $k \neq i$, $k \neq j$, and T is the kinetic energy. Let us be specific to illustrate and take $\alpha = 1$, $\lambda = 2$. Then V_1 equals the sum of V^2 and V^3 while V_2 equals the sum of V^1 and V^3. We then have:

$$T_{21} = V^2 + V^3 + (V^1 + V^3)\, G_2^+\, (V^2 + V^3) + \ldots \tag{170}$$

Now in the first two terms, there is one particle which does not interact. The third term contains four contributions and in three of them, all three particles interact (e.g., in the term $V^3 G_2 V^2$, particles 1 and 2 interact and particles 1 and 3 interact so there is no ''spectator'' particle). However, in the term $V^3 G_2 V^3$, only particles 1 and 2 interact while 3 is a spectator. The Green's function always contains terms which correspond to free propagation and in fact this is reflected in the continuum portion of the Green's operator. The term $V^3 G_2 V^3$ corresponds to a so-called disconnected diagram[4] and one may easily verify that no matter how high an iterate is taken in Equation 170, there will always be terms involving such disconnected diagrams. In the specific case of Equation 170, these all involve powers that contain only V^3 so that particle 3 is always a spectator. As a result, the kernel of the integral equation is not compact,[4] and this may cause problems in its numerical solution. It is the presence of the continuum portion of G_2 that is of concern and so we see that the basic source of difficulty in solving Equations 163 and 164 is the fact that they are not connected upon iteration. The solution of this difficulty may be achieved in a variety of ways. In part, it arises because we wanted to write what may be called a global solution. That is, the LS equation is an attempt to write the solution to the Schrödinger equation which also incorporates the asymptotic boundary conditions. For nonreactive scattering, this is straightforward because there is only a single asymptotic region possible. That is, in a nonreactive atom-diatom problem, the diatom is assumed not to dissociate or react and there is no continuum. For reactive systems, there is more than one asymptotic limit and concomitant boundary condition which must be described by the integral equation. In the case of the LS equation this is achieved only when the continuum portion of the Green's operator is included. One way to proceed is to solve the Schrödinger equation in a piecewise fashion and match the solutions in various regions with the appropriate asymptotic boundary conditions. This is the procedure used in most of the reactive studies carried out in the past decade.[90-91] It is not basically an integral equation method and we shall not consider it here. Another alternative is to introduce new types of coordinates which allow one to separate into all the asymptotic regions smoothly.[17] Again, this is not essentially an integral equation method in character and we shall not pursue it. Instead, in the following section, we describe a method for developing more general integral equations for reactive scattering than the LS equation. These equations avoid the problems alluded to herein.

IV. MATRIX INTEGRAL EQUATION FORMALISMS

The first integral equations which avoid the problems of the LS equation for reactive systems were those developed by Faddeev.[14] Subsequently, there has been an enormous amount of research devoted to these and other similar types of integral equations which are capable of treating reactive systems.[19-75] In our discussion we shall assume that the collision energy will be well below the dissociation threshold. Thus, we shall consider only systems in which there are two clusters in each partition (one an atom and the other a diatom). This does *not* necessarily mean we are truncating the Green's operator, however. In a part of our discussion, it will be convenient to consider the three-particle system with pairwise interactions but our primary concern will be three-body systems with general interactions. We shall not be concerned with electronically nonadiabatic systems so that each particle will be effectively structureless.

A. Generalized Lippmann-Schwinger Equations for Green's and Transition Operators

We recall that in Section III one could write an LS equation for the full Green's operators (G^+) in terms of any one of the partition Green's operators (G^+_λ). Each partition Green's operator is most convenient for describing the asymptotic behavior of the system in its own particular partition. What is desired is to obtain equations for the transition operators which incorporate explicitly all the various possible asymptotic limits. To do this, we follow Baer and Kouri[30] in introducing the channel or partition coupling array **W** whose elements satisfy the conditions:

$$\sum_{\lambda} W_{\lambda\lambda'} = \sum_{\lambda'} W_{\bar{\lambda}\lambda'} = 1 \tag{171}$$

The elements $W_{\lambda\lambda'}$ are themselves ordinary numbers rather than operators. We now multiply the equation:

$$G^+ = G^+_{\lambda''} + G^+_{\lambda''} V_{\lambda''} G^+ \tag{172}$$

by $W_{\lambda\,\lambda''}$, sum over λ'', and use Equation 171 to write:

$$G^+ = \sum_{\lambda''} W_{\bar{\lambda}\lambda'} (G^+_{\lambda''} + G^+_{\lambda''} V_{\lambda''} G^+) \tag{173}$$

We now consider the basic equation for an arbitrary transition operator: $\tilde{T}_{\lambda\lambda'}$,

$$T_{\lambda\lambda'} = V_\lambda + V_\lambda G^+ V_{\lambda'} \tag{174}$$

and substitute Equation 173 into 174, obtaining:

$$T_{\lambda\lambda'} = V_\lambda + V_\lambda \sum_{\lambda''} W_{\bar{\lambda}\lambda''} (G^+_{\lambda''} + G^+_{\lambda''} V_{\lambda''} G^+) V_{\lambda'} \tag{175}$$

$$= V_\lambda + V_\lambda \sum_{\lambda''} W_{\bar{\lambda}\lambda''} G^+_{\lambda''} (V_{\lambda'} + V_{\lambda''} G^+ V_{\lambda'}) \tag{176}$$

This in turn can be expressed as:

$$\tilde{T}_{\lambda\lambda'} = V_\lambda + V_\lambda \sum_{\lambda''} W_{\bar{\lambda}\lambda''} G^+_{\lambda''} (\tilde{T}_{\lambda''\lambda'} + V_{\lambda'} - V_{\lambda''}) \tag{177}$$

Let us consider the term:

$$V_\lambda \sum_{\lambda''} W_{\bar{\lambda}\lambda''} G^+_{\lambda''} (V_{\lambda'} - V_{\lambda''}) = V_\lambda \sum_{\lambda''} W_{\bar{\lambda}\lambda''} G^+_{\lambda''} (H_{\lambda''} - H_{\lambda'}) \tag{178}$$

We add and subtract $E + i\epsilon$ and rewrite this as:

$$= V_\lambda \sum_{\lambda''} W_{\bar{\lambda}\lambda''} G^+_{\lambda''} (G^+_{\lambda'})^{-1} - V_\lambda \sum_{\lambda''} W_{\bar{\lambda}\lambda''} G^+_{\lambda''} (G^+_{\lambda''})^{-1} \tag{179}$$

Now by Equation 171 this becomes:

$$V_\lambda \sum_{\lambda''} W_{\bar{\lambda}\lambda''} G^+_{\lambda''} (V_{\lambda'} - V_{\lambda''}) = V_\lambda \left(\sum_{\lambda''} W_{\bar{\lambda}\lambda''} G^+_{\lambda''} (G^+_{\lambda'})^{-1} - 1 \right) \tag{180}$$

so that our equation for $T_{\lambda\lambda'}$ is

$$T_{\lambda\lambda'} = V_{\lambda} \sum_{\lambda''} W_{\bar{\lambda}\lambda''} G^{+}_{\lambda''} (G^{+}_{\lambda'})^{-1} + V_{\lambda} \sum_{\lambda''} W_{\bar{\lambda}\lambda''} G^{+}_{\lambda''} T_{\lambda''\lambda'} \tag{181}$$

This is the new integral equation for $T_{\lambda\lambda'}$ which will replace the uncoupled equations considered in Section III. We note that the structure of the equation depends on how the elements $W_{\lambda\lambda'}$ are assigned.

One can also obtain similar integral equations for the operators $T_{\lambda\lambda'}$, and the result there is

$$T_{\lambda\lambda'} = V_{\lambda'} + V_{\lambda} \sum_{\lambda''} W_{\bar{\lambda}\lambda''} G^{+}_{\lambda''} T_{\lambda''\lambda'} \tag{182}$$

In this case, the analysis of Equations 178 to 180 is not necessary. For reasons which will soon be made clear, we shall work primarily with the form of Equation 181.

One may also examine the effect of $\tilde{T}_{\lambda\lambda'}$ on a particular initial state $|\phi(\lambda' i)\rangle$ with energy E. We note that:

$$T_{\lambda\lambda'} |\phi(\lambda' i)\rangle = V_{\lambda} \sum_{\lambda''} W_{\bar{\lambda}\lambda''} G^{+}_{\lambda''} (G^{+}_{\lambda'})^{-1} |\phi(\lambda' i)\rangle$$

$$+ V_{\lambda} \sum_{\lambda''} W_{\bar{\lambda}\lambda''} T_{\lambda''\lambda'} |\phi(\lambda' i)\rangle \tag{183}$$

and it is clear that in the first term on the right,

$$(G^{+}_{\lambda'})^{-1} |\phi(\lambda' i)\rangle = i\epsilon |\phi(\lambda' i)\rangle \tag{184}$$

except when $\lambda'' = \lambda'$ in the sum. Thus, Equation 183 becomes:

$$T_{\lambda\lambda'} |\phi(\lambda' i)\rangle = V_{\lambda} W_{\bar{\lambda}\lambda'} |\phi(\lambda' i)\rangle$$

$$+ i\epsilon V_{\lambda} \sum_{\lambda'' \neq \lambda'} W_{\bar{\lambda}\lambda'} G^{+}_{\lambda''} |\phi(\lambda' i)\rangle$$

$$+ V_{\lambda} \sum_{\lambda''} W_{\bar{\lambda}\lambda'} G^{+}_{\lambda''} T_{\lambda''\lambda'} |\phi(\lambda' i)\rangle \tag{185}$$

Now the terms proportional to ϵ on the r.h.s. of the above equation tend to zero as $\epsilon \to 0_{+}$ provided only that the state $|\phi(\lambda' i)\rangle$ is not a breakup state. However, at energies below the dissociation threshold, one never has to consider initial states corresponding to three free particles and thus, this term goes to zero with ϵ. In fact, we can replace the operators $T_{\lambda''\lambda'}$ by new operators $T_{\lambda''\lambda'}$ defined by the equations:

$$T_{\lambda\lambda'} = V_{\lambda} W_{\bar{\lambda}\lambda'} + \sum_{\lambda''} (V_{\lambda} W_{\bar{\lambda}\lambda''}) G^{+}_{\lambda''} T_{\lambda''\lambda'} \tag{186}$$

These operators have *exactly* the same physical matrix elements as do the $T_{\lambda\lambda'}$ once ϵ has been taken to zero (which must be done in order to get the physical matrix element). These various T operators correspond to what are known as different "off-shell" extensions of the transition operator. This simply means that as operators, they are not equal because their nonphysical matrix elements need not be equal even though their physical matrix elements are. That is, the operator is defined in terms of its matrix elements with respect to the *entire* complete sets $\{<\phi(\lambda,j)|\}$ and $\{|\phi(\lambda',t)>\}$ and not all these states have the same energy as that at which the collision is occurring. Equation 186 is the basic general equation that will replace the LS equation when one deals with reactive scattering.

B. Faddeev-Type Integral Equations

In order to proceed further, it is necessary to be more specific as to the choice of the elements of the channel or partition coupling array **W**. We begin by showing how one may obtain equations which are essentially those due to Faddeev. In fact, there are numerous alternative versions of these equations which again correspond to various possible off-shell extensions of the transition operator. We begin by noting that in substituting Equation 173 into 174 the choice of the index $\bar{\lambda}$ was completely free. We shall consider again the special situation of pairwise interactions. Then, one can write the perturbation (V_λ) as

$$V_\lambda = \sum_{\tilde{\lambda}=1}^{3} (1 - \delta_{\lambda\tilde{\lambda}})V^{\tilde{\lambda}} \tag{187}$$

Then Equation 159 can be written as:

$$T_{\lambda\lambda'} = V_\lambda + \sum_{\tilde{\lambda}=1}^{3} (1 - \delta_{\lambda\tilde{\lambda}})V^{\tilde{\lambda}}\, G^+ V_{\lambda'} \tag{188}$$

and a different form of Equation 173 for G^+ can be used in each term of the sum over $\tilde{\lambda}$. In particular, we shall choose $\bar{\lambda}$ to be equal to $\tilde{\lambda}$ so that our general Equation 171 becomes:

$$T_{\lambda\lambda'} = \sum_{\tilde{\lambda}=1}^{3} (1 - \delta_{\lambda\tilde{\lambda}})V^{\tilde{\lambda}}\, W_{\tilde{\lambda}\lambda'} + \sum_{\lambda''}\sum_{\tilde{\lambda}=1}^{3} (1 - \delta_{\lambda\tilde{\lambda}})V^{\tilde{\lambda}}\, W_{\tilde{\lambda}\lambda''}\, G^+_{\lambda''}\, T_{\lambda''\lambda'} \tag{189}$$

Finally, we choose for $W_{\tilde{\lambda}\lambda''}$ a Kronnecker delta,

$$W_{\tilde{\lambda}\lambda''} = \delta_{\tilde{\lambda}\lambda''} \tag{190}$$

so that our equation reads:

$$T_{\lambda\lambda'} = (1 - \delta_{\lambda\lambda'})V^{\lambda'} + \sum_{\lambda''} (1 - \delta_{\lambda\lambda''})V^{\lambda''}\, G^+_{\lambda''}\, T_{\lambda''\lambda'} \tag{191}$$

This is one form of Faddeev's equations. It is customary to write it slightly differently. We note that:

$$G^+_{\lambda''} = 1/(E - T - V^{\lambda''} + i\epsilon) \quad (192)$$

and we note that $V^{\lambda''}$ is the interaction potential between the two particles which form the bound cluster in arrangement λ'' (i.e., if $\lambda''=1$, then $V^{\lambda''}$ is the interaction between particles 2 and 3). Thus, $G^+_{\lambda''}$ is the Green's operator that would arise if the two particles (other than particle λ'') were colliding in the presence of a third spectator particle. Then by Equation 64 we may write:

$$V^{\lambda''} G^+_{\lambda''} = T_{\lambda''} G^+_o \quad (193)$$

where G^+_o is the totally free Green's operator,

$$G^+_o = (E - T + i\epsilon)^{-1} \quad (194)$$

and T_λ'' is a two-particle (nonreactive) T-operator embedded in a three-body space (with particle λ'' being the extra spectator particle). It satisfies the equation:

$$T_{\lambda''} = V^{\lambda''} + V^{\lambda''} G^+_{\lambda''} V^{\lambda''} \quad (195)$$

It is helpful in visualizing the equations to write them out more explicitly. Let us take the initial arrangement to be 1 (so particle 1 is free and (2,3) are bound initially). Then our equations read:

$$T_{11} = T_2 G^+_o T_{21} + T_3 G^+_o T_{31} \quad (196)$$

$$T_{21} = V^1 + T_1 G^+_o T_{11} + T_3 G^+_o T_{31} \quad (197)$$

$$T_{31} = V^1 + T_1 G^+_o T_{11} + T_2 G^+_o T_{21} \quad (198)$$

and

$$V^1 = V_{23} \quad (199)$$

with v_{23} the interaction between particles 2 and 3. These equations can also be iterated and the result can be understood if one substitutes, e.g., Equations 196 and 198 into 197. This yields:

$$T_{21} = V^1 + T_1 G^+_o (T_2 G^+_o T_{21} + T_3 G^+_o T_{31})$$

$$+ T_3 G^+_o (V^1 + T_1 G^+_o T_{11} + T_2 G^+_o T_{21}) \quad (200)$$

It is clear that none of the terms has just a single type two-body T operator occurring. Instead, there are products like:

$$P = T_1 G^+_o T_2;\ \ T_1 G^+_o T_3;\ \ T_3 G^+_o T_1;\ \ T_3 G^+_o T_2;\ \ T_3 G^+_o V^1 \quad (201)$$

In order to see the general pattern, we consider $T_1 G_o{}^+ T_2$. In T_1, there occurs the interaction V^1 to infinite order while in T_2, V^2 occurs. Thus, each term in $T_1 G^+_o T_2$ involves products in each of which particles 2 and 3 interact and also particles 1 and 3 interact. Thus, this quantity is completely connected since there are no spectator particles. Similarly, all of the other products do not have any spectator particles so that after a single iteration Equations 196 to 198 become connected. This greatly ameliorates the role played by the continuum in these equations, and makes the Faddeev equations amenable to solution by techniques not feasible with the LS equation. We remark that although the discussion of the connectivity of the Faddeev equation is usually couched in terms of

$$T_{\lambda\lambda'} = (1 - \delta_{\lambda\lambda'})V^{\lambda'} + \sum_{\lambda''} (1 - \delta_{\lambda\lambda''})T_{\lambda''}G_o^+ T_{\lambda''\lambda'} \tag{202}$$

there is no difference if one instead uses the form of Equation 191.

To illustrate another off-shell extension of the Faddeev equations (known as the AGS equations[20]) we note that:

$$(E - H_{\lambda'} + i\epsilon)|\phi(\lambda'i)> = 0 \tag{203}$$

in the limit $\epsilon \rightarrow 0_+$. But we then note that:

$$(E - H_{\lambda'} + i\epsilon)|\phi(\lambda'i)> = (E - T + i\epsilon)|\phi(\lambda'i)> - V^{\lambda'}|\phi(\lambda'i)> \tag{204}$$

so that in the limit that $\epsilon \rightarrow 0_+$,

$$(E - T + i\epsilon)|\phi(\lambda'i)> = V^{\lambda'}|\phi(\lambda'i)> \tag{205}$$

$$= (G_o^+)^{-1}|\phi(\lambda'i)> \tag{206}$$

As a result, we may define:

$$T^{AGS}_{\lambda\lambda'} = (1 - \delta_{\lambda\lambda'})(G_o^+)^{-1} + \sum_{\lambda''} (1 - \delta_{\lambda\lambda''})T_{\lambda''}G_o^+ \, T^{AGS}_{\lambda''\lambda'} \tag{207}$$

as an alternative off-shell extension. It is clear from Equations 205 and 206 that physical matrix elements of $T^{AGS}_{\lambda\lambda'}$ and $T_{\lambda\lambda'}$ will be equal.

It is possible to write the Faddeev equations in a matrix form. If we use Equation 191, then define matrixes of arrangement channel operators:

$$(\mathcal{V})_{\lambda\lambda'} = (1 - \delta_{\lambda\lambda'})V^{\lambda'} \tag{208}$$

$$(\mathcal{G}_o^+)_{\lambda\lambda'} = \delta_{\lambda\lambda'}\, G_{\lambda'}^+ \tag{209}$$

and

$$(\mathcal{T})_{\lambda\lambda'} = T_{\lambda\lambda'} \tag{210}$$

our coupled equations can be written as:

$$\mathcal{T} = \mathcal{V} + \mathcal{V}\mathcal{G}_o^+\mathcal{T} \tag{211}$$

The analogy with the form of the simple nonreactive LS equation (Equation 61) is striking. We shall examine this in more detail later.

C. Baer-Kouri-Levin-Tobocman Equations

Another set of coupled integral equations which has received much attention are those known now as the BKLT equations.[23,30,34,35,36,38] They arise from the general form of Equation 186 in which $\bar{\lambda}$ is chosen to equal λ, so that one has:

$$T_{\lambda\lambda'} = V_\lambda W_{\lambda\lambda'} + \sum_{\lambda''} V_\lambda W_{\lambda\lambda''} G^+_{\lambda''} T_{\lambda''\lambda'} \qquad (212)$$

Of course, there are a nondenumerably infinite number of ways to choose the $W_{\lambda\lambda'}$ since the only constraints are that they sum to one. The particular choice which has been most widely used is that of the channel or partition permuting choice. For the present three-body problem, there are two such choices,[35,36]

$$\mathbf{W} = \begin{pmatrix} 0 & 1 & 0 \\ 0 & 0 & 1 \\ 1 & 0 & 0 \end{pmatrix} \qquad (213)$$

and

$$\mathbf{W} = \begin{pmatrix} 0 & 0 & 1 \\ 1 & 0 & 0 \\ 0 & 1 & 0 \end{pmatrix} \qquad (214)$$

However, the channel or partition permuting scheme was actually first utilized in studies of piecewise constant potential models (e.g., the Hulburt-Hirschfelder model[92]) for a collinear atom-diatom reactive system.[23,24] In that case, there are only two arrangements and a single permuting array:

$$\mathbf{W} = \begin{pmatrix} 0 & 1 \\ 1 & 0 \end{pmatrix} \qquad (215)$$

In this section, we shall concern ourselves with the more general three-partition case. We shall arbitrarily choose one of the permuting schemes, in particular that of Equation 213. It is again helpful to consider the specific example used in the Faddeev case so we assign λ' equal to 1. Then our equations read:

$$T_{11} = V_1 G_2^+ T_{21} \qquad (216)$$

$$T_{21} = V_2 G_3^+ T_{31} \qquad (217)$$

and

$$T_{31} = V_3 + V_3 G_1^+ T_{11} \qquad (218)$$

since W_{11} and W_{21} are zero while W_{31} equals one. These equations have a very interesting structure. We first try to iterate them by substituting Equation 218 into Equation 217 followed by substitution into 216. The result is

$$T_{11} = V_1 G_2^+ V_2 G_3^+ V_3 (1 + G_1^+ T_{11}) \qquad (219)$$

The first point we note is that the equations have decoupled upon iteration! However, in the process of decoupling the equations, the kernel of the final integral equation is now much more complicated. In particular, we note that the quantity $V_1G_2^+V_2G_3^+V_3$ appears as an effective potential so that one has the form:

$$T_{11} = \mathcal{U}_1 + \mathcal{U}_1G_1^+ T_{11} \tag{220}$$

Because of the presence of the G_2^+ and G_3^+ in U_1, it is *nonlocal* and *energy dependent*. This reminds one of an optical potential. In fact, these features of U_1 are a reflection of the presence of other arrangements into which flux can flow. Thus, unlike Equation 164, even if one writes the related reactance operator equation:

$$R_{11} = \mathcal{U}_1 + \mathcal{U}_1G_1^P R_{11} \tag{221}$$

one will not conserve all the flux in arrangement 1 because U_1 is complex and R_{11} will therefore not produce an S_{11} which is unitary by itself.

Another feature of interest in the iterated equations is the question of connectivity. We note that for pairwise interactions,

$$V_1 = V_{12} + V_{13} \tag{222}$$

$$V_2 = V_{12} + V_{23} \tag{223}$$

and

$$V_3 = V_{13} + V_{23} \tag{224}$$

We then note that the connectivity structure of the kernel of Equation 219 is determined solely by the product of V_1, V_2, and V_3, which is

$$V_1V_2V_3 = (V_{12} + V_{13})(V_{12} + V_{23})(V_{13} + V_{23}) \tag{225}$$

$$= (V_{12}^2 + V_{12}V_{23} + V_{13}V_{12} + V_{13}V_{23})(V_{13} + V_{23}) \tag{226}$$

Now in the first factor in Equation 226, only the term V_{12}^2 is not already completely connected. It corresponds to the successive interaction of particle 1 and 2 with 3 present only as a spectator. However, the other factor contains two interactions, *each of which involves particle 3*. Thus, the product V_{12}^2 $(V_{13} + V_{23})$ is also completely connected. We see then that our BKLT equations with the partition permuting choice of **W** also has a kernel which becomes connected (after two iterations) and therefore again the problems of the continuum portions of the G_λ^+ should be ameliorated just as they are in the Faddeev equations.

We now consider the matrix form of the BKLT equations. We define:

$$(\mathcal{V})_{\lambda\lambda'} = V_\lambda W_{\lambda\lambda'} \tag{227}$$

$$(\mathcal{G}_o^+)_{\lambda\lambda'} = \delta_{\lambda\lambda'}\, G_\lambda^+ \tag{228}$$

and

$$(\mathcal{T})_{\lambda\lambda'} = T_{\lambda\lambda'} \tag{229}$$

Equation 212 becomes:

$$\mathcal{T} = \mathcal{V} + \mathcal{V}\mathcal{G}_o^+\mathcal{T} \tag{230}$$

which is identical in form to the matrix Faddeev equations. We define a matrix generalized full Green's operator $\mathcal{G}^+$ by:

$$\mathcal{G}^+ = \mathcal{G}_o^+ + \mathcal{G}_o^+\mathcal{V}\mathcal{G}^+ \tag{231}$$

$$= \mathcal{G}_o^+ + \mathcal{G}^+\mathcal{V}\mathcal{G}_o^+ \tag{232}$$

If we apply $(\mathcal{G}^+)^{-1}$ from the right to Equation 231 and $(\mathcal{G}_o^+)^{-1}$ from the left, we find that:

$$(\mathcal{G}_o^+)^{-1} = (\mathcal{G}^+)^{-1} + \mathcal{V} \tag{233}$$

But $\mathcal{G}_o^+$ certainly exists and is given by:

$$[(\mathcal{G}_o^+)^{-1}]_{\lambda\lambda'} = \delta_{\lambda\lambda'}(E - H_\lambda + i\epsilon) \tag{234}$$

so that:

$$(\mathcal{G}^+)^{-1} = E - \mathcal{H}_o - \mathcal{V} + i\epsilon \tag{235}$$

where

$$(\mathcal{H}_o)_{\lambda\lambda'} = \delta_{\lambda\lambda'}H_\lambda \tag{236}$$

We therefore can use the matrix generalized LS equation for $\mathcal{G}^+$ to write Equation 230 as:

$$\mathcal{T} = \mathcal{V} + \mathcal{V}(\mathcal{G}^+ - \mathcal{G}^+\mathcal{V}\mathcal{G}_o^+)\mathcal{T} \tag{237}$$

$$= \mathcal{V} + \mathcal{V}\mathcal{G}^+\mathcal{T} - \mathcal{V}\mathcal{G}^+(\mathcal{V}\mathcal{G}_o^+\mathcal{T}) \tag{238}$$

Finally, the quantity $\mathcal{V}\mathcal{G}_o^+\mathcal{T}$ can be replaced using Equation 233 to arrive at:

$$\mathcal{T} = \mathcal{V} + \mathcal{V}\mathcal{G}^+\mathcal{V} \tag{239}$$

This is the arrangement channel space matrix of operators generalization of Equation 59 for nonreactive scattering and illustrates how the structure of the CIE formalism is the same. However, there *are* important differences and by far the most profound is the fact that the potential operator matrix $\mathcal{V}$ is not, in general, Hermitian. This is obvious for the channel or partition permuting choice of W-array, where, e.g., in our sample three-body problem:

$$\mathcal{V} = \begin{pmatrix} 0 & V_1 & 0 \\ 0 & 0 & V_2 \\ V_3 & 0 & 0 \end{pmatrix} \tag{240}$$

This introduces complications in the structure of the time-dependent formulation but one can still formulate the general scattering theory in this case.[33,61,70] We shall not discuss that here however.

There are other forms that the BKLT equations can take. One approach to these equations is to derive them not by starting with the basic $\tilde{T}_{\lambda\lambda'}$ or $\widetilde{T}_{\lambda\lambda'}$ operators at all but rather to begin with a matrix generalization of the Schrödinger equation and we shall briefly describe that approach now.[33,54,61,70] We note first that by Equations 133 and 171 we can write:

$$H = \sum_{\lambda} W'_{\lambda\lambda'} H_{\lambda'} + \sum_{\lambda} W_{\lambda\lambda'} V_{\lambda'} \tag{241}$$

for *any* λ'. Now we decompose the full state vector $|\Psi(\alpha i)>$ into partition components such that:

$$|\psi(\alpha i) > = \sum_{\lambda'} |\psi_{\lambda'}(\alpha i)> \tag{242}$$

We then write the usual Schrödinger equation in the form:

$$\sum_{\lambda'} \{E|\psi_{\lambda'}(\alpha i) > - \left(\sum_{\lambda} W'_{\lambda\bar{\lambda}} H_{\bar{\lambda}} + \sum_{\lambda} W_{\lambda\bar{\lambda}} V_{\bar{\lambda}}\right)|\psi_{\lambda'}(\alpha i)>\} = 0 \tag{243}$$

and recall that the term in parenthesis $(\sum_{\lambda} W_{\lambda\bar{\lambda}} H_{\bar{\lambda}} + \sum_{\lambda} W_{\lambda\bar{\lambda}}{}^{v}{}_{\bar{\lambda}}$ is independent of $\bar{\lambda}$ so that we can set $\bar{\lambda}$ equal to the summation index λ'. We then have:

$$\sum_{\lambda'} \{E|\psi_{\lambda'}(\alpha i)> - \left(\sum_{\lambda} W'_{\lambda\lambda'} H_{\lambda'} + \sum_{\lambda} W_{\lambda\lambda'} V_{\lambda'}\right)|\psi_{\lambda'}(\alpha i)>\} = 0 \tag{244}$$

We next note that:

$$\begin{aligned} &\sum_{\lambda}\sum_{\lambda'} (W'_{\lambda\lambda'} H_{\lambda'} + W_{\lambda\lambda'} V_{\lambda'})|\psi_{\lambda'}(\alpha i)> \\ &= \sum_{\lambda}\sum_{\lambda'} (W'_{\lambda'\lambda} H_{\lambda} + W_{\lambda'\lambda} V_{\lambda})|\psi_{\lambda}(\alpha i)> \end{aligned} \tag{245}$$

and thus Equation 244 can be written as:

$$\sum_{\lambda'} \{E|\psi_{\lambda'}(\alpha i)> - \left(\sum_{\lambda} W'_{\lambda'\lambda} H_{\lambda} + \sum_{\lambda} W_{\lambda'\lambda} V_{\lambda}\right)|\psi_{\lambda}(\alpha i)>\} = 0 \tag{246}$$

We then require that each term in the sum over λ' vanish separately, so that:

$$E|\psi_{\lambda}(\alpha i)> - \left(\sum_{\lambda'} W'_{\lambda\lambda'} H_{\lambda'} + \sum_{\lambda'} W_{\lambda\lambda'} V_{\lambda'}\right)|\psi_{\lambda'}(\alpha i)> = 0 \tag{247}$$

Finally, we shall take $W_{\lambda\lambda'}$ to equal a Kronnecker delta and our partition component Schrödinger equation is then:

$$(E - H_{\lambda})|\psi_{\lambda}(\alpha i)> - \sum_{\lambda'} W_{\lambda\lambda'} V_{\lambda'}|\psi_{\lambda'}(\alpha i)> = 0 \tag{248}$$

We can write this equation in the arrangement channel space matrix form by defining:

$$(|\Psi(\alpha i)>)_{\lambda} = |\psi_{\lambda}(\alpha i)> \tag{249}$$

$$(\mathcal{H}_o)_{\lambda\lambda'} = \delta_{\lambda\lambda'}H_\lambda \tag{250}$$

$$(\mathcal{E})_{\lambda\lambda'} = \delta_{\lambda\lambda'}E \tag{251}$$

$$(\mathcal{W})_{\lambda\lambda'} = W_{\lambda\lambda'} \tag{252}$$

$$\overline{\mathcal{V}} = \delta_{\lambda\lambda'}V_\lambda \tag{253}$$

Then we have:

$$(\mathcal{E} - \mathcal{H}_o)|\Psi(\alpha i)> = \mathcal{W}\overline{\mathcal{V}}|\Psi(\alpha i)> \tag{254}$$

and we note also by Equations 252, 253, and 227 that for a suitable choice of $W_{\lambda\lambda'}$ in Equation 252,

$$\mathcal{W}\overline{\mathcal{V}} = \mathcal{V}^T \tag{255}$$

We now may write an arrangement channel space matrix generalized LS equation for $|\Psi(\alpha i)>$,

$$|\Psi^+(\alpha i)> = |\Phi(\alpha i)> + \mathcal{G}_o^+\mathcal{V}^T|\Psi(\alpha i)> \tag{256}$$

where

$$(|\Phi(\alpha i)>)_\lambda = \delta_{\lambda\alpha}|\phi\ (\alpha i)> \tag{257}$$

It is also straightforward to define a matrix generalized T operator by:

$$\mathcal{T}|\Phi(\alpha i)> = \mathcal{V}^T|\Psi^+(\alpha i)> \tag{258}$$

This, combined with Equation 256 then yields:

$$\mathcal{T} = \mathcal{V}^T(1 + \mathcal{G}_o^+\mathcal{T}) \tag{259}$$

However, using an analog of Equation 232 we arrive at the result:

$$\mathcal{T} = \mathcal{V}^T + \mathcal{V}^T\mathcal{G}^+\mathcal{V}^T \tag{260}$$

where

$$(\mathcal{G}^+)^{-1} = \mathcal{E} + i\epsilon - \mathcal{H}_o - \mathcal{V}^T \tag{261}$$

In fact, it is clear that:

$$\mathcal{G}^+ = (\mathcal{G}^+)^T \tag{262}$$

and also,

$$\mathcal{T} = \mathcal{T}^T \tag{263}$$

Thus, we find that a formulation based on the matrix generalized Schrödinger equation leads to a transpose of the generalized matrix $\mathcal{T}$-operator obtained earlier. It is interesting at this point to note that the post and prior definitions of the ordinary reactive T operator satisfy the equations:

$$T_{\lambda\lambda'} = V_{\lambda'} + V_{\lambda} G^{+} V_{\lambda'} \tag{264}$$

$$T_{\lambda\lambda'} = V_{\lambda} + V_{\lambda} G^{+} V_{\lambda'} \tag{265}$$

Now we see that

$$T_{\lambda'\lambda} = (T_{\lambda\lambda'})^{T} \tag{266}$$

which is very similar to Equation 263.

D. Generalized Reactance and Scattering Operators

It is extremely simple to derive matrix generalized reactance operator equations at this point because we have seen that the structure of the BKLT (and Faddeev) matrix equations is precisely analogous to ordinary nonreactive scattering. Thus, by analogy to the discussion in Section II, we define $\mathcal{R}$ by:

$$\mathcal{R} = \mathcal{V} + \mathcal{V}\, \mathcal{G}_{o}^{P} \mathcal{R} \tag{267}$$

where

$$(\mathcal{G}_{o}^{P})_{\lambda\lambda'} = \delta_{\lambda\lambda'} \frac{P}{E - H_{\lambda}} \tag{268}$$

We next note that:

$$(\mathcal{G}_{o}^{+})_{\lambda\lambda'} = \delta_{\lambda\lambda'} \frac{1}{E - H_{\lambda} + i\epsilon} \tag{269}$$

$$= \delta_{\lambda\lambda'} \left[\frac{P}{E - H_{\lambda}} - i\pi\delta(E - H_{\lambda})\right] \tag{270}$$

Thus, we can write:

$$\mathcal{G}_{o}^{+} = \mathcal{G}_{o}^{P} - i\pi\delta(\mathcal{E} - \mathcal{H}_{o}) \tag{271}$$

where

$$(\delta(\mathcal{E} - \mathcal{H}_{o}))_{\lambda\lambda'} = \delta_{\lambda\lambda'}\, \delta(E - H_{\lambda}) \tag{272}$$

We now subtract Equation 267 from Equation 230 to obtain:

$$\mathcal{T} - \mathcal{R} = \mathcal{G}_{o}^{P} (\mathcal{T} - \mathcal{R}) - i\pi\mathcal{V}\, \delta(\mathcal{E} - \mathcal{H}_{o})\mathcal{T} \tag{273}$$

We note that by Equations 267 and 273,

$$\mathcal{R} = (\mathcal{I} - \mathcal{V}\, \mathcal{G}_{o}^{P})^{-1}\mathcal{V} \tag{274}$$

and

$$\mathcal{T} - \mathcal{R} = -i\pi(\mathcal{I} - \mathcal{V}\mathcal{G}_o^P)^{-1}\mathcal{V}\,\delta(\mathcal{E} - \mathcal{H}_o)\mathcal{T} \tag{275}$$

so that:

$$\mathcal{T} = \mathcal{R} - i\pi\,\mathcal{R}\,\delta(\mathcal{E} - \mathcal{H}_o)\mathcal{T} \tag{276}$$

This is the channel or partition matrix generalization of the Heitler damping equation.[34,86] The equations for $\mathcal{R}$ have the computational advantage that they only involve real arithmetic. We note that the structure of the equations is independent of whether one chooses to use the BKLT or Faddeev-type coupled equations. The alternative form of BKLT equations which results from the use of Equation 182 does not lead to such a simple generalization of the damping equation.[33] For this reason, that set of equations is not currently used in numerical applications to realistic chemical reaction systems. It is also straightforward to introduce the partition matrix generalization of the scattering operator $\mathcal{S}$. We define it by:

$$\mathcal{S} = \mathcal{I} - 2\pi i\,\delta(\mathcal{E} - \mathcal{H}_o)\mathcal{T} \tag{277}$$

Then solving Equation 276 for $\mathcal{T}$ and substituting the result yields the expression:

$$\mathcal{S} = (\mathcal{I} - i\pi\delta(\mathcal{E} - \mathcal{H}_o)\mathcal{R})(\mathcal{I} + i\pi\delta(\mathcal{E} - \mathcal{H}_o)\mathcal{R})^{-1} \tag{278}$$

If the actual matrix elements of $\mathcal{R}$ satisfy the necessary symmetry conditions, the $\mathcal{S}$ is guaranteed to satisfy unitarity.

E. Distorted Wave Green's Operators and Partition Matrix Generalized Equations

In actual applications of the BKLT equations to realistic reactive systems, it is essential to make use of distortion potentials in defining the various unperturbed Green's functions. The reason for this has to do with the fact that the totally unperturbed translational functions in each arrangement do not tend to zero rapidly enough in regions where the interactions can be quite large. Although this does not invalidate the equations, it does create problems with respect to the rate of convergence of numerical solutions of the equations and thus, basis sets are required that are impractically large. By introducing distortion potentials appropriate to each arrangement λ, the radial portions of the Green's operators can be made to go to zero in highly nonclassical regions, and the result is that many fewer expansion functions are required. The distorted wave version of the BLKT equations was first presented by Tobocman.[36] It causes almost no complication in the formalism except that now, the states between which matrix elements of the $T_{\lambda\lambda'}$ operators are taken must satisfy certain boundary conditions. In particular, the matrix elements arising from the solution of Equation 230 (or equivalently, Equation 212) is of the form:

$$\langle\chi^{(-)}(\lambda f)|T_{\lambda\lambda'}|\chi^{(+)}(\lambda' i)\rangle \tag{279}$$

where $\chi^{(-)}$ satisfies incoming scattered wave boundary conditions and $\chi^{(+)}$ has outgoing scattered wave boundary conditions. The full T matrix is given by:

$$\begin{aligned}\langle\phi(\lambda f)|T^{\text{full}}_{\lambda\lambda'}|\phi(\lambda' i)\rangle &= \delta_{\lambda\lambda'}\langle\phi(\lambda f)|T_o|\phi(\lambda' i)\rangle \\ &\quad + \langle\chi^{(-)}(\lambda f)|T_{\lambda\lambda'}|\chi^{(+)}(\lambda' i)\rangle\end{aligned} \tag{280}$$

where $<\phi(\lambda f)|T_o|\phi(\lambda' i)>$ gives the scattering due just to the distortion potential and the states $|\phi(\lambda f)>$ and $|\phi(\lambda' i)>$ are noninteracting. Also, we have taken the distortion potential to be such that it cannot produce reaction or internal excitation of the target molecules. In such a case $<\phi(\lambda f)|T_o|\phi(\lambda' i)>$ is diagonal in λ,λ' and also in all the internal molecular quantum numbers. Tobocman[75] has shown that the full S matrix elements are given by:

$$<\phi(\lambda f)|S^{full}_{\lambda\lambda'}|\phi(\lambda' i)> = \exp(i\delta_\lambda(f))<\chi^{(-)}(\lambda f)|S_{\lambda\lambda''}|\chi^{(+)}(\lambda' i)> \exp(i\delta_{\lambda'}(i)) \quad (281)$$

where the distorted wave S matrix element on the right is obtained from the finite basis version of Equation 263. The phase shifts $\delta_\lambda(f)$ and $\delta_{\lambda'}(i)$ are elastic scattering phase shifts in arrangements λ and λ' with the system in state f and i, respectively. The analogy of this expression for reactive systems with Equation 111 for nonreactive ones is evident. We now turn to consider how one goes about solving these CIEs.

V. NUMERICAL ASPECTS OF THE BKLT EQUATIONS

In this section we shall discuss the details of how one can solve the BKLT equations for reaction. Although initially we shall restrict our consideration to collinear atom-diatom reactive systems, the basic methods and concepts readily generalize to three-dimensional systems. Of course, such systems will be more complicated. One may choose to solve either the equations for $T_{\lambda'\lambda}$ or $R_{\lambda'\lambda}$. If one chooses the former, then it turns out that one can obtain results for a single initial condition of the system. This is in marked contrast to the situation with propagative methods where one must solve for all possible initial conditions. The reason this is necessary has to do with the imposition of boundary conditions. One must form linear combinations of the linearly independent solutions of the Schrödinger equation in order to obtain solutions that satisfy the physical boundary conditions because the numerically generated solutions do not automatically satisfy these conditions. By contrast, the boundary conditions are explicitly contained in the BKLT integral equations and as a result, one does not have to obtain all the linearly independent solutions. On the other hand, the $T_{\lambda'\lambda}$ equations involve complex arithmetic while the $R_{\lambda'\lambda}$ are real. If one solves for the $R_{\lambda'\lambda}$ matrix elements and generates the $T_{\lambda'\lambda}$ or $S_{\lambda'\lambda}$ matrix elements via the generalized Heitler damping equation, then it is necessary to obtain the matrix elements for all initial and final states in order to solve the damping equation for the scattering matrix elements. We shall choose to solve the real $R_{\lambda'\lambda}$ equations here.

Another feature of the propagative methods is that one must carry out so-called stabilization transformations on the matrix of solutions in order to maintain their linear independence. Although there are propagation techniques which do not stabilize in the classic sense,[93] these methods essentially generate the R-matrix in various subregions and then join the solutions in various regions by a connection formula. The calculation of the R-matrix in the various subregions essentially stabilizes them. In the BKLT-type formalism, there is no propagation of linearly independent solutions and as a result, nothing corresponding to stabilization is required. We shall begin by discussing the solution of the BKLT equations as Fredholm integral equations. Next we shall examine them to see if we can obtain Volterra-type equations. The approach taken in the Fredholm discussion is patterned after that developed by Shima et al.[66]

A. Coordinate Representation of the BKLT Equations

In order to solve the equations, it is necessary to (1) put them into the coordinate representation and (2) make the various interactions, Green's functions, reference states, expansion bases, etc. explicit. We consider the case of a collinear reaction with the coordinates shown in Figure 1.

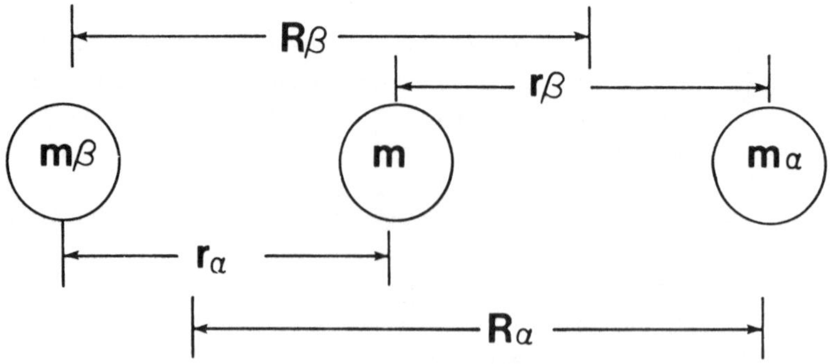

FIGURE 1. Coordinate system for the reaction system. R_γ is the center of mass scattering variable and r_γ the internal vibrational variable for arrangement γ. (From Shima, Y., Baer, M., and Kouri, D. J., *J. Chem. Phys.*, 78, 6666, 1983. With permission).

Thus, $\bar{R}_\alpha$ is the actual separation between the projectile in arrangement α, having mass m_α, and the center of mass of the diatom and $\bar{r}_\alpha$ is the separation distance for the α-arrangement diatom. We use the convention that the central atom has mass m and the end atom of the α-arrangement diatom has mass m_β. There are analogous coordinates $\bar{R}_\beta$, $\bar{r}_\beta$ for the β-arrangement. We then define mass-scaled coordinates as usual by:

$$r_\gamma = f_\gamma \bar{r}_\gamma \tag{282}$$

$$R_\gamma = \bar{R}_\gamma / f_\gamma \tag{283}$$

where

$$f_\gamma = \left[\frac{m m_{\gamma'}(m_\gamma + m + m_{\gamma'})}{m_\gamma(m + m_{\gamma'})^2}\right]^{1/2}, \quad \gamma \neq \gamma' \tag{284}$$

$$= (\mu_\gamma/\mu)^{1/2} \tag{285}$$

Here

$$\mu_\gamma = m m_\gamma/(m + m_\gamma) \tag{286}$$

and

$$\mu = \left(\frac{m m_{\gamma'} m_\gamma}{m + m_\gamma + m_{\gamma'}}\right)^{1/2} \tag{287}$$

It is clear that the system reduced mass (μ) is independent of which arrangement is being considered. The coordinates for different arrangements are related via a unitary transformation so that:

$$R_\beta = R_\alpha \cos\chi + r_\alpha \sin\chi \tag{288}$$

$$r_\beta = R_\alpha \sin\chi - r_\alpha \cos\chi \tag{289}$$

$$R_\alpha = R_\beta \cos\chi + r_\beta \sin\chi \tag{290}$$

$$r_\alpha = R_\beta \sin\chi - r_\beta \cos\chi \quad (291)$$

with

$$\cos\chi = [m_\alpha m_\beta/(m + m_\alpha)(m + m_\beta)]^{1/2} \quad (292)$$

$$\sin\chi = [m(m_\alpha + m + m_\beta)/(m + m_\alpha)(m + m_\beta)]^{1/2} \quad (293)$$

$$\tan\chi = [m(m_\alpha + m + m_\beta)/m_\alpha m_\beta]^{1/2} \quad (294)$$

$$= m/\mu \quad (295)$$

The angle χ is the skewing angle and it places a restriction on how large r_α can be for a fixed value of R_α:

$$r_\alpha < R_\alpha \tan\chi \quad (296)$$

For systems like HFH, χ is about 87.13° and r_α can be large for given R_α while for heavy-light-heavy systems such as IHI, χ is very small. The H_3 system has $\chi = 60°$. The possible processes that can occur in our model collinear system are:

$$A + BC \rightarrow AB + C \quad (297)$$

$$A + BC \rightarrow A + BC \quad (298)$$

$$AB + C \rightarrow A + BC \quad (299)$$

$$AB + C \rightarrow AB + C \quad (300)$$

Thus, there are four possible processes that must be described. We shall employ the BKLT equations with the channel or partition permuting choice of the W-array so our basic equations read:

$$R_{\alpha\alpha} = V_\alpha G_\beta^P R_{\beta\alpha} \quad (301)$$

$$R_{\beta\alpha} = V_\beta + V_\beta G_\alpha^P R_{\alpha\alpha} \quad (302)$$

$$R_{\beta\beta} = V_\beta G_\alpha^P R_{\alpha\beta} \quad (303)$$

$$R_{\alpha\beta} = V_\alpha + V_\alpha G_\beta^P R_{\beta\beta} \quad (304)$$

It is clear that there is no coupling between the various initial states and we therefore concentrate on the equations with α as the initial arrangement. We now wish to apply these operators to a particular initial state, $|\theta(\alpha n_o k_{n0})>$, to obtain:

$$|\zeta(\alpha|\alpha n_o k_{n_o})> = V_\alpha G_\beta^P|\zeta(\beta|\alpha n_o k_{n_o})> \quad (305)$$

$$|\zeta(\beta|\alpha n_o k_{n_o})> = V_\beta|\theta(\alpha n_o k_{n_o})> + V_\beta G_\alpha^P|\zeta(\alpha|\alpha n_o k_{n_o})> \quad (306)$$

where

$$|\zeta(\gamma|\alpha n_o k_{n_o})> = R_{\gamma\alpha}\,|\theta(\alpha n_o k_{n_o})> \tag{307}$$

We now project with the coordinate representative bra $<R_\alpha r_\alpha|$ on Equation 290 and $<R_\beta r_\beta|$ on Equation 291 to obtain:

$$\zeta(\alpha|\alpha n_o k_{n_o}|R_\alpha r_\alpha) = <r_\alpha R_\alpha|V_\alpha G_\beta^p|\zeta(\beta|\alpha n_o k_{n_o})> \tag{308}$$

$$\zeta(\beta|\alpha\ n_o k_{n_o}|R_\beta r_\beta) = <r_\beta R_\beta|V_\beta|\theta(\alpha n_o k_{n_o})> + <r_\beta R_\beta|V_\beta G_\alpha^p|\zeta(\alpha|\alpha n_o k_{n_o})> \tag{309}$$

Here R_γ is the distance from the atom to the diatom center of mass in arrangement γ and r_γ is the separation distance for the α-arrangement diatom. We then insert resolutions of the identity in the appropriate coordinate representation to convert Equations 308 and 309 into the integral equations:

$$\zeta(\alpha|\alpha n_o k_{n_o}|r_\alpha R_\alpha) = \int dr'_\alpha\, dR'_\alpha dr'_\beta\, dR'_\beta\, dr_\beta dR_\beta <r_\alpha R_\alpha|V_\alpha|r'_\alpha R'_\alpha> \times <r'_\alpha R'_\alpha|r_\beta R_\beta> G_\beta^p(r_\beta R_\beta|r'_\beta R'_\beta)\zeta(\beta|\alpha n_o k_{n_o}|r'_\beta R'_\beta) \tag{310}$$

and

$$\begin{aligned}\zeta(\beta|\alpha n_o k_{n_o}|r_\beta R_\beta) = &\int dr'_\beta\, dR'_\beta\, dr_\alpha dR_\alpha < r_\beta R_\beta|V_\beta|r'_\beta R'_\beta> \\ &\times < r'_\beta R'_\beta|r_\alpha R_\alpha> \theta\,(\alpha n_o k_{n_o}|r_\alpha R_\alpha) \\ &+ \int dr'_\beta\, dR'_\beta\, dr_\alpha\, dR_\alpha dr'_\alpha\, dR'_\alpha < r_\beta R_\beta|V_\beta|r'_\beta R'_\beta >< r'_\beta R'_\beta|r_\alpha R_\alpha > \\ &\times G_\alpha^p(r_\alpha R_\alpha|r'_\alpha R'_\alpha)\zeta(\alpha|\alpha n_o k_{n_o}|r'_\alpha R'_\alpha)\end{aligned} \tag{311}$$

where

$$\zeta(\gamma|\alpha n_o k_{n_o}|r_\gamma R_\gamma) = < r_\gamma R_\gamma|\zeta(\gamma|\alpha n_o k_{n_o})> \tag{312}$$

Now the potentials V_α and V_β involve just the usual Born-Oppenheimer potential energy surfaces and these are local, i.e.,

$$<r_\gamma R_\gamma|V_\gamma|r'_\gamma R'_\gamma> = \delta(r_\gamma - r'_\gamma)\delta(R_\gamma - R'_\gamma)V_\gamma(r_\gamma R_\gamma) \tag{313}$$

In addition, the inner products $<r_\beta' R_\beta'|r_\alpha R_\alpha>$ and $<r_\alpha' R_\alpha'|r_\beta R_\beta>$ are easily evaluated since a given configuration of the ABC collinear system can be described either by r_α, R_α, or r_β, R_β. Thus, $<r_\alpha' R_\alpha'|r_\beta R_\beta>$ is given by

$$<r'_\alpha R'_\alpha|r_\beta R_\beta> = \delta((r'_\beta R'_\beta) - (r_\alpha R_\alpha)) \tag{314}$$

where the notation signifies that one expresses r_α as a function of r_β', R_β' and also expresses R_α as a function of r_β', R_β', or vice versa. We then can write:

$$\zeta(\alpha|\alpha n_o k_{n_o}|r_\alpha R_\alpha) = V_\alpha(r_\alpha R_\alpha)\int_0^\infty dR'_\beta \int_0^{\bar{r}_\beta} dr'_\beta\; G_\beta^p\,(r_\beta R_\beta|r'_\beta R'_\beta) \times \zeta(\beta|\alpha n_o k_{n_o}|r'_\beta R'_\beta) \tag{315}$$

with r_β, R_β describing the *same* configuration as r_α, R_α via Equations 288 and 289, and

$$\zeta(\beta|\alpha n_o k_{n_o}|r_\beta R_\beta) = V_\beta(r_\beta R_\beta)\theta(\alpha n_o k_{n_o}|r_\alpha R_\alpha) + V_\beta(r_\beta R_\beta)\int_0^\infty dR'_\alpha \int_0^{\bar{r}_\alpha} dr'_\alpha G^P_\alpha(r_\alpha R_\alpha|r'_\alpha R'_\alpha)\zeta(\alpha|\alpha n_o k_{n_o}|r'_\alpha R'_\alpha) \tag{316}$$

with r_α, R_α describing the same point as r_β, R_β (via Equations 290 and 291). The limits on the integrals over the r_γ' ($\bar{r}_\alpha$ and $\bar{r}_\beta$) reflect the fact that r_γ' cannot exceed R_γ' tan χ. Now in order to proceed further, we must specify more about the coordinate representation Green's functions, the unperturbed states $\theta(\gamma n k_n|r_\gamma R_\gamma)$, and the perturbations. The perturbation in the α-arrangement is given by:

$$V_\alpha(r_\alpha R_\alpha) = \bar{V}_\alpha(r_\alpha R_\alpha) - \bar{V}_\alpha(r_{\alpha o} R_\alpha) - \bar{V}_\alpha(r_\alpha R_{\alpha L}) - D_\alpha \tag{317}$$

where $\bar{V}_\alpha(r_\alpha R_\alpha)$ is the full potential for the system (subscripted by α *only* because we write it as a function of the α-arrangement variables), $\bar{V}_\alpha(r_{\alpha o} R_\alpha)$ is a distortion potential (which is taken to be the full potential evaluated at a specific value of the initial diatom coordinate, namely the equilibrium separation $r_{\alpha 0}$), $\bar{V}_\alpha(r_\alpha R_{\alpha L})$ is the value of the potential in the limit of large $R_\alpha \equiv R_{\alpha L}$ (this is just the diatom binding potential in arrangement α) and because the zero of the potential is defined to be $\bar{V}_\alpha = 0$, r_α and $R_\alpha \rightarrow \infty$, we also subtract the depth (D_α) of the diatom binding potential. This latter point implies that $\bar{V}_\alpha(r_\alpha R_{\alpha L})$ tends to zero as r_α becomes large. Next we must know the coordinate representations of the principal value Green's operators G^P_γ, $\gamma = \alpha$ or β. In addition, the calculation of the physical probability amplitudes will require the matrix elements of the $R_{\gamma\alpha}$, $\gamma = \alpha$, β between the eigenstates of the reference Hamiltonians for the α and β arrangements. Essentially, we have that:

$$G^P_o = P\left(\frac{1}{E - H_{o\gamma} - \bar{V}_\gamma(r_{\gamma o} R_\gamma) - \bar{V}_\gamma(r_\gamma R_{\gamma L})}\right) \tag{318}$$

and we require the solutions of

$$[E - H_{o\gamma} - \bar{V}_\gamma(r_{\gamma o} R_\gamma) - \bar{V}_\gamma(r_\gamma R_{\gamma L})]\ \theta(\gamma n k_n|r_\gamma R_\gamma) = 0 \tag{319}$$

both to expand the G^P_γ and also to serve as possible initial and final states in matrix elements of the $R_{\gamma\alpha}$. That is, we must calculate:

$$R(\gamma m|\alpha n) = \langle\theta(\gamma m k_m)|R_{\gamma\alpha}|\theta(\alpha n k_n)\rangle \tag{320}$$

and also expand $G^P_\gamma(r_\gamma R_\gamma|r_\gamma' R_\gamma')$ in terms of the solutions of Equation 319. We note that Equation 319 is separable so that the solutions $\theta(\gamma n k_n|r_\gamma R_\gamma)$ can be expressed as products of the solutions of the equations:

$$\left[-\frac{\hbar^2}{2\mu}\frac{d^2}{dr_\gamma^2} + \bar{V}_\gamma(r_\gamma R_{\gamma L})\right]\phi(\gamma n|r_\gamma) = (\epsilon^\gamma_n - D_\gamma)\phi(\gamma n|r_\gamma) \tag{321}$$

$$\left[-\frac{\hbar^2}{2\mu}\frac{d^2}{dR_\alpha^2} + \bar{V}_\alpha(r_{\alpha o} R_\alpha)\right]\psi(\alpha k_n|R_\alpha) = (E - \epsilon^\alpha_n - D_\alpha)\psi(\alpha k_n|R_\alpha) \tag{322}$$

and

$$\left[-\frac{\hbar^2}{2\mu}\frac{d^2}{dR_\beta^2} + \overline{V}_\beta(r_{\beta o}R_\beta)\right]\psi(\beta k_n|R_\beta) = (E - \epsilon_n^\beta - D_\alpha)\psi(\beta k_n|R_\beta) \tag{323}$$

In the above, ϵ_n^α is the (positive) eigenvalue associated with $\theta(\alpha n|r_\alpha)$ and $E - D_\alpha$ is the total energy of the system. One must solve Equations 322 and 323 for both the regular and irregular solutions in order to construct the principal value Green's function. We require the regular solution to satisfy the asymptotic boundary condition:

$$\psi^{reg}(\gamma k_n|R_\gamma) \sim \frac{1}{k_n^{1/2}}\sin(k_n R_\gamma - \delta_\gamma(n)) \tag{324}$$

where

$$\hbar^2 k_n^2/2\mu = E - \epsilon_n^\gamma - D_\alpha \tag{325}$$

and the derivative is

$$\frac{d}{dR_\gamma}\psi^{reg}(\gamma k_n|R_\gamma) \sim k_n^{1/2}\cos(k_n R_\gamma - \delta_\gamma(n)) \tag{326}$$

The irregular solutions are obtained by integrating Equations 322 and 323 inward toward the origin with the initial conditions

$$\psi^{irr}(\gamma k_n|R_\gamma) = -\frac{1}{k_n^{1/2}}\cos(k_n R_\gamma - \delta_\gamma(n)) \tag{327}$$

and

$$\frac{d\psi^{irr}}{dR_\gamma}(\gamma k_n|R_\gamma) = k_n^{1/2}\sin(k_n R_\gamma - \delta_\gamma(n)) \tag{328}$$

This gives a Wronskian which equals one. The procedure for closed channels is similar except that the regular solution has the asymptotic form exp $(\kappa_m R_\gamma)$, where

$$\frac{\hbar^2\kappa_m^2}{2\mu} = D_\alpha + \epsilon_n^\gamma - E > 0 \tag{329}$$

and the irregular solution tends to zero for large R_γ. We then express $G_\gamma^P(r_\gamma R_\gamma|r_\gamma' R_\gamma')$ as:[88]

$$G_\gamma^P(r_\gamma R_\gamma|r_\gamma' R_\gamma') = \sum_n G_n^\gamma(R_\gamma|R_\gamma')\phi(\gamma n|r_\gamma)\phi(\gamma n|r_\gamma') \tag{330}$$

where

$$G_n^\gamma(R_\gamma|R_\gamma') = \frac{2\mu}{\hbar^2}\psi^{reg}(\gamma k_n|R_\gamma^<)\psi^{irr}(\gamma k_n|R_\gamma^>) \tag{331}$$

and

$$\theta(\gamma n k_n|r_\gamma R_\gamma) = \psi^{reg}(\gamma k_n|R_\gamma)\phi(\gamma n|r_\gamma) \tag{332}$$

We next substitute these results into Equations 315 and 316 to obtain:

$$\zeta(\alpha|\alpha n_o k_{n_o}|r_\alpha R_\alpha) = V_\alpha(r_\alpha R_\alpha) \sum_{n=1}^{P\beta} \phi(\beta n|r_\beta) \int dR'_\beta dr'_\beta G_n^\beta(R_\beta|R'_\beta) \times \phi(\beta n|r'_\beta)\zeta(\beta|\alpha n_o k_{n_o}|r'_\beta R'_\beta) \tag{333}$$

$$\zeta(\beta|\alpha n_o k_{n_o}|r_\beta R_\beta) = V_\beta(r_\beta R_\beta)\phi(\alpha n_o|r_\alpha)\, \psi^{reg}(\alpha k_{n_o}|R_\alpha) + V_\beta(r_\beta R_\beta) \sum_{n=1}^{P\alpha} \phi(\alpha n|r_\alpha) \int dR'_\alpha dr'_\alpha G_n^\alpha (R_\alpha|R'_\alpha)\phi(\alpha n|r'_\alpha) \times \zeta(\alpha|\alpha n_o k_{n_o}|r'_\alpha R'_\alpha) \tag{334}$$

Here, P_γ is the number of expansion terms retained in the Green's function for arrangement γ. It is now possible to examine how the amplitude densities behave as either r_α or R_α becomes large. The behavior with respect to R_α is very simple since in the above equations, it is clear that $\zeta(\gamma|\alpha n_0 k_{no}|r_\gamma R_\gamma)$ is proportional to V_γ and it is defined (cf. Equation 302) so that it goes to zero as R_γ gets large. Thus, we have that

$$\lim_{R\gamma \to \text{large}} \zeta(\gamma|an_o k_{n_o}|r_\gamma R_\gamma) = 0 \tag{335}$$

This is analogous to the behavior of the nonreactive amplitude density. The behavior with respect to r_γ is more subtle. First we note by Equation 317 that V_γ does *not* tend to zero as r_γ gets large. However, we note that by Equation 296, the only way that r_γ can become large is if R_γ is also large. In this case, the potential V_γ *does* tend to zero so that we also have:

$$\lim_{r\gamma \to \text{large}} \zeta(\gamma|\alpha n_o k_{n_o}|r_\gamma R_\gamma) = 0 \tag{336}$$

As a consequence, we see that the amplitude density is confined and we can use quadratically integrable (L^2) functions as an expansion basis. This fact is quite interesting since it implies that the philosophy underlying the solution of the present equations is much more akin to traditional quantum chemistry that to close-coupling methods. It ultimately will be important to examine the use of a variety of basis functions, including perhaps some used in the study of electronic structure. Whether these will be efficient for heavy particle reactions remains to be seen. (Of course, quantum chemists are generally interested in eigenvalue problems, whereas the reactive scattering problem will consist of linear *inhomogeneous* integral equations.) At any rate, this feature of the BKLT equations is novel and may have important practical consequences so far as the efficiency of this approach to reactive scattering is concerned. This feature also causes the present BKLT approach to bear some similarities to R-matrix theory. The strongest similarity arises in the types of integrals that are calculated. The connection with R-matrix theory has been discussed recently in some detail by Tobocman.[75] However, his approach is somewhat different from that which we consider here. For example, at no point have we divided up configuration space into an inner region for which the solutions to the ordinary Schrödinger equation are L^2. Nor do we then carry out a matching of such interior solutions to scattering solutions on various ''channel entrance surfaces'' which connect the inner region to the respective asymptotic regions associated with the different possible arrangements. Instead, we make use of the fact that just like in ordinary structureless particle scattering, the BKLT amplitude density function belongs to the L^2 class of functions (for suitable potential functions). As we noted earlier, the amplitude density associated with the ordinary LS equation does *not* have this property for reactive systems due to the fact that there is now more than one asymptotic region due to the occurrence of more than one arrangement. By introducing CIEs, one essentially obtains an amplitude density for each arrangement and each one of these turns out to be L^2 in the scattering variable for its respective arrangement. Thus, the analogy with the R-matrix approach primarily arises from the fact that they both generate L^2 functions as unknowns but they do so in quite different ways.

B. Solution of the BKLT Equations as Fredholm Integral Equations

Now in order to proceed further, we must expand the $\zeta(\gamma|\alpha n_o k_{no}|r_\gamma R_\gamma)$ in a basis set. There are, of course, a great multitude of possibilities. The simplest is to expand the r_γ dependence in terms of the asymptotic vibrational states $\theta(\gamma n|r_\gamma)$, satisfying Equation 321. A more general procedure would be to use a basis set that changes as R_γ does (essentially an adiabatic-type basis). We shall first derive the equations assuming such a general basis set, which we denote by $\{\xi^\gamma_m(r_\gamma R_\gamma)\}$. They satisfy the condition:

$$\int_0^{\tilde{r}_\gamma} dr_\gamma \; \xi^\gamma_m(r_\gamma R_\gamma)\xi^\gamma_{m'}(r_\gamma R_\gamma) = \delta_{mm'} \tag{337}$$

for any given value of R_γ. Note again that $\tilde{r}_\gamma$ is $R_\gamma \tan\chi$. Then we can write:

$$\zeta(\gamma|\alpha n_o k_{n_o}|r_\gamma R_\gamma) = \sum_{m=1}^{N_\gamma} Z(\gamma m|\alpha n_o k_{n_o}|R_\gamma)\xi^\gamma_m(r_\gamma R_\gamma) \tag{338}$$

For any fixed R_γ, this expression can be inverted to yield the coefficients Z in terms of the ξ. The reactance matrix elements are given by (cf. Equations 319 and 332):

$$R(\gamma n'|\alpha n_o) = \frac{2\mu}{\hbar^2}\int_0^\infty dR_\gamma \int_0^{\tilde{r}_\gamma} dr_\gamma \; \psi^{reg}(\gamma k_{n'}|R_\gamma)\phi(\gamma n'|r_\gamma) \sum_{m=1}^{N_\gamma} \xi^\gamma_m(r_\gamma R_\gamma) \times Z(\gamma m|\alpha n_o k_{n_o}|R_\gamma) \tag{339}$$

Now the integral over r_γ will *only* give a Kronnecker delta for large R_γ values (where the expansion eigenstates $\xi^\gamma_m(r_\gamma R_\gamma)$ are just the asymptotic vibrational states of the diatom, $\phi(\gamma m|r_\gamma)$ and so we introduce the overlap:

$$S^\gamma(n'm|R_\gamma) = \int_0^{\tilde{r}_\gamma} dr_\gamma \; \phi(\gamma n'|r_\gamma)\, \xi^\gamma_m(r_\gamma R_\gamma) \tag{340}$$

Then we have:

$$R(\gamma n'|\alpha n_o) = \frac{2\mu}{\hbar^2}\sum_{m=1}^{N_\gamma}\int_0^\infty dR_\gamma \; \psi^{reg}(\gamma k_{n'}|R_\gamma) \times Z(\gamma m|\alpha n_o k_{n_o}|R_\gamma)S^\gamma(n'm|R_\gamma) \tag{341}$$

Now we substitute Equation 338 into Equations 333 and 334 to obtain:

$$\sum_{m=1}^{N_\alpha} Z(\alpha m|\alpha n_o k_{n_o}|R_\alpha)\xi^\alpha_m(r_\alpha R_\alpha) = V_\alpha(r_\alpha R_\alpha)\sum_{n=1}^{P_\beta}\phi(\beta n|r_\beta) \times \int dR'_\beta\, dr'_\beta\, G^\beta_n(R_\beta|R'_\beta)\phi(\beta n|r'_\beta)\sum_{m'=1}^{N_\beta} Z(\beta m'|\alpha n_o k_{n_o}|R'_\beta) \times \xi^\beta_{m'}(R'_\beta\, r'_\beta) \tag{342}$$

and a similar equation for the other arrangement. We multiply by a particular $\xi^{\alpha}_{m}(r_\alpha R_\alpha)$ and integrate over r_α to obtain:

$$Z(\alpha m|\alpha n_o k_{n_o}|R_\alpha) = \int_0^{r_\alpha} dr_\alpha\, \xi^{\alpha}_{m}(r_\alpha R_\alpha) V_\alpha(r_\alpha R_\alpha) \sum_{n=1}^{P_\beta} \phi(\beta n|r_\beta) \int_0^{\infty} dR'_\beta\, G^\beta_n (R_\beta|R'_\beta) \times \sum_{m'=1}^{N_\beta} Z(\beta m'|\alpha n_o k_{n_o}|R'_\beta) S^\beta(nm'|R'_\beta) \tag{343}$$

and for the other arrangement, we have:

$$Z(\beta m'|\alpha n_o k_{n_o}|R_\beta) = \int_0^{r_\beta} dr_\beta\, \xi^{\beta}_{m'}(r_\beta R_\beta) V_\beta(r_\beta R_\beta) \phi(\alpha n_o|r_\alpha) \psi^{reg}(\alpha k_{n_o}|R_\alpha) + \int_0^{r_\beta} dr_\beta\, \xi^{\beta}_{m'}(r_\beta R_\beta) V_\beta(r_\beta R_\beta) \sum_{n=1}^{P_\alpha} \phi(\alpha n|r_\alpha) \int_0^{\infty} dR'_\alpha\, G^\alpha_n (R_\alpha|R'_\alpha) \times \sum_{m=1}^{N_\alpha} Z(\alpha m|\alpha n_o k_{n_o}|R'_\alpha)\, S^\alpha(nm|R'_\beta) \tag{344}$$

Here N_γ, $\gamma = \alpha$ or β is the number of expansion functions used for the r_γ dependence of $\zeta(\gamma|\alpha n_o k_{no}|r_\gamma R_\gamma)$ and P_γ is the number of asymptotic vibrational states taken in expanding the Green's function G^P_γ. The final step is to expand the R_γ dependence of $Z(\gamma m|\alpha n_o k_{no}|R_\gamma)$ in an L^2 basis of orthonormal functions $\lambda(\gamma t|R_\gamma)$:

$$Z(\gamma m|\alpha n_o k_{n_o}|R_\gamma) = \sum_{t=1}^{M_\gamma} a(\gamma mt|\alpha n_o k_{n_o}) \lambda(\gamma t|R_\gamma) \tag{345}$$

It then follows that the $R(\gamma n|\alpha n_o)$ are given by:

$$R(\gamma n|\alpha n_o) = \frac{2\mu}{\hbar^2} \sum_{m=1}^{N_\gamma} \sum_{t=1}^{M_\gamma} a(\gamma mt|\alpha n_o k_{n_o}) \times \int_0^{\infty} dR_\gamma \psi^{reg}(\gamma k_n|R_\gamma) \lambda(\gamma t|R_\gamma) S^\gamma(nm|R_\gamma) \tag{346}$$

The coefficients $a\,(\gamma mt|\alpha n_o k_{no})$ satisfy the coupled equations:

$$a(\alpha mt|\alpha n_o k_{n_o}) = \int_0^{\infty} dR_\alpha \int_0^{r_\alpha} dr_\alpha \lambda(\alpha t|R_\alpha) \xi^{\alpha}_{m}(r_\alpha R_\alpha) V_\alpha(r_\alpha R_\alpha) \times \sum_{n=1}^{P_\beta} \phi(\beta n|r_\beta) \int_0^{\infty} dR'_\beta\, G^\beta_n (R_\beta|R'_\beta) \sum_{m'=1}^{N_\beta} \sum_{t'=1}^{M_\beta} S^\beta(nm'|R'_\beta) \times \lambda(\beta t'|R'_\beta) a(\beta m't'|\alpha n_o k_{n_o}) \tag{347}$$

and

$$a(\beta m't'|\alpha n_o k_{n_o}) = \int_0^\infty dR_\beta \int_0^{r_\beta} dr_\beta\ \lambda(\beta t'|R_\beta)\ \xi^\beta_{m'}\ (r_\beta R_\beta)V_\beta(r_\beta R_\beta)\phi(\alpha n_o|r_\alpha)$$

$$\times\ \psi^{reg}(\alpha k_{n_o}|R_\alpha) + \int_0^\infty dR_\beta \int_0^{r_\beta} dr_\beta\ \lambda(\beta t'|R_\beta)\xi^\beta_{m'}\ (r_\beta R_\beta)V_\beta(r_\beta R_\beta) \sum_{n=1}^{P_\alpha} \phi(\alpha n|r_\alpha)$$

$$\times \int_0^\infty dR'_\alpha\ G^\alpha_n\ (R_\alpha|R'_\alpha) \sum_{m=1}^{N_\alpha} \sum_{t=1}^{M_\alpha} S^\alpha(nm|R'_\alpha)\lambda(\alpha t|R'_\alpha) \tag{348}$$

$$\times\ a(\alpha mt|\alpha n_o k_{n_o})$$

These are the algebraic equations which one must solve (they are the BKLT reactive analogs of Equation 115). These equations simplify slightly if one takes as the $\xi^\gamma_m(r_\gamma R_\gamma)$ basis the asymptotic vibrational states $\phi(\gamma m|r_\gamma)$. Then the $S^\gamma(nm|R_\gamma)$ become equal to Kronnecker deltas provided R_γ is not so small that the upper limit $(\bar{r}_\gamma)$ in Equation 340 is too small. If the system is not too highly skewed, this should not pose problems. If the system is highly skewed and the potential such that the projectile and target can get close to one another, then the general form of Equations 347 and 348 must be used. In the cases where the minimum R_γ that contributes (say $R_{\gamma o}$) is not too small so that $S^\gamma\ (nm|R_\gamma)$ is always a Kronnecker delta (assuming we have chosen $\xi^\gamma_m(r_\gamma R_\gamma) \equiv \phi(\gamma m|r_\gamma)$), then the R-matrix is given by:

$$R(\gamma n|\alpha n_o) = \frac{2\mu}{\hbar^2} \sum_{t=1}^{M_\gamma} a(\gamma nt|\alpha n_o k_{n_o}) \int_0^\infty dR_\gamma\ \psi^{reg}(\gamma k_n|R_\gamma)\lambda(\gamma t|R_\gamma) \tag{349}$$

$$\equiv \frac{2\mu}{\hbar^2} \sum_{t=1}^{M_\gamma} a(\gamma nt|\alpha n_o k_{n_o})\omega(\gamma nt) \tag{350}$$

We now introduce a more compact notation and write our equations in the form:

$$a(\alpha n't'|\alpha n_o k_{n_o}) = \sum_{n=1}^{N_\beta} \sum_{t=1}^{M_\beta} C(\alpha n't'|\beta nt)a(\beta nt|\alpha n_o k_{n_o}) \tag{351}$$

and

$$a(\beta nt|\alpha n_o k_{n_o}) = b(\beta nt|\alpha n_o k_{n_o})$$

$$+ \sum_{n'=1}^{N_\alpha} \sum_{t'=1}^{M_\alpha} C(\beta nt|\alpha n't')a(\alpha n't'|\alpha n_o k_{n_o}) \tag{352}$$

where

$$b(\beta nt|\alpha n_o k_{n_o}) = \int_{R_{\beta 0}}^\infty dR_\beta \int_0^{r_\beta} dr_\beta\lambda(\beta t|R_\beta)\xi^\beta_n(r_\beta R_\beta)V_\beta(r_\beta R_\beta)\phi(\alpha n_o|r_\alpha)\psi^{reg}(\alpha k_{n_o}|R_\alpha) \tag{353}$$

$$C(\alpha n't'|\beta nt) = \sum_{m=1}^{P_\beta} \int_{R_{\alpha 0}}^\infty dR_\alpha \int_0^{r_\alpha} dr_\alpha\ \xi^\alpha_{n'}\ (r_\alpha R_\alpha)\lambda(\alpha t'|R_\alpha)V_\alpha(r_\alpha R_\alpha)\phi(\beta m|r_\beta)$$

$$\times \int_{R_{\beta 0}}^\infty dR'_\beta\ G^\beta_m\ (R_\beta|R'_\beta)S^\beta(mn|R'_\beta)\lambda(\beta t|R'_\beta) \tag{354}$$

$$C(\beta nt|\alpha n't') = \sum_{m=1}^{P_\alpha} \int_{R_{\beta 0}}^{\infty} dR_\beta \int_0^{r_\beta} dr_\beta\, \xi_n^\beta\,(r_\beta R_\beta)\lambda(\beta t|R_\beta)V_\beta(r_\beta R_\beta)$$

$$\times\ \phi(\alpha m|r_\alpha) \int_{R_{\alpha 0}}^{\infty} dR'_\alpha\, G_m^\alpha\,(R_\alpha|R'_\alpha)S^\alpha(mn'|R'_\alpha)\lambda(\alpha t'|R'_\alpha) \tag{355}$$

We remark that for an asymmetric reaction in general, the coefficient matrixes need not be square. These quantities may be cast into a more convenient form for the purpose of computation. We note that

$$\int_{R_{\alpha 0}}^{\infty} dR_\alpha \int_0^{r_\alpha} dr_\alpha\, f = \int_{R_{\beta 0}}^{\infty} dR_\beta \int_0^{r_\beta} dr_\beta\, f \tag{356}$$

i.e., the double integral over r_α, R_α can be transformed into one over the entire range of r_β, R_β with a unit Jacobian. Then we write:

$$C(\alpha n't'|\beta nt)$$

$$= \sum_{m=1}^{P_\beta} \int_{R_{\beta 0}}^{\infty} dR_\beta \int_{R_{\beta 0}}^{\infty} dR'_\beta\, G_m^\beta\,(R_\beta|R'_\beta)S^\beta(mn|R'_\beta)\lambda(\beta t|R'_\beta) \tag{357}$$

$$\times\ E_{t'}\,(\alpha n'|\beta m|R_\beta)$$

where

$$E_{t'}\,(\alpha n'|\beta m|R_\beta) \equiv \int_0^{r_\beta} dr_\beta\, \xi_{n'}^\alpha\,(r_\alpha R_\alpha)\lambda(\alpha t'|R_\alpha)V_\alpha(r_\alpha R_\alpha)\phi(\beta m|r_\beta) \tag{358}$$

with

$$n' = 1, \ldots, N_\alpha \tag{359}$$

$$t' = 1, \ldots, M_\alpha \tag{360}$$

$$m = 1, \ldots, P_\beta \tag{361}$$

$$n = 1, \ldots, N_\beta \tag{362}$$

$$t = 1, \ldots, M_\beta \tag{363}$$

Similarly, we define

$$C(\beta nt|\alpha n't') = \sum_{m=1}^{P_\alpha} \int_{R_{\alpha 0}}^{\infty} dR_\alpha \int_{R_{\alpha 0}}^{\infty} dR'_\alpha\, G_m^\alpha\,(R_\alpha|R'_\alpha)S^\alpha(mn'|R'_\alpha)$$
$$\times\ \lambda(\alpha t'|R'_\alpha)E_t(\beta n|\alpha m|R_\alpha) \tag{364}$$

and

$$E_t(\beta n|\alpha m|R_\alpha) = \int_0^{r_\alpha} dr_\alpha\, \xi_n^\beta\,(r_\beta R_\beta)\lambda(\beta t|R_\beta)V_\beta(r_\beta R_\beta)\phi(\alpha m|r_\alpha) \tag{365}$$

Finally, the same procedure applies to $b(\beta nt|\alpha n_o k_{n_o})$ so that

$$b(\beta nt|\alpha n_o k_{n_o}) = \int_{R_{\alpha 0}}^{\infty} dR_\alpha \; \psi^{reg}(\alpha k_{n_o}|R_\alpha) E_t(\beta n|\alpha n_o|R_\alpha) \tag{366}$$

These forms are especially useful computationally because they enable one to separate a large amount of calculation which is independent of collision energy. Thus, the $E_{t'}(\alpha n'|\beta m|R_\beta)$ and $E_t(\beta n|\alpha m|R_\alpha)$ are independent of energy and are calculated once and stored, thereby reducing the time required for generating the coefficient matrix at other energies.

It is convenient to look now at the structure of the algebraic equations (Equations 351 and 352). The a coefficients constitute a vector and we can write the equations in the form:

$$\boldsymbol{a}^\alpha - \mathbf{C}^{\alpha\beta}\, \boldsymbol{a}^\beta = \mathbf{0} \tag{367}$$

$$-\mathbf{C}^{\beta\alpha}\boldsymbol{a}^\alpha + \boldsymbol{a}^\beta = \mathbf{b}^\beta \tag{368}$$

It is therefore necessary to invert the matrix:

$$\mathbf{C} = \begin{pmatrix} \mathbf{1}^{\alpha\alpha} & -\mathbf{C}^{\alpha\beta} \\ -\mathbf{C}^{\beta\alpha} & \mathbf{1}^{\beta\beta} \end{pmatrix} \tag{369}$$

The matrixes $1^{\gamma\gamma}$ have elements $\delta_{nn'}\delta_{tt'}$ and dimension $(N_\gamma \times M_\gamma)$ by $(N_\gamma \times M_\gamma)$. The matrix $\mathbf{C}_{\lambda\lambda'}$ has dimension $(N_\gamma \times M_\gamma)$ by $(N_\gamma' \times M_\gamma')$. Thus, the overall dimension of $\mathbf{C}$ is $(N_\alpha \times M_\alpha + N_\beta \times M_\beta)$ by $(N_\alpha \times M_\alpha + N_\beta \times M_\beta)$, i.e., $\mathbf{C}$ is a square matrix even when the $\mathbf{C}^{\gamma\gamma'}$ are not. Now it is clear from Equations 367 and 368 that back substitution decouples the algebraic equations so far as the arrangement channel index is concerned. If one substitutes Equation 367 into 368, the coefficient matrix becomes $\mathbf{1}^{\beta\beta} - \mathbf{C}^{\alpha\beta}\mathbf{C}^{\beta\alpha}$ and it is of dimension $(N_\beta \times M_\beta)$ by $(N_\beta \times M_\beta)$; if one substitutes Equation 368 into 367, then the coefficient matrix is $1^{\alpha\alpha} - C^{\alpha\beta}C^{\beta\alpha}$ and its dimension is $(N_\alpha \times M_\alpha)$ by $(N_\alpha \times M_\alpha)$. We are free to choose which back substitution is performed and in a general asymmetric system, the dimension of one may be much smaller than the other. Thus, if $(N_\alpha \times M_\alpha << (N_\beta \times M_\beta)$, one would invert $\mathbf{1}^{\alpha\alpha} - \mathbf{C}^{\alpha\beta}\mathbf{C}^{\beta\alpha}$ to solve for the $\mathbf{a}^\alpha$ coefficients. Then the $\mathbf{a}^\beta$ coefficients can be arrived at by a simple matrix multiplication. This could be a very important feature of these equations. However, even for a symmetric reaction, it is computationally better to do the back substitution and invert the resulting smaller coefficient matrix.

An alternative to matrix inversion is to try and solve the equations by iteration (indeed, the back substitution is the first step of an iteration). This requires multiplying matrixes, but, if it converges rapidly, may be less work than inverting the coefficient matrix.

Now it is also necessary to carry out calculations (in asymmetric systems) for the other possible initial condition. In the above equations, we took the initial arrangement to be α. The equations governing the situation where β is the initial condition are Equations 303 and 304. We shall write them in reversed order,

$$R_{\alpha\beta} = V_\alpha + V_\alpha G_\beta^P R_{\beta\beta} \tag{370}$$

and

$$R_{\beta\beta} = V_\beta G_\alpha^P R_{\alpha\beta} \tag{371}$$

If this is compared to the equations with α as the initial condition,

$$R_{\alpha\alpha} = V_\alpha G^P_\beta R_{\beta\alpha} \tag{372}$$

and

$$R_{\beta\alpha} = V_\beta + V_\beta G^P_\alpha R_{\alpha\alpha} \tag{373}$$

it should be clear that the kernels of the two sets are identical. As a result, when the same sort of analysis is applied to Equations 370 and 371, one finds the exact same coefficient matrix **C** as in Equation 369. The only difference in the two sets of equations occurs in the inhomogeneity, which we can denote as $\mathbf{b}^\alpha$. Thus, once the coefficient matrix has been inverted for one of the two sets of algebraic equations, the other set is also solved without any additional matrix inversion.

As a final point, once one has obtained the matrix elements of the R operator, the S matrix is determined (for the normalization we employ) via:

$$S(\gamma n|\alpha n_o) = \sum_{n'} \sum_{\gamma'} \{(\mathbf{1} + i\mathbf{R})^{-1}\}^{\gamma\gamma'}_{nn'} \{\mathbf{1} - i\mathbf{R}\}^{\gamma'\alpha}_{n'n_o} \tag{374}$$

and the physical S matrix is then obtained via Equation 281.

C. Volterra Equation Form of the BKLT Equations

We now wish to examine the BKLT equations along the same lines as the analysis of nonreactive scattering given in Equations 121 to 132. In that discussion it was shown how one can obtain Volterra equations in place of Fredholm equations. We begin with Equations 333 and 334 for the amplitude density functions and we explicitly eliminate the lesser and greater variables in the Green's functions $G^\alpha_n(R_\gamma|R'_\gamma)$. Then one has:

$$\begin{aligned}\zeta(\alpha|\alpha n_o k_{n_o}|r_\alpha R_\alpha) &= V_\alpha(r_\alpha R_\alpha) \sum_{n=1}^{P_\beta} \phi(\beta n|r_\beta) \int_0^{R_\beta} dR'_\beta \, [\psi^{irr}(\beta k_n|R_\beta) \\ &\times \psi^{reg}(\beta k_n|R'_\beta) - \psi^{reg}(\beta k_n|R_\beta)\psi^{irr}(\beta k_n|R'_\beta)] \int_0^{r'_\beta} dr'_\beta \, \phi(\beta n|r'_\beta) \\ &\times \zeta(\beta|\alpha n_o k_{n_o}|r'_\beta R'_\beta) + V_\alpha(r_\alpha R_\alpha) \sum_{n=1}^{P_\beta} \phi(\beta n|r_\beta) \, \psi^{reg}(\beta n|R_\beta) \\ &\times \int_0^{\infty} dR'_\beta \int_0^{r'_\beta} dr'_\beta \, \psi^{irr}(\beta k_n|R'_\beta)\phi(\beta n|r'_\beta)\zeta(\beta|\alpha n_o k_{n_o}|r'_\beta R'_\beta)\end{aligned} \tag{375}$$

and a similar equation for $\zeta(\beta|\alpha n_o k_{no}|r_\beta R_\beta)$, where

$$r'_\beta = R'_\beta \tan \chi \tag{376}$$

Now we define the constant:

$$\begin{aligned}B(\beta n|\alpha n_o) &= \int_0^{\infty} dR'_\beta \int_0^{r'_\beta} dr'_\beta \, \psi^{irr}(\beta k_n|R'_\beta)\phi(\beta n|r'_\beta) \\ &\times \zeta(\beta|\alpha n_o k_{n_o}|r'_\beta R'_\beta)\end{aligned} \tag{377}$$

Then Equation 375 becomes:

$$\zeta(\alpha|\alpha n_o k_{n_o}|r_\alpha R_\alpha) = V_\alpha(r_\alpha R_\alpha) \sum_{n=1}^{P_\beta} \phi(\beta n|r_\beta)\psi^{reg}(\beta k_n|R_\beta)B(\beta n|\alpha n_o)$$

$$+ \sum_{n=1}^{P_\beta} \int_0^{R_\beta} dR'_\beta \int_0^{r_\beta} dr'_\beta\, K(\alpha|\beta n|r_\beta R_\beta, r'_\beta R'_\beta)\zeta(\beta|\alpha n_o k_{n_o}|R'_\beta r'_\beta) \tag{378}$$

where

$$K(\alpha|\beta n|r_\beta R_\beta, r'_\beta R'_\beta) = V_\alpha(r_\alpha R_\alpha)\phi(\beta n|r_\beta)[\psi^{irr}(\beta k_n|R_\beta)\psi^{reg}(\beta k_n|R'_\beta)$$

$$- \psi^{reg}(\beta k_n|R_\beta)\psi^{irr}(\beta k_n|R'_\beta)]\ \phi(\beta n|R'_\beta) \tag{379}$$

Similarly, one has:

$$\zeta(\beta|\alpha n_o k_{n_o}|r_\beta R_\beta) = V_\beta(r_\beta R_\beta)\phi(\alpha n_o|r_\alpha)\psi^{reg}(\alpha k_{n_o}|R_\alpha)$$

$$+ V_\beta(r_\beta R_\beta) \sum_{n=1}^{P_\alpha} \phi(\alpha n|r_\alpha)\psi^{reg}(\alpha k_n|R_\alpha)B(\alpha n|\alpha n_o) \tag{380}$$

$$+ \sum_{n=1}^{P_\alpha} \int_0^{R_\alpha} dR'_\alpha \int_0^{r'_\alpha} dr'_\alpha\, K(\beta|\alpha n|r_\alpha R_\alpha, r'_\alpha R'_\alpha)\zeta(\alpha|\alpha n_o k_{n_o}|r'_\alpha R'_\alpha)$$

We can define the new constants:

$$D(\gamma n|\alpha n_o) = \delta_{\alpha\gamma}\delta_{nn_o} + B(\gamma n|\alpha n_o) \tag{381}$$

and write our equations as:

$$\zeta(\alpha|\alpha n_o k_{n_o}|r_\alpha R_\alpha) = V_\alpha(r_\alpha R_\alpha) \sum_{n=1}^{P_\beta} \phi(\beta n|r_\beta)\ \psi^{reg}(\beta k_n|R_\beta)D(\beta n|\alpha n_o)$$

$$+ \sum_{n=1}^{P_\beta} \int_0^{R_\beta} dR'_\beta \int_0^{r'_\beta} dr'_\beta\, K(\alpha|\beta n|r_\beta R_\beta r'_\beta R'_\beta)\zeta(\beta|\alpha n_o k_{n_o}|r'_\beta R'_\beta) \tag{382}$$

and

$$\zeta(\beta|\alpha n_o k_{n_o}|r_\alpha R_\alpha) = V_\beta(r_\beta R_\beta) \sum_{n=1}^{P_\alpha} \phi(\alpha n|r_\alpha)\ \psi^{reg}(\alpha n|R_\alpha)D(\alpha n|\alpha n_o)$$

$$+ \sum_{n=1}^{P_\alpha} \int_0^{R_\alpha} dR'_\alpha \int_0^{r'_\alpha} dr'_\alpha\, K(\beta|\alpha n|r_\alpha R_\alpha, r'_\alpha R'_\alpha)\zeta(\alpha|\alpha n_o k_{n_o}|r'_\alpha R'_\alpha) \tag{383}$$

We next define the matrixes:

$$(\boldsymbol{\zeta})_{\gamma n,\gamma' n'} = \zeta(\gamma|\gamma' n' k_{n'}|r_\gamma R_\gamma) \tag{384}$$

$$(\mathbf{V})_{\gamma n,\gamma' n'} = \delta_{\gamma\gamma'} V_\gamma(r_\gamma R_\gamma) \tag{385}$$

$$(\mathbf{K})_{\gamma n,\gamma' n'} = (1 - \delta_{\gamma\gamma'}) \int_0^{R_{\gamma'}} dR'_{\gamma'} \int_0^{r'_{\gamma'}} dr'_{\gamma'}\, K(\gamma|\gamma' n'|r_{\gamma'} R_{\gamma'}, r'_{\gamma'} R'_{\gamma'}) \tag{386}$$

$$(\mathbf{I})_{\gamma n,\gamma' n'} = (1 - \delta_{\gamma\gamma'})\phi(\gamma' n|r_{\gamma'})\ \psi^{reg}(\gamma' n|R_{\gamma'})\delta_{nn'} \tag{387}$$

$$(\mathbf{D})_{\gamma n,\gamma' n'} = \delta_{\gamma\gamma'}\ \delta_{nn'} + (\mathbf{B})_{\gamma n,\gamma' n'} \tag{388}$$

and

$$(\mathbf{B})_{\gamma n,\gamma' n'} = \int_0^{\infty} dR_{\gamma} \int_0^{r_{\gamma}} dr_{\gamma}\ \psi^{irr}(\gamma k_n|R_{\gamma})\phi(\gamma n|r_{\gamma})\zeta(\gamma|\gamma' n' k_{n'}|r_{\gamma}R_{\gamma}) \tag{389}$$

Then the equations can be written in matrix form as:

$$\zeta = \mathbf{V} \cdot \mathbf{I} \cdot \mathbf{D} + \mathbf{K} \cdot \zeta \tag{390}$$

where it is understood that $\hat{\mathbf{K}}$ is a matrix of integral operators. Now we can factor out the matrix of contants **D** by defining the new matrix of functions $\boldsymbol{\xi}$ by:

$$\zeta = \boldsymbol{\xi} \cdot \mathbf{D} \tag{391}$$

It follows that $\boldsymbol{\xi}$ satisfies the equation:

$$\boldsymbol{\xi} = \mathbf{V} \cdot \mathbf{I} + \mathbf{K} \cdot \boldsymbol{\xi} \tag{392}$$

Now one notes that

$$\mathbf{D} = \mathbf{1} + \mathbf{B} \tag{393}$$

and

$$\mathbf{B} = \mathbf{B} \cdot \mathbf{D} \tag{394}$$

where $\hat{\mathbf{B}}$ is given by Equation 377 except with ζ replaced by ξ. Now we note that:

$$\zeta(\gamma|\gamma' n' k_{n'}|r_{\gamma}R_{\gamma}) = \sum_{\lambda m} \xi(\gamma n|\lambda m k_m|r_{\gamma}R_{\gamma})D(\gamma m|\gamma' n') \tag{395}$$

and since ζ is independent of n, so also must ξ be. In addition, we note that $(\hat{\mathbf{K}})_{\gamma n,\gamma' n'}$ is independent of the index n. Thus, the equations satisfied by the elements of ξ are explicitly given by

$$\begin{aligned} \xi(\gamma|\gamma' n' k_{n'}|r_{\gamma}R_{\gamma}) &= V_{\gamma}(r_{\gamma}R_{\gamma})(1 - \delta_{\gamma\gamma'})\phi(\gamma' n'|r_{\gamma'})\ \psi^{reg}(\gamma' k_{n'}|R_{\gamma'}) \\ &+ \sum_{m=1}^{P_{\gamma''}} \int_0^{R_{\gamma''}} dR'_{\gamma''} \int_0^{r'_{\gamma''}} dr'_{\gamma''}\ K(\gamma|\gamma'' m|r_{\gamma''}R_{\gamma''},r'_{\gamma''}R'_{\gamma''}) \\ &\times\ \xi(\gamma''|\gamma' n' k_{n'}|r'_{\gamma''}R'_{\gamma''}) \end{aligned} \tag{396}$$

where γ'' is not equal to γ. It is clear from this equation that the ξ functions are not dependent on n. Now if one uses a Newton-Cotes quadrature on these equations, it turns out that one does not get a simple recursion formula. The reason for this is that specifying $r_{\gamma}R_{\gamma}$ variables does not produce the same value of $r_{\gamma''}$, $R_{\gamma''}$ (see the transformation Equations 288 to 291). As a result, the values of $\xi(\alpha|\alpha n k_n|r_{\alpha}R_{\alpha})$ occurring in the integral portion of the $\xi(\beta|\alpha n k_n|r_{\beta}R_{\beta})$

equation are not the ones generated by the $\xi(\alpha|\alpha n k_n|r_\alpha R_\alpha)$ equation. One can, however, transform the double integral from $R'_{\alpha''}$, $r'_{\alpha''}$ to a double integral over R'_α. These Volterra equations have never been tested by actual applications.

D. Computational Tests of the BKLT Equations

The first applications of the BKLT equations were carried out by Baer and Kouri[23-24] for collinear reactive scattering and used piecewise constant potentials in the spirit of the Hulburt-Hirschfelder model.[92] The models studied all contained the assumption that the middle atom was infinitely massive. This greatly simplifies the transformation between α and β arrangement coordinates. The method of solution was via the Fredholm technique discussed in Section V.B. In addition to collinear models, Baer and Kouri[25] also considered a three-dimensional model for an atom-heteronuclear diatom collision having only two arrangements:

$$A + BC \begin{array}{l} \nearrow \ AB + C \\ \searrow \ A + BC \end{array} \tag{397}$$

The atom B common to both diatoms was taken to be infinitely massive. Both integral and differential cross-sections were computed for a model (piecewise constant) potential. Subsequently, Baer and Kouri,[30] Kouri et al.,[32,33] and Eccles and Secrest[57,58] applied the BKLT formalism to low energy electron-H atom scattering and got good results. In this instance, the BKLT equations were formulated as Volterra equations (but not in the same way as discussed in Section V.C). Most recently, the BKLT equations have been successfully applied to a number of collinear atom-diatom systems with realistic smooth potentials and all three atoms having finite mass. The systems studied were the H + FH exchange reaction,[63,68] the H + H_2 exchange reaction[65,66] with the Porter-Karplus potential, and the D + $H_2 \rightarrow$ DH + H exchange reaction[95] with the potential surface of Siegbahn et al.[96-97] In all cases the results were in satisfactory agreement with those obtained by close coupling methods. In the latter two instances, the expansion functions $\xi^\gamma_m(r_\gamma R_\gamma)$ were taken to be the asymptotic diatom vibrational states for arrangement γ, namely $\phi(\gamma m|r_\gamma)$. These are independent of R_γ and the overlap matrix $S^\gamma(n'm|R_\gamma)$ taken to the Kronnecker delta $\delta_{n'm}$. This will be true only so long as the upper limit $(\tilde{r}_\gamma)$ in the integral in Equation 340 is large enough. This is guaranteed if the system does not reach too small a minimum relative distance $(R_{\gamma o})$. The potential surface in the small R_γ region is highly repulsive and the minimum value $(R_{\gamma o})$ for all three of the above systems is large enough that there is no problem with taking $S^\gamma(n'm|R_\gamma)$ to be a Kronnecker delta over the entire range of R_γ. The choice employed for the $\lambda(\gamma t|R_\gamma)$ is not as straightforward. It is necessary to employ functions that cover the region where the $Z(\gamma m|\alpha n_o k_{no}|R_\gamma)$ are nonzero. This region is determined by the potential surface. In the calculations for H + H_2, and D + H_2, the $\lambda(\gamma t|R_\gamma)$ basis was taken to be:

$$\lambda(\gamma t|R_\gamma) = \frac{2}{(L_2 - L_1)} \sin\left[\frac{t\pi(R_\gamma - L_1)}{L_2 - L_1}\right] \tag{398}$$

where L_1 is the minimum value of R_γ and L_2 its maximum value. These functions have the property that they remain orthonormal if one discretizes the range $L_2 - L_1$ with equally spaced points. In that case,

$$R_{\gamma i} = L_1 + i\Delta \tag{399}$$

$$\Delta = (L_2 - L_1)/J \tag{400}$$

where J is the number of grid points used. One then has

Table 1
REACTIVE PROBABILITIES FOR THE H + H_2 EXCHANGE REACTION

E^a	$0 \rightarrow 0^b$	$0 \rightarrow 1^c$	$1 \rightarrow 1$
0.3116	$2.2(-8)^{d,e}$		
	$2.0(-8)$		
0.6486	0.999		
	0.999		
0.8976	0.73	0.14	0.26
	0.67	0.18	0.29
0.9806	0.69	0.21	0.45
	0.62	0.22	0.51
1.0916	0.43	0.30	0.36
	0.39	0.33	0.41
1.2026	0.23	0.34	0.19
	0.20	0.38	0.26

[a] This is the total energy in electron volts.
[b] This is for reaction from H_2 in vibrational state 0 to final product in state 0.
[c] The reactive probability for the reverse reaction $1 \rightarrow 0$ is equal to that for this process.
[d] The upper entry is the BKLT result and the lower entry is the close coupling result of Diestler.
[e] This means 10^{-8}.

$$\Delta \sum_{i=1}^{J} \lambda(\gamma t|R_{\gamma i})\lambda(\gamma t'|R_{\gamma i}) = \delta_{tt'} \tag{401}$$

provided that t, t′ are both less than J. In the calculations for these systems, only about 10 of the $\lambda(t|R_\gamma)$ were needed (so the maximum value of t was 10) while the number of grid points (J) was about 200. The vibrational basis states $\phi(\gamma n|r_\gamma)$ were generated by numerical solution of Equation 321 at the same grid points as were used in the quadrature over the r_γ integrals. As a result, the numerical $\phi(\gamma n|r_\gamma)$ were also orthonormal when a quadrature evaluation of the orthogonality integral was performed. In the H + H_2 calculations, 127 points were used in the quadrature over r_γ and a total of 9 vibrational states were used to expand the r_γ dependence of the Green's functions and the amplitude density functions. Thus, in a typical calculation, the coefficient matrix was 180 × 180. In Table 1, we give a comparison of the results obtained for H + H_2 using the BKLT equations with those obtained by Diestler using a close-coupling method.[98]

E. Application of the BKLT Equations to Three Physical Dimensional Reaction Systems

The BKLT equations are now being applied by Shima et al.[99] to treat real three-dimensional (3D) systems. The first studies focus on the D + H_2 and H + H_2 systems using the most accurate available potential surface due to Siegbahn and Liu[96] as fitted by Truhlar and Horowitz.[97] In this section we now wish to show briefly how the BKLT formalism may be applied to such 3D systems. In fact, most of the techniques employed in the collinear studies can be generalized to three physical dimensions without any difficulty. This includes the separation of the energy-dependent and independent parts of the problem. The first discussion of the BKLT equations for 3D systems having two arrangements was given earlier by Baer and Kouri.[25,30,100,101] The present discussion is based on the more recent work of Shima et al.[99] and applies to a reaction in which there are three possible arrangements. Before going into details of the coordinate representation of the equations, it is useful to comment on the structure of the solution of the equations when one has three arrangements possible. The equations for the case where arrangement 1 is the initial one are given again as:

$$R_{11} = V_1 G_2^p R_{21} \tag{402}$$

$$R_{21} = V_2 G_3^p R_{31} \tag{403}$$

$$R_{31} = V_3 + V_3 G_1^p R_{11} \tag{404}$$

The equations corresponding to the other two possible initial arrangements are written as:

$$R_{12} = V_1 + V_1 G_2^p R_{22} \tag{405}$$

$$R_{22} = V_2 G_3^p R_{32} \tag{406}$$

$$R_{32} = V_3 G_1^p R_{12} \tag{407}$$

and

$$R_{13} = V_1 G_2^p R_{23} \tag{408}$$

$$R_{23} = V_2 + V_2 G_3^p R_{33} \tag{409}$$

$$R_{33} = V_3 G_1^p R_{13} \tag{410}$$

Our purpose in writing the equations in this particular sequence is that now the kernels of the three sets of three CIEs will be equal and need be computed only once, just as in the case of the two-arrangement, collinear example discussed earlier. The inhomogeneity in each equation will change. In the case of a symmetric exchange reaction (such as $H + H_2 \rightarrow H_2 + H$), even greater simplifications occur. For example, the nonzero portions of the inhomogeneities in the three sets of equations are equal to one another and, in addition, there are simplifications in the three pieces of the matrix kernel (i.e., $V_1 G_2^+$, $V_3 G_1^+$ and $V_2 G_3^+$). In particular, when the equations are converted into simultaneous algebraic equations, their structure will look like this:

$$\mathbf{R}_{11} = \mathbf{C}_{12} \mathbf{R}_{21} \tag{411}$$

$$\mathbf{R}_{21} = \mathbf{C}_{23} \mathbf{R}_{31} \tag{412}$$

$$\mathbf{R}_{31} = \mathbf{I}_{31} + \mathbf{C}_{31} \mathbf{R}_{11} \tag{413}$$

$$\mathbf{R}_{12} = \mathbf{I}_{12} + \mathbf{C}_{12} \mathbf{R}_{22} \tag{414}$$

$$\mathbf{R}_{22} = \mathbf{C}_{23} \mathbf{R}_{32} \tag{415}$$

$$\mathbf{R}_{32} = \mathbf{C}_{31} \mathbf{R}_{12} \tag{416}$$

and

$$\mathbf{R}_{13} = \mathbf{C}_{12} \mathbf{R}_{23} \tag{417}$$

$$\mathbf{R}_{23} = \mathbf{I}_{23} + \mathbf{C}_{23}\mathbf{R}_{33} \tag{418}$$

$$\mathbf{R}_{33} = \mathbf{C}_{31}\mathbf{R}_{13} \tag{419}$$

These are most easily solved by back substitution. The first set of equations (Equations 411 to 413) yields:

$$\mathbf{R}_{31} = (\mathbf{1} - \mathbf{C}_{31}\mathbf{C}_{12}\mathbf{C}_{23})^{-1}\,\mathbf{I}_{31} \tag{420}$$

$$\mathbf{R}_{21} = \mathbf{C}_{23}(\mathbf{1} - \mathbf{C}_{31}\mathbf{C}_{12}\mathbf{C}_{23})^{-1}\,\mathbf{I}_{31} \tag{421}$$

$$\mathbf{R}_{11} = \mathbf{C}_{12}\mathbf{C}_{23}(\mathbf{1} - \mathbf{C}_{31}\mathbf{C}_{12}\mathbf{C}_{23})^{-1}\,\mathbf{I}_{31} \tag{422}$$

where the unit matrix **1** in Equation 420 has dimensions determined by the number of basis functions ocurring in arrangement 3. Similarly, Equations 414 to 416 yield:

$$\mathbf{R}_{12} = (\mathbf{1} - \mathbf{C}_{12}\mathbf{C}_{23}\mathbf{C}_{31})^{-1}\,\mathbf{I}_{12} \tag{423}$$

$$\mathbf{R}_{32} = \mathbf{C}_{31}(\mathbf{1} - \mathbf{C}_{12}\mathbf{C}_{23}\mathbf{C}_{31})^{-1}\,\mathbf{I}_{12} \tag{424}$$

$$\mathbf{R}_{22} = \mathbf{C}_{23}\mathbf{C}_{31}(\mathbf{1} - \mathbf{C}_{12}\mathbf{C}_{23}\mathbf{C}_{31})^{-1}\,\mathbf{I}_{12} \tag{425}$$

where now the unit matrix **1** in Equation 423 has dimensions determined by the number of basis functions in arrangement 1. Finally, solution of Equations 417 to 419 yields:

$$\mathbf{R}_{23} = (\mathbf{1} - \mathbf{C}_{23}\mathbf{C}_{31}\mathbf{C}_{12})^{-1}\,\mathbf{I}_{23} \tag{426}$$

$$\mathbf{R}_{13} = \mathbf{C}_{12}(\mathbf{1} - \mathbf{C}_{23}\mathbf{C}_{31}\mathbf{C}_{12})^{-1}\,\mathbf{I}_{23} \tag{427}$$

$$\mathbf{R}_{33} = \mathbf{C}_{31}\mathbf{C}_{12}(\mathbf{1} - \mathbf{C}_{23}\mathbf{C}_{31}\mathbf{C}_{12})^{-1}\,\mathbf{I}_{23} \tag{428}$$

and the identity **1** in these equations has dimensions determined by the number of basis functions in arrangement 2. In the case where all three atoms are identical, the structure of the equations simplifies since then:

$$\mathbf{C}_{12} = \mathbf{C}_{23} = \mathbf{C}_{31} = \mathbf{C} \tag{429}$$

and

$$\mathbf{I}_{31} = \mathbf{I}_{12} = \mathbf{I}_{23} = \mathbf{I} \tag{430}$$

Then we find that

$$\mathbf{R}_{11} = \mathbf{R}_{22} \tag{431}$$

$$= \mathbf{R}_{33} \tag{432}$$

$$= \mathbf{C}^2(\mathbf{1} - \mathbf{C}^3)^{-1}\,\mathbf{I} \tag{433}$$

$$\mathbf{R}_{31} = \mathbf{R}_{12} \tag{434}$$

$$= \mathbf{R}_{23} \tag{435}$$

$$= (\mathbf{1} - \mathbf{C}^3)^{-1}\mathbf{I} \tag{436}$$

and

$$\mathbf{R}_{21} = \mathbf{R}_{13} \tag{437}$$

$$= \mathbf{R}_{32} \tag{438}$$

$$= \mathbf{C}(\mathbf{1} - \mathbf{C}^3)^{-1}\mathbf{I} \tag{439}$$

We remark that in solving the algebraic equations, one need not make the back substitution in the sequence given above. For example, in place of Equations 423 to 425, we may also solve Equations 414 to 416 by the sequence of substituting Equation 414 into Equation 416 followed by the substitution of Equation 415, thereby yielding:

$$\mathbf{R}_{32} = \mathbf{C}_{31}\mathbf{I}_{12} + \mathbf{C}_{31}\mathbf{C}_{12}\mathbf{C}_{23}\mathbf{R}_{32} \tag{440}$$

or

$$\mathbf{R}_{32} = (\mathbf{1} - \mathbf{C}_{31}\mathbf{C}_{12}\mathbf{C}_{23})^{-1}\mathbf{C}_{31}\mathbf{I}_{12} \tag{441}$$

The point is that in the above solution, one does not have to compute a different matrix inverse from the one occurring in the solutions to Equations 411 to 413 (i.e., Equations 420 to 422) even when one is dealing with a system for which all three arrangements are distinct.

The final point which we wish to illustrate is with regard to the fact that just as for the collinear reaction systems, we can divide the work up into energy-independent and energy-dependent parts. Furthermore, as we shall see, a major additional complication of going to three physical dimensions is that in place of the triple integrals encountered in the collinear case, one now must evaluate fourfold integrals. However, we show that the additional integral occurs in the energy-independent part of the problem, so its effect on the calculation is minimized. We shall concentrate on the set of Equations 402 to 404, for which arrangement 1 is the initial arrangement. We then introduce amplitude density functions:

$$\zeta(1|1n_oj_om_oJM|\mathbf{R}_1\mathbf{r}_1) = <\mathbf{R}_1\mathbf{r}_1|R_{11}|1n_oj_om_oJM> \tag{442}$$

$$\zeta(2|1n_oj_om_oJM|\mathbf{R}_2\mathbf{r}_1) = <\mathbf{R}_2\mathbf{r}_2|R_{21}|1n_oj_om_oJM> \tag{443}$$

and

$$\zeta(3|1n_oj_om_oJM|\mathbf{R}_3\mathbf{r}_3) = <\mathbf{R}_3\mathbf{r}_3|R_{31}|1n_oj_om_oJM> \tag{444}$$

Here $(1n_oj_om_o)$ denotes the initial vibration-rotational quantum numbers in arrangement 1, J is the total angular momentum quantum number, and $M\hbar$ is the projection of total angular momentum along an arbitrary space-fixed axis. The rotor projection of angular momentum $m_o\hbar$ is taken along the scattering vector in arrangement 1 (i.e., it is a helicity quantum number) $\mathbf{R}_1$. The vector $\mathbf{R}_i$ is the vector from the diatom center of mass to the atom in arrangement i and $\mathbf{r}_i$ is the internuclear vector for the diatom in arrangement i. All the vectors are taken to be mass-scaled, just as in the collinear case. The form of the initial state is given by:

$$\theta(1n_\alpha j_o m_o JM|\mathbf{R}_1\mathbf{r}_1)$$

$$= \frac{2\mu}{\hbar^2} \sum_m \chi_{n_o j_o}(r_1)\, D^J_{mM}(\phi_1\theta_1\psi_1) Y_{j_o m}(\gamma_1,0)\, \psi^{reg}_{J j_o m m_o}(R_1) \tag{445}$$

where the function $\psi^{reg}_{Jjmm_o}(R_1)$ is a solution of a distorted wave radial equation[99] expressed in the body frame,[102] with distortion potential given by $V(R_1, r_{1,o}, \gamma_{1,o})$. V is the *full* potential and $r_{1,o}$ is a fixed value of the diatom separation in arrangement 1 and $\gamma_{1,o}$ is a fixed value of the angle between $\mathbf{R}_1$ and $\mathbf{r}_1$ corresponding to the minimum energy path (for H + H_2, $\gamma_{1,o}$ will be taken to be zero). In addition, ϕ_1, θ_1, and ψ_1 are the Euler angles associated with the orientation of the three-atom system in arrangement 1, the rotation matrix is assumed to be properly normalized, and the vibrational states are defined as the solutions of the equation:

$$\left\{ -\frac{\hbar^2}{2\mu}\left[\frac{1}{r_1^2}\frac{\partial}{\partial r_1}\left(r_1^2\frac{\partial}{\partial r_1}\right) - \frac{j(j+1)}{r_1^2}\right] + V(R_{1,0}, r_1)\right\} \chi_{nj}(r_1)$$
$$= \epsilon_{nj}\, \chi_{nj}(r_1) \tag{446}$$

where $R_{1,o}$ is a value of R_1 large enough that there is no longer any interaction between the atom and the diatom.[102]

The coordinate representation of the equations is then:

$$\zeta(i|1n_\alpha j_o m_o JM|\mathbf{R}_1\mathbf{r}_1)$$
$$= \delta_{i3}\, V_3(\mathbf{R}_3,\mathbf{r}_3)\theta(1n_\alpha j_o m_o JM|\mathbf{R}_1\mathbf{r}_1) \tag{447}$$
$$+ \frac{2\mu}{\hbar^2}\int d\mathbf{R}'_{i+1}\int d\mathbf{r}'_{i+1}\, V_i(\mathbf{R}_i,\mathbf{r}_i)$$
$$G^p_{i+1}\,(\mathbf{R}_{i+1}(\mathbf{R}_i\mathbf{r}_i)\mathbf{r}_{i+1}(\mathbf{R}_i\mathbf{r}_i)|\mathbf{R}'_{i+1}\,\mathbf{r}'_{i+1})$$
$$\times\ \zeta(i+1|1n_\alpha j_o m_o JM|\mathbf{R}'_{i+1}\mathbf{r}_{i+1})$$

where if i equals 3, then by convention, i + 1 is taken to be 1. The amplitude densities are expanded in basis sets according to:

$$\zeta(i|1n_\alpha j_o m_o JM|\mathbf{R}_i\mathbf{r}_i)$$
$$= \sum_{\substack{njm\\J'M'}} \chi_{nj}(r_i)\, D^{J'}_{mM'}(\phi_i\theta_i\psi_i)Y_{jm}(\gamma_i,0)\zeta(injmJ'M'|1n_\alpha j_o m_o JM|R_i) \tag{448}$$

and in addition, the Green's functions are expanded as:

$$G^p_i(\mathbf{R}_i\mathbf{r}_i|\mathbf{R}'_i\mathbf{r}'_i) = \sum_{\substack{Jnj\\Mmm}} D^J_{mM}(\phi_i\theta_i\psi_i)D^{J*}_{m'M}(\phi'_i\theta'_i\psi'_i)Y_{jm}(\gamma_i,0)Y^*_{jm'}(\gamma'_i,0)$$
$$\times\ g_i^{Jnjmm'}(R_i|R'_i)\chi_{nj}(r_i)\chi^*_{nj}(r'_i) \tag{449}$$

The radial portion of the Green's function is expressed as:

$$g_i^{Jnjmm'}(R_i|R'_i) = \sum_{m''} \psi^{reg}_{Jnjm''m}(R_i^<)\, \psi^{irreg}_{Jnjm''m'}(R_i^>) \tag{450}$$

where $\psi^{irreg}_{Jnjm''m'}(R_i)$ is the irregular solution of the body frame distorted wave radial equation and $R_i^<$, $R_i^>$ are the lesser and greater of (R_i, R'_i), respectively. We substitute Equations 448 to 450 appropriately into Equation 447 and obtain:

$$\sum_{\substack{njm\\J'M'}} \chi_{nj}(r_i)\, D^{J'}_{mM'}(\phi_i\theta_i\psi_i)Y_{jm}(\gamma_i,0)\zeta(injmJ'M'|ln_oj_om_oJM|R_i)$$

$$= \delta_{i3}\, V_3(R_3r_3\gamma_3)\theta(ln_oj_om_oJM|\mathbf{R}_1\mathbf{r}_1)$$

$$+ \frac{2\mu}{\hbar^2}\sum_{\substack{J''M''\\j''n''m''\\m}} \int_0^\infty dR'_{i+1}\, R'^2_{i+1}\, D^{J''}_{mM''}(\phi_{i+1}\theta_{i+1}\psi_{i+1})Y_{j''m}(\gamma_{i+1},0)\chi_{n''j''}(r_{i+1})$$

$$\times\ V_i(R_ir_i\gamma_i)g^{J''n''j''mm''}_{i+1}(R_{i+1}|R'_{i+1})\zeta(i+1n''j''m''J''M''|ln_oj_oJM|R'_{i+1}) \qquad (451)$$

We next multiply both sides of Equation 451 by particular i-th arrangement basis functions and integrate over $(r_i,\varphi_i,\theta_i,\psi_i,\gamma_i)$, making use of orthogonality and the relation:[100]

$$D^J_{mM}(\phi_{i+1},\theta_{i+1},\psi_{i+1}) = \sum_{\bar{m}} d^J_{m\bar{m}}(\Delta_{i,i+1})D^J_{\bar{m}M}(\phi_i,\theta_i,\psi_i) \qquad (452)$$

to obtain:

$$\zeta(injmJ'M'|ln_oj_om_oJM|R_i) = \frac{2\mu}{\hbar^2}\delta_{i3}\,\delta_{J'J}\,\delta_{M'M}\sum_{\bar{m}}\int dr_3\, r_3^2\int d(\cos\gamma_3)\chi^*_{nj}(r_3)$$

$$\times\ Y^*_{jm}(\gamma_3,0)V_3(R_3r_3\gamma_3)\,\chi_{n_oj_o}(r_1)Y_{j_o\bar{m}}(\gamma_1,0)\, d^J_{\bar{m}m}(\Delta_{31})\psi^{reg}_{Jj_o\bar{m}m_o}(R_1)$$

$$+ \frac{2\mu}{\hbar^2}\int\sum_{\substack{n''j''\\m''\bar{m}}} dr_i\, r_i^2\int d(\cos\gamma_i)\int dR'_{i+1}\, R'^2_{i+1}\,\chi^*_{nj}(r_i)Y^*_{jm}(\gamma_i,0)d^{J'}_{\bar{m}m}(\Delta_{i,i+1}) \qquad (453)$$

$$\times\ V_i(R_ir_i\gamma_i)Y_{j''\bar{m}}(\gamma_{i+1},0)\,\chi_{n''j''}(r_{i+1})g^{J'n''j''\bar{m}m''}_{i+1}(R_{i+1}|R'_{i+1})$$

$$\times\ \zeta(i+1n''j''m''J'M'|ln_oj_om_oJM|R'_{i+1})$$

Now we see that the equations are completely uncoupled in J',M', and further for $J' \neq J$, and/or $M' \neq M$, the equations are completely homogeneous. Since we are looking at positive (collision) energies, it follows that the amplitude density functions must be proportional to $\delta_{JJ'}\,\delta_{MM'}$ and we need only consider equations for which J equals J' and M equals M'. This is just the conservation of total angular momentum. Furthermore, the resulting equations are independent of the total angular momentum projection ($M\hbar$) along an arbitrary space-fixed axis, as indeed is well known from the Wigner-Eckart theorem.[103] We thus write the equations as:

$$\zeta^J(injm|ln_oj_om_o|R_i)$$

$$= \frac{2\mu}{\hbar^2}\delta_{i3}\sum_{\bar{m}}\int dr_3\, r_3^2\int d(\cos\gamma_3)\chi^*_{nj}(r_3)Y^*_{jm}(\gamma_3,0)V_3(R_3r_3\gamma_3)\chi_{n_oj_o}(r_1)$$

$$\times\ Y_{j_om_o}(\gamma_1,0)d^J_{\bar{m}m}(\Delta_{13})\,\psi^{reg}_{Jj_o\bar{m}m_o}(R_1) + \frac{2\mu}{\hbar^2}\sum_{\substack{n''j''\\m''\bar{m}}}\int dr_ir_i^2\int d(\cos\gamma_i) \qquad (454)$$

$$\times \int dR'_{i+1}\, R'^2_{i+1}\,\chi^*_{nj}(r_i)Y^*_{jm}(\gamma_i,0)V_i(R_ir_i\gamma_i)\, d^J_{\bar{m}m}(\Delta_{i,i+1})Y_{j''\bar{m}}(\gamma_{i+1},0)$$

$$\times\ \chi_{n''j''}(r_{i+1})g^{Jn''j''\bar{m}m''}_{i+1}(R_{i+1}|R'_{i+1})$$

$$\times\ \zeta^J(i+1n''j''m''|ln_oj_om_o\, R'_{i+1})$$

At this point, the analysis of the equations now becomes completely parallel to that of the collinear case. We must expand the amplitude densities in terms of basis functions in the appropriate scattering distances, taking advantage again that the amplitude density is contained so that L^2-type basis functions can be employed. Thus, we write:

$$\zeta^J(\text{injm}|\text{ln}_\alpha\text{j}_o\text{m}_o|R_i) = \sum_t a^J(\text{injmt}|\text{ln}_\alpha\text{j}_o\text{m}_o)\lambda(\text{it}|R_i) \tag{455}$$

substitute into Equation 454, multiply by a particular basis function, and integrate over R_i, to obtain:

$$\begin{aligned}
&a^J(\text{injmt}|\text{ln}_\alpha\text{j}_o\text{m}_o) \\
&= \frac{2\mu}{\hbar^2}\delta_{i3}\sum_{\bar{m}}\int dR_3R_3^2\int dr_3r_3^2\int d(\cos\gamma_3)\lambda^*(3t|R_3)\chi^*_{nj}(r_3)Y^*_{jm}(\gamma_3,0) \\
&\times V_3(R_3r_3\gamma_3)\chi_{n_oj_o}(r_1)Y_{j_o\bar{m}}(\gamma_1,0)\,d^J_{\bar{m}m}(\Delta_{13})\,\psi^{reg}_{Jj_o\bar{m}m_o}(R_1) \\
&+ \frac{2\mu}{\hbar^2}\sum_{\substack{n''j''\\ m''\bar{m}\,t'}}\int dR_iR_i^2\int dr_ir_i^2\int d(\cos\gamma_i)\int dR'_{i+1}\,R'^2_{i+1}\,\lambda^*(\text{it}|R_i)\chi^*_{nj}(r_i) \\
&\times Y^*_{jm}(\gamma_i,0)\,d^J_{\bar{m}m}(\Delta_{i,i+1})\,Y_{j''\bar{m}}(\gamma_{i+1},0)\chi_{n''j''}(r_{i+1})V_i(R_ir_i\gamma_i) \\
&\times g^{Jn''j''\bar{m}m''}_{i+1}(R_{i+1}|R'_{i+1})\lambda(i+1t'|R'_{i+1})a^J(i+1n''j''m''t'|\text{ln}_\alpha\text{j}_o\text{m}_o)
\end{aligned} \tag{456}$$

The next step is to change the variables of integration in Equation 456 from R_i, r_i, cos γ_i to R_{i+1}, r_{i+1}, cos γ_{i+1}, and we then note that the following energy-independent quantities can be defined:

$$\begin{aligned}
&E^J(\text{injmt}|i+1n''j''\bar{m}|R_{i+1}) \\
&= \frac{2\mu}{\hbar^2}\int dr_{i+1}\,r_{i+1}^2\int d(\cos\gamma_{i+1})\lambda^*(\text{it}|R_i)\chi^*_{nj}(r_i)Y^*_{jm}(\gamma_i,0) \\
&\times d^J_{\bar{m}m}(\Delta_{i,i+1})\,Y_{j''\bar{m}}(\gamma_{i+1},0)\,\chi_{n''j''}(r_{i+1})V_i(R_ir_i\gamma_i)
\end{aligned} \tag{457}$$

In terms of these, we may write our equation finally as:

$$\begin{aligned}
a^J(\text{injmt}|\text{ln}_\alpha\text{j}_o\text{m}_o) &= \delta_{i3}\sum_{\bar{m}}\int dR_1R_1^2\,E^J(3\text{njmt}|\text{ln}_\alpha\text{j}_o\bar{m}|R_1) \\
&\times \psi^{reg}_{Jj_o\bar{m}m_o}(R_1) + \sum_{\substack{n''j''\\ m''\bar{m}t'}}\int dR_{i+1}R_{i+1}^2\int dR'_{i+1}\,R'^2_{i+1}\,E^J(\text{injmt}|i+1n''j''\bar{m}|R'_{i+1}) \\
&\times \lambda\,(i+1t'|R'_{i+1})\,g^{Jn''j''\bar{m}m''}_{i+1}(R_{i+1}|R'_{i+1})a^J(i+1n''j''m''t'|\text{ln}_\alpha\text{j}_o\text{m}_o)
\end{aligned} \tag{458}$$

This confirms our earlier statement that the additional integral (over cos γ_i), in fact can be isolated into the energy-independent portion of the calculation so that it does not introduce too great an additional computational burden compared to the collinear case discussed earlier. At this point, the solution of the simultaneous algebraic equations proceeds just as in the collinear case. This completes our discussion of the 3D, three-arrangment case of the BKLT equations.

VI. FADDEEV EQUATIONS FOR CHEMICAL REACTIONS

The application of the Faddeev equations to chemical systems has been made primarily by Micha and co-workers,[76-82] and Brumer and Shapiro.[83,84] As was noted in Section IV, these equations require that the interaction be expressed as a sum of pairwise interactions. This is not accurate for triatomic systems in general, if one uses pair potentials which depend only on position variables. Micha[76] has shown that one can express the triatomic potential as a sum of pairwise interactions provided that one uses spin-dependent potentials. Then if the potential is diagonalized in spin space, the resulting diagonal elements are simply potential surfaces of the London-Eyring-Polany-Sato (LEPS) type in which the Coulomb and exchange integrals are approximated by the Heitler-London expressions. Thus, the approach essentially involves a blending of the DIM approach with the Faddeev equations. Micha[76] employs a different off-shell extension of the Faddeev equations than was given in Section IV. However, the essential step is to note that the two-body T-operators (T_λ) defined by:

$$T_\lambda = V^\lambda + V^\lambda G_o^+ T_\lambda \tag{459}$$

now depend on spin through the fact that the V^λ are spin dependent. For a system of spin $^1/_2$ atoms, the total spin can be either $^1/_2$ or $^3/_2$. Micha[76] restricts his discussion to total spin equal to $^1/_2$. The interaction is expressed as:

$$V^\lambda = {}^1E(\lambda',\lambda'')\,{}^1O_\lambda + {}^3E(\lambda',\lambda'')\,{}^3O_\lambda \tag{460}$$

where ${}^{2j+1}O_\lambda$ is a projection operator for spin j. Since the two atoms λ',λ'' each have spin $^1/_2$, they produce a singlet and triplet contribution to the pair potential. In Equation 460 above, $\lambda' \neq \lambda'' \neq \lambda \neq \lambda'$. We note that

$${}^1O_\lambda = {}^1/_4 - \mathbf{S}_{\lambda'} \cdot \mathbf{S}_{\lambda''} \tag{461}$$

$${}^3O_\lambda = {}^3/_4 - \mathbf{S}_{\lambda'} \cdot \mathbf{S}_{\lambda''} \tag{462}$$

so that one may also write:

$$V^\lambda = V^{(\lambda,c)} + V^{(\lambda,s)}\,\mathbf{S}_{\lambda'} \cdot \mathbf{S}_{\lambda''} \tag{463}$$

with

$$V^{(\lambda,c)} = [{}^1E(\lambda',\lambda'') + 3\,{}^3E(\lambda',\lambda'')]/4 \tag{464}$$

and

$$V^{\lambda,s} = {}^3E(\lambda',\lambda'') - {}^1E(\lambda',\lambda'') \tag{465}$$

The result of using the spin-dependent V^γ in Equation 459 is that the two-body T-operators also depend on spin and can be expressed as:

$$T_\lambda = {}^1O_\lambda\,{}^1T_\lambda + {}^3O_\lambda\,{}^3T_\lambda \tag{466}$$

or

$$T_\lambda = T_\lambda^{(c)} + T_\lambda^{(s)}\,\mathbf{S}_{\lambda'} \cdot \mathbf{S}_{\lambda''} \tag{467}$$

with

$$T_\lambda^{(c)} = ({}^1T_\lambda + 3\,{}^3T_\lambda)/4 \tag{468}$$

and

$$T_\lambda^{(s)} = {}^3T_\lambda - {}^1T_\lambda \tag{469}$$

We note that $T_\lambda^{(c)}$ is *not* obtained using $V^{(\lambda,c)}$. It is the spin-dependent T_λ which are used by Micha[76] in the Faddeev equations. Essentially then the singlet ${}^1T_\lambda$ is determined using as its potential ${}^1E(\lambda',\lambda'')$ and the triplet ${}^3T_\lambda$ is determined using ${}^3E(\lambda',\lambda'')$ as its potential:

$${}^1T_\lambda = {}^1E(\lambda',\lambda'') + {}^1E(\lambda',\lambda'')G_o^+\,{}^1T_\lambda \tag{470}$$

and

$${}^3T_\lambda = {}^3E(\lambda',\lambda'') + {}^3E(\lambda',\lambda'')G_o^+\,{}^3T_\lambda \tag{471}$$

The full interaction (T_λ) is then constructed out of these according to Equation 466, or equivalently, Equations 467 to 469. This interaction is then used in the Faddeev equations. The solution of Equations 470 and 471 is nontrivial because it is necessary to obtain not only on-shell matrix elements (i.e., matrix elements between states whose momentum and energy quantum numbers satisfy energy conservation) but also off-shell matrix elements. Several techniques have been used by Micha and co-workers.[82] One has been to expand the ${}^1T_\lambda$ and ${}^3T_\lambda$ in terms of the eigenstates of $G_o^+\,{}^1E(\lambda',\lambda'')$ and $G_o^+\,{}^3E(\lambda',\lambda'')$, respectively. Other methods include direct solution of the inhomogeneous radial equation. From a computation standpoint, methods which represent the two-body T matrixes as separable expansions are convenient because these reduce the Faddeev equations to simultaneous algebraic equations. At that point, the solution of the Faddeev equations is then analogous to the BKLT equations. At the present time, there exist no converged calculations of the Faddeev equations for a chemical reaction. However, it is hoped that results will soon be available for comparison with other methods.

VII. FUTURE AVENUES FOR STUDY

The numerical application of integral equations for nonperturbative studies of chemical reactions has lagged behind that of the Schrödinger equation. However, it now appears that new progress is forthcoming. The integral equations approach attempts a global representation of the solution including all boundary conditions for a reactive system, and this has been both its major attraction and source of difficulty. The difficulties have stemmed from the fact that it has not been easy to deduce optimum basis sets to use in expanding the amplitude density functions. In addition, the unperturbed radial functions appearing in the various Green's functions did not tend to zero rapidly in highly nonclassical regions. The result was that convergence of the solutions was very slow, making the solution of the equations impractical. The use of distortion potentials and elucidation of how to treat systems with all particles having finite mass has significantly opened up these equations to exploitation. In addition, the version of the BKLT method discussed herein admits the possiblity of using a vibrational expansion basis which adjusts to the collision in various regions. This should significantly enhance the convergence of the expansion of the r_γ dependence of the amplitude density. Future studies will focus on several aspects of the BKLT equations. First and most important is the need to test the method on a real 3D system. Second is the exploration of other methods of solving the equations. Some of these include the finite element method, iteration, and the solution of the Volterra form of the equations. Third, it is important to study some highly asymmetric reaction systems in order to test the utility of the back substitution procedure for reducing the dimension of the coefficient matrix in order that the smallest possible matrix is the one which is inverted. Fourth, because the BKLT method can be used to solve for a single initial condition (in the T-operator form rather than the reactance operator form), it is important to try and develop reliable factorization relations which can be used to predict the full transition matrix from a subset of T-matrix elements. In the case of the Faddeev equation, it is important that converged results be obtained for some simple three-dimensional reactive system (e.g., the H_3 system).

ACKNOWLEDGMENT

Acknowledgment is made to the donors of the Petroleum Research Fund, administered by the American Chemical Society, for partial support of this work. Partial support was also provided by the National Science Foundation under grant CHE-8215317.

REFERENCES

1. **Lippmann, B. A. and Schwinger, J.,** Variational principles for scattering processes. I, *Phys. Rev.*, 79, 469, 1950.
2. **Gell-Mann, M. and Goldberger, M. L.,** The formal theory of scattering, *Phys. Rev.*, 91, 398, 1953.
3. **Secrest, D.,** Amplitude densities in molecular scattering, in *Methods in Computational Physics*, Vol. 10, Alder, B., Fernbach, S., and Rotenberg, M., Eds., Academic Press, N.Y., 1971, 243.
4. **Newton, R. G.,** *Scattering Theory of Waves and Particles*, 2nd ed., Springer-Verlag, N.Y., 1982.
5. **Kouri, D. J.,** Quantum mechanical theory of reactive scattering, in *Energy, Structure and Reactivity*, Smith, D. W. and McRae, W. B., Eds., John Wiley & Sons, N. Y., 1973, 26.
6. **Baer, M.,** A review of quantum-mechanical approximate treatments of three-body reactive systems, in *Advances in Chemical Physics*, Vol. 49, Rice, S. A. and Prigogine, I., Eds., John Wiley & Sons, N.Y., 1982, 191.
7. **Johnson, B. R. and Secrest, D.,** The solution of the nonrelativistic quantum scattering problem without exchange, *J. Math. Phys.*, 7, 2187, 1966.
8. **Johnson, B. R. and Secrest, D.,** Quantum-mechanical calculations of the inelastic cross sections for rotational excitation of para and ortho H_2 upon collision with He, *J. Chem. Phys.*, 48, 4682, 1968.
9. **Sams, W. N. and Kouri, D. J.,** Noniterative solutions of integral equations for scattering. I. Single channels, *J. Chem. Phys.*, 51, 4809, 1969.
10. **Sams, W. N. and Kouri, D. J.,** Noniterative solutions of integral equations for scattering. II. Coupled channels, *J. Chem. Phys.*, 51, 4815, 1969.
11. **White, R. A. and Hayes, E. F.,** Quantum mechanical studies of the vibrational excitation of H_2 by Li^+, *J. Chem. Phys.*, 57, 2985, 1972.
12. **Smith, E. R. and Henry, R. J. W.,** Noniterative integral-equation approach to scattering problems, *Phys. Rev. A*, 7, 1585, 1973.
13. **Morrison, M. A.,** The coupled-channels integral-equations method in the theory of low-energy electron-molecule scattering, in *Electron-Molecule and Photon-Molecule Collisions*, Rescigno, T., McKoy, V., and Schneider, B., Eds., Plenum Press, N.Y., 1978, 15.
14. **Faddeev, L. D.,** *Mathematical Aspects of the Three Body Problem in Quantum Scattering Theory*, D., Davey, N.Y., 1965.
15. **Foldy, L. L. and Tobocman, W.,** Application of formal scattering theory to many-body problems, *Phys. Rev.*, 105, 1099, 1956.
16. **Epstein, S.,** Theory of rearrangement collisions, *Phys. Rev.*, 106, 598, 1957.
17. **Marcus, R. A.,** Analytical mechanics of chemical reactions. III. National collision coordinates, *J. Chem. Phys.*, 49, 2610, 1968.
18. **Eyges, L.,** Quantum-mechanical three body problem, *Phys. Rev.*, 115, 1643, 1959.

19. **Lovelace, C.,** Practical theory of three-particle states. I. Nonrelativistic, *Phys. Rev.*, 135, B1225, 1964.
20. **Alt, E. O., Grassberger, P., and Sandhas, W.,** Reduction of the three particles collision problem to multi-channel two-particle Lippmann-Schwinger equations, *Nucl. Phys. B*, 2, 167, 1967.
21. **Hahn, Y.,** Asymptotic channel states and matrix equations in the unified reaction theory, *Phys. Rev.*, 169, 794, 1968.
22. **Glockle, W.,** A new approach to the three body problem, *Nucl. Phys. A*, 141, 620, 1970.
23. **Baer, M. and Kouri, D. J.,** Rearrangement channel operator approach to models for three body reactions. I, *Phys. Rev. A*, 4, 1924, 1971.
24. **Baer, M. and Kouri, D. J.,** Theory of reactive scattering. II. Application of T-operator formalism to a linear model for three body rearrangements, *J. Chem. Phys.*, 56, 4840, 1972.
25. **Baer, M. and Kouri, D. J.,** Theory of reactive scattering. IV. Exact quantum mechanical study of angular independent and angular dependent models for three dimensional rearrangement collisions, *J. Chem. Phys.*, 56, 1758, 1972.
26. **Sloan, I. H.,** Equations for four-particle scattering, *Phys. Rev. C*, 6, 1945, 1972.
27. **Hahn, Y. and Watson, K. M.,** Reduction method and distortion potentials for many particle scattering equations, *Phys. Rev. A*, 5, 1718, 1972.
28. **Chandler, C. and Gibson, A.,** Transition from time dependent to time independent multichannel quantum scattering theory, *J. Math. Phys.*, 14, 1328, 1973.
29. **Bencze, Gy.,** Integral equations for N-particle scattering, *Nucl. Phys. A*, 210, 568, 1973.
30. **Baer, M. and Kouri, D. J.,** Coupled channel approach to e-H scattering, *J. Math. Phys.*, 14, 1637, 1973.
31. **Redish, E. F.,** Connected kernel methods in nuclear reactions, *Nucl. Phys. A*, 225, 16, 1974.
32. **Kouri, D. J., Craigie, M., and Secrest, D.,** Coupled channel operators and rearrangement scattering. I. Comparative study of onesate s-wave e-H scattering, *J. Chem. Phys.*, 60, 1851, 1974.
33. **Kouri, D. J., Levin, F. S., Craigie, M., and Secrest, D.,** A general relationship between alternative equations for rearrangement channel T operators, *J. Chem. Phys.*, 61, 17, 1974.
34. **Kouri, D. J. and Levin, F. S.,** On channel T and K operators and the Heitler damping equation for identical particle scattering, *Phys. Rev. A*, 10, 1616, 1974.
35. **Kouri, D. J. and Levin, F. S.,** Coupled channel T operator equations with connected kernels. I. Three body problem with pairwise interactions, *Nucl. Phys. A*, 250, 127, 1975.
36. **Tobocman, W.,** New integral equations for the transition operators of many-body systems, *Phys. Rev. C*, 9, 2466, 1974.
37. **Tobocman, W.,** Variational many-channel nuclear reaction formalism, *Phys. Rev. C*, 10, 60, 1974.
38. **Kouri, D. J. and Levin, F. S.,** A new method for multiparticle scattering, *Phys. Lett., B*, 50, 421, 1974.
39. **Hahn, Y., Kouri, D. J., and Levin, F. S.,** Coupled channel arrays and the reduction method in many body scattering, *Phys. Rev. C*, 10, 1615, 1974.
40. **Goldflam, R. and Kouri, D. J.,** Some simple remarks on variational bounds in Faddeev-type formalisms, *Chem. Phys. Lett.*, 34, 594, 1975.
41. **Kouri, D. J. and Levin, F. S.,** Coupled channel T operator equations with connected kernels. II. N-coupled two-body channels, *Nucl. Phys. A*, 253, 395, 1975.
42. **Tobocman, W.,** New coupled-reaction channels formalism for nuclear reactions, *Phys. Rev. C*, 11, 43, 1975.
43. **Kouri, D. J. and Levin, F. S.,** General many channel variational principle for nuclear reactions, *Phys. Rev. C*, 11, 352, 1975.
44. **Tobocman, W.,** Exchange effects in the coupled-equations nuclear-reaction formalism, *Phys. Rev. C*, 12, 1146, 1957.
45. **Kouri, D. J., Levin, F. S., and Sandhas, W.,** On K operators and unitary approximations for the three body problem, *Phys. Rev. C*, 13, 1825, 1976.
46. **Kouri, D. J., Kruger, H., and Levin, F. S.,** Arrangement channel quantum mechanics: a general time dependent formalism for multiparticle systems, *Phys. Rev. D*, 15, 1156, 1977.
47. **Kruger, H. and Levin, F. S.,** On the validity of the simplest molecular orbital approximation for H_2^+ bonding, *Chem. Phys. Lett.*, 46, 95, 1977.
48. **Rabitz, S. and Rabitz, H.,** Decomposition theory of chemical reactions, *J. Chem. Phys.*, 67, 2965, 1977.
49. **Levin, F. S. and Kruger, H.,** Many-body scattering theory methods as a means for solving bound-state problems: applications of arrangement-channel quantum mechanics, *Phys. Rev. A*, 15, 2147, 1977.
50. **Levin, F. S. and Kruger, H.,** Channel-coupling theory of covalent bonding in H_2: a further application of arrangement-channel quantum mechanics, *Phys. Rev. A*, 16, 836, 1977.
51. **Goldflam, R. and Tobocman, W.,** Exhange symmetry and preservation of kernel connectivity in the Baer-Kouri-Levin-Tobocman coupled-equations nuclear reaction formalism, *Phys. Rev. C*, 17, 1914, 1978.
52. **Goldflam, R. and Tobocman, W.,** Alternative treatment of exchange effects in theory of radioactive decay, *Phys. Rev. C*, 18, 1857, 1978.
53. **Levin, F. S. and Greben, J. M.,** Validity of neglecting continuum contributions in two-body rearrangement collisions, *Phys. Rev. Lett.*, 41, 1447, 1978.

54. **Hoffmann, D. K., Kouri, D. J., and Top, Z. H.,** Kinetic theory of reacting fluid mixtures I. The BBGKY hierarchy in arrangement channel quantum mechanics, *J. Chem. Phys.*, 70, 4640, 1979.
55. **Goldflam, R. and Tobocman, W.,** Feshbach projection operator method and few-body reaction models in Baer-Kouri-Levin-Tobocman many-body scattering theory, *Phys. Rev. C*, 20, 904, 1979.
56. **Baer, M. and Kouri, D. J.,** in a proposal to the U.S.-Israel Binational Science Foundation, unpublished.
57. **Eccles, J. and Secrest, D.,** personal communication, 1980.
58. **Eccles, J.,** Reactive Scattering of an Atom by a Diatom: The Coupled Arrangement Wavefunction Method, Ph. D. thesis, University of Illinois, Urbana-Champaign, 1980.
59. **Goldflam, R., Kowalski, K. L., and Tobocman, W.,** Partition permuting array approach to few body Hamiltonian models of nuclear reactions, *J. Math. Phys.*, 21, 1888, 1980.
60. **Levin, F. S.,** Wave-function formalisms in the channel coupling array theory of many body scattering, *Phys. Rev. C*, 21, 2199, 1980.
61. **Evans, J. W.,** The mathematical structure of arrangement channel quantum mechanics, *J. Math. Phys.*, 22, 1672, 1981.
62. **Evans, J. W. and Hoffman, D. K.,** Faddeev's equations in differential form: completeness of physical and spurious solutions and spectral properties, *J. Math. Phys.*, 22, 2858, 1981.
63. **Shima, Y. and Baer, M.,** Quantum mechanical reactive transition probability. Application of the arrangement channel approach, *Chem. Phys. Lett.*, 91, 43, 1982.
64. **Top, Z. H. and Shapiro, M.,** A single coordinate, arrangement channels approach to reactive scattering, *J. Chem. Phys.*, 77, 5009, 1983.
65. **Shima, Y., Baer, M., and Kouri, D. J.,** BKLT equations for reactive scattering: a successful application to $H + H_2$, *Chem. Phys. Lett.*, 94, 321, 1983.
66. **Shima, Y., Baer, M., and Kouri, D. J.,** BKLT equations for reactive scattering I. Theory and application to 3 finite mass atoms system, *J. Chem. Phys.*, 78, 6666, 1983.
67. **Evans, J. W., Hoffmann, D. K., and Kouri, D. J.,** The reactive quantum Boltzmann equations: a derivation from an arrangement channel space representation and BBGKY hierarchy, *J. Chem. Phys.*, 78, 2665, 1983.
68. **Shima, Y. and Baer, M.,** Arrangement channel approach to reactive systems, *J. Phys. B.*, in press.
69. **Ford, W. K. and Levin, F. S.,** Channel coupling theory of molecular structure. IV. Finite element method solution for H_2^+, *Chem. Phys. Lett.*, in press.
70. **Evans, J. W., Hoffman, D. K., and Kouri, D J.,** Reactive scattering theory in arrangement channel quantum mechanics, *J. Math. Phys.*, 24, 576, 1983.
71. **Sandhas, W.,** Dynamical equations and approximation methods, in *Few Body Dynamics,* Mitra, A. N., Slaus, I., Bhasin, V. S., and Gupta, V. K., Eds., Elsevier/North-Holland, Amsterdam, 1976, 540.
72. **Kowalski, K. L.,** N-body systems, in *Lecture Notes in Physics,* Vol. 87, 1978, 393.
73. **Vanzani, V.,** The N-body problem. Wave-function decomposition, integral equations and structural properties, in *Few-Body Nuclear Physics,* IAEA, Vienna, 1978, 57.
74. **Levin, F. S.,** Some recent developments in n-particle scattering theory, in *Proc. 9th Int. Conf. Few Body Problem,* Levin, F. S., Eds., Elsevier/North-Holland, Amsterdam, 1981, 143c.
75. **Barrett, R. F., Robson, B. A., and Tobocman, W.,** Calculable methods for many-body scattering, *Rev. Mod. Phys.*, 55, 155, 1983.
76. **Micha, D. A.,** Collision dynamics of three interacting atoms: the Faddeev equations, *J. Chem. Phys.*, 57, 2184, 1972.
77. **Micha, D. A.,** Collision dynamics of three interacting atoms: permutational symmetry for identical nuclei, *J. Chem. Phys.*, 60, 2480, 1974.
78. **Micha, D. A. and Yuan, J.-M.,** Collision dynamics of three interacting atoms: the multiple collision expansion, *J. Chem. Phys.*, 63, 5462, 1975.
79. **Micha, D. A.,** Operator formalisms of reactive molecular scattering, *Int. J. Quantum Chem.*, 10, 259, 1976.
80. **Kuruoglu, Z. C. and Micha, D. A.,** Diatomic transition operators: results of L^2 basis expansions, *J. Chem. Phys.*, 72, 3327, 1980.
81. **Beard, L. H. and Micha, D. A.,** Collision dynamics of three interacting atoms: energy transfer and dissociation in collinear motions, *J. Chem. Phys.*, 73, 1193, 1980.
82. **Micha, D. A.,** Few-body processes in atom-diatom collisions, in *Proc. 9th Int. Conf. Few Body Problems,* Levin, F. S., Ed., Elsevier/North-Holland, Amsterdam, 1981, 39c.
83. **Brumer, P. and Shapiro, M.,** Multiple scattering theory. I. Off-shell t matrix elements and momentum space wavefunctions for local two body potentials, *J. Chem. Phys.*, 63, 427, 1974.
84. **Shapiro, M. and Brumer, P.,** Multiple scattering theory. II. Collision induced dissociation in first order, *J. Chem. Phys.*, in press.
85. **Gioumousis, G. and Curtiss, C. F.,** Molecular collisions. I. Formal theory and the Pauli principle, *J. Chem. Phys.*, 29, 996, 1958.

86. **Heitler, W.,** The influence of radiation damping on the scattering of light and mesons by free particles. I, *Camb. Phil. Soc.*, 37, 291, 1941.
87. **Margenau, H. and Murphy, G. M.,** *The Mathematics of Physics and Chemistry*, 2nd ed., D Van Nostrand, Princeton, 1962, 520.
88. **Morse, P. M. and Feshbach, H.,** *Methods of Theoretical Physics*, Vol. 1, 2, McGraw-Hill, N.Y., 1953.
89. **Kouri, D. J.,** Theory of reactive scattering. I. Homogeneous integral solution formalism for rearrangement T operator, *J. Chem. Phys.*, 51, 5204, 1969.
90. **Wyatt, R. E.,** Reactive scattering cross sections. II. Approximate quantual treatments, in *Atom-Molecule Collision Theory. A Guide for the Experimentalist*, Bernstein, R. B., Ed., Plenum Press, N.Y., 1979, 477.
91. **Wyatt, R. E.,** Direct-mode chemical reactions. I. Methodology for accurate quantal calculations, in *Atom-Molecule Collision Theory. A Guide for the Experimentalist*, Bernstein, R. B., Ed., Plenum Press, N.Y., 1979, 567.
92. **Hulburt, H. M. and Hirschfelder, J. O.,** The transmission coefficient in the theory of absolute reaction rates, *J. Chem. Phys.*, 11, 276, 1943.
93. **Light, J. C. and Walker, R. B.,** An R matrix approach to the solution of coupled equations for atom-molecule reactive scattering, *J. Chem. Phys.*, 65, 4272, 1976.
94. **Porter, R. N. and Karplus, M.,** Potential energy surface for H_3, *J. Chem. Phys.*, 40, 1105, 1964.
95. **Abu-Salbi, N., Kouri, D. J., Shima, Y., and Baer, M.,** Application of the BKLT equations to collinear asymmetric reactions with realistic potentials, to be published.
96. **Siegbahn, P. and Liu, B.,** An accurate three-dimensional potential energy surface for H_3, *J. Chem. Phys.*, 68, 2457, 1978.
97. **Truhlar, D. G. and Horowitz, C. J.,** Functional representation of Liu and Siegbahn's accurate *ab initio* potential energy calculations for $H + H_2$, *J. Chem. Phys.*, 68, 2466, 1978.
98. **Diestler, D. J.,** Close-coupling technique for chemical exchange reaction of the type $A + BC \rightarrow AB + C$. $H + H_2 \rightarrow H_2 + H^*$, *J. Chem. Phys.*, 54, 4547, 1971.
99. **Shima, Y., Baer, M., and Kouri, D. J.,** private communication.
100. **Baer, M. and Kouri, D. J.,** Theory of reactive scattering. III. Exact quantum mechanical calculations for a three dimensional model for three-body rearrangements, *J. Chem. Phys.*, 57, 3441, 1972.
101. **Baer, M. and Kouri, D. J.,** Exact quantum cross sections for a three body reactions, *Chem. Phys. Lett.*, 11, 238, 1971.
102. **Kouri, D. J., Heil, T. G., and Shimoni, Y.,** On the Lippman-Schwinger equation for atom-diatom collisions: a rotating frame treatment, *J. Chem. Phys.*, 65, 226, 1976.
103. **Tinkham, M.,** *Group Theory and Quantum Mechanics*, McGraw-Hill, N.Y., 1964.

INDEX

A

B

C

D

E

F

G

H

I

J

K

L

M

N

O

P

T

U

V

W

Z